Lecture Notes in Economics and Mathematical Systems

For information about Vols. 1–156, please contact your bookseller or Springer-Verlag

Vol. 157: Optimization and Operations Research. Proceedings 1977. Edited by R. Henn, B. Korte, and W. Oettli. VI, 270 pages. 1978.

Vol. 158: L. J. Cherene, Set Valued Dynamical Systems and Economic Flow. VIII, 83 pages. 1978.

Vol. 159: Some Aspects of the Foundations of General Equilibrium Theory: The Posthumous Papers of Peter J. Kalman. Edited by J. Green. VI, 167 pages. 1978.

Vol. 160: Integer Programming and Related Areas. A Classified Bibliography. Edited by D. Hausmann. XIV, 314 pages. 1978.

Vol. 161: M. J. Beckmann, Rank in Organizations. VIII, 164 pages. 1978.

Vol. 162: Recent Developments in Variable Structure Systems, Economics and Biology. Proceedings 1977. Edited by R. R. Mohler and A. Ruberti. VI, 326 pages. 1978.

Vol. 163: G. Fandel, Optimale Entscheidungen in Organisationen. VI, 143 Seiten. 1979.

Vol. 164: C. L. Hwang and A. S. M. Masud, Multiple Objective Decision Making – Methods and Applications. A State-of-the-Art Survey. XII, 351 pages. 1979.

Vol. 165: A. Maravall, Identification in Dynamic Shock-Error Models. VIII, 158 pages. 1979.

Vol. 166: R. Cuninghame-Green, Minimax Algebra. XI, 258 pages. 1979.

Vol. 167: M. Faber, Introduction to Modern Austrian Capital Theory. X, 196 pages. 1979.

Vol. 168: Convex Analysis and Mathematical Economics. Proceedings 1978. Edited by J. Kriens. V, 136 pages. 1979.

Vol. 169: A. Rapoport et al., Coalition Formation by Sophisticated Players. VII, 170 pages. 1979.

Vol. 170: A. E. Roth, Axiomatic Models of Bargaining. V, 121 pages. 1979.

Vol. 171: G. F. Newell, Approximate Behavior of Tandem Queues. XI, 410 pages. 1979.

Vol. 172: K. Neumann and U. Steinhardt, GERT Networks and the Time-Oriented Evaluation of Projects. 268 pages. 1979.

Vol. 173: S. Erlander, Optimal Spatial Interaction and the Gravity Model. VII, 107 pages. 1980.

Vol. 174: Extremal Methods and Systems Analysis. Edited by A. V. Fiacco and K. O. Kortanek. XI, 545 pages. 1980.

Vol. 175: S. K. Srinivasan and R. Subramanian, Probabilistic Analysis of Redundant Systems. VII, 356 pages. 1980.

Vol. 176: R. Färe, Laws of Diminishing Returns. VIII, 97 pages. 1980.

Vol. 177: Multiple Criteria Decision Making-Theory and Application. Proceedings, 1979. Edited by G. Fandel and T. Gal. XVI, 570 pages. 1980.

Vol. 178: M. N. Bhattacharyya, Comparison of Box-Jenkins and Bonn Monetary Model Prediction Performance. VII, 146 pages. 1980.

Vol. 179: Recent Results in Stochastic Programming. Proceedings, 1979. Edited by P. Kall and A. Prékopa. IX, 237 pages. 1980.

Vol. 180: J. F. Brotchie, J. W. Dickey and R. Sharpe, TOPAZ – General Planning Technique and its Applications at the Regional, Urban, and Facility Planning Levels. VII, 356 pages. 1980.

Vol. 181: H. D. Sherali and C. M. Shetty, Optimization with Disjunctive Constraints. VIII, 156 pages. 1980.

Vol. 182: J. Wolters, Stochastic Dynamic Properties of Linear Econometric Models. VIII, 154 pages. 1980.

Vol. 183: K. Schittkowski, Nonlinear Programming Codes. VIII, 242 pages. 1980.

Vol. 184: R. E. Burkard and U. Derigs, Assignment and Matching Problems: Solution Methods with FORTRAN-Programs. VIII, 148 pages. 1980.

Vol. 185: C. C. von Weizsäcker, Barriers to Entry. VI, 220 pages. 1980.

Vol. 186: Ch.-L. Hwang and K. Yoon, Multiple Attribute Decision Making – Methods and Applications. A State-of-the-Art-Survey. XI, 259 pages. 1981.

Vol. 187: W. Hock, K. Schittkowski, Test Examples for Nonlinear Programming Codes. V. 178 pages. 1981.

Vol. 188: D. Bös, Economic Theory of Public Enterprise. VII, 142 pages. 1981.

Vol. 189: A. P. Lüthi, Messung wirtschaftlicher Ungleichheit. IX, 287 pages. 1981.

Vol. 190: J. N. Morse, Organizations: Multiple Agents with Multiple Criteria. Proceedings, 1980. VI, 509 pages. 1981.

Vol. 191: H. R. Sneessens, Theory and Estimation of Macroeconomic Rationing Models. VII, 138 pages. 1981.

Vol. 192: H. J. Bierens: Robust Methods and Asymptotic Theory in Nonlinear Econometrics. IX, 198 pages. 1981.

Vol. 193: J. K. Sengupta, Optimal Decisions under Uncertainty. VII, 156 pages. 1981.

Vol. 194: R. W. Shephard, Cost and Production Functions. XI, 104 pages. 1981.

Vol. 195: H. W. Ursprung, Die elementare Katastrophentheorie. Eine Darstellung aus der Sicht der Ökonomie. VII, 332 pages. 1982.

Vol. 196: M. Nermuth, Information Structures in Economics. VIII, 236 pages. 1982.

Vol. 197: Integer Programming and Related Areas. A Classified Bibliography. 1978 – 1981. Edited by R. von Randow. XIV, 338 pages. 1982.

Vol. 198: P. Zweifel, Ein ökonomisches Modell des Arztverhaltens. XIX, 392 Seiten. 1982.

Vol. 199: Evaluating Mathematical Programming Techniques. Proceedings, 1981. Edited by J.M. Mulvey. XI, 379 pages. 1982.

Vol. 200: The Resource Sector in an Open Economy. Edited by H. Siebert. IX, 161 pages. 1984.

Vol. 201: P. M. C. de Boer, Price Effects in Input-Output-Relations: A Theoretical and Empirical Study for the Netherlands 1949–1967. X, 140 pages. 1982.

Vol. 202: U. Witt, J. Perske, SMS – A Program Package for Simulation and Gaming of Stochastic Market Processes and Learning Behavior. VII, 266 pages. 1982.

Vol. 203: Compilation of Input-Output Tables. Proceedings, 1981. Edited by J. V. Skolka. VII, 307 pages. 1982.

Vol. 204: K. C. Mosler, Entscheidungsregeln bei Risiko: Multivariate stochastische Dominanz. VII, 172 Seiten. 1982.

Vol. 205: R. Ramanathan, Introduction to the Theory of Economic Growth. IX, 347 pages. 1982.

Vol. 206: M. H. Karwan, V. Lotfi, J. Telgen, and S. Zionts, Redundancy in Mathematical Programming. VII, 286 pages. 1983.

Vol. 207: Y. Fujimori, Modern Analysis of Value Theory. X, 165 pages. 1982.

Vol. 208: Econometric Decision Models. Proceedings, 1981. Edited by J. Gruber. VI, 364 pages. 1983.

Vol. 209: Essays and Surveys on Multiple Criteria Decision Making. Proceedings, 1982. Edited by P. Hansen. VII, 441 pages. 1983.

Vol. 210: Technology, Organization and Economic Structure. Edited by R. Sato and M. J. Beckmann. VIII, 195 pages. 1983.

continuation on page 361

Lecture Notes
in Economics and
Mathematical Systems

Managing Editors: M. Beckmann and W. Krelle

312

Jayalakshmi Krishnakumar

Estimation of Simultaneous Equation Models with Error Components Structure

Springer-Verlag
Berlin Heidelberg GmbH

Author

Dr. Jayalakshmi Krishnakumar
Dept. of Econometrics, University of Geneva
2, rue Dancet, CH-1211 Geneva 4, Switzerland

ISBN 978-3-540-50031-5 ISBN 978-3-642-45647-3 (eBook)
DOI 10.1007/978-3-642-45647-3

2142/3140-543210

To my daughter Raga

ACKNOWLEDGEMENTS

This book is my doctoral thesis in Econometrics accepted at the Faculty of Economic and Social Sciences, University of Geneva, in December 1986.

I wish to express my profound gratitude to my thesis supervisor, Professor Pietro Balestra, for his extremely valuable guidance and advice at all stages of this research. He had been a source of constant encouragement and never hesitated to spare his time for long hours of useful discussion.

I am also greatly indebted to Professor Fabrizio Carlevaro for his unfailing support and interest in the course of my work. I am particularly grateful to him for his constructive criticisms and pertinent suggestions concerning the chapter on the practical application.

A special word of thanks is due to Professor A.L.Nagar of the Delhi School of Economics, who was my guide during my two years of research work in the University of Delhi. His useful and timely advice on the subject of finite sample properties was a key factor in the progress of this work.

I also benefitted substantially from the helpful comments and suggestions of Professor Alberto Holly of the University of Lausanne, on numerous occasions.

I would also like to thank Professor Elvezio Ronchetti for readily accepting to be a member of the thesis jury.

My sincere acknowledgements are due to Dr.Manfred Gilli and Ms.Bernadette Laplanche for their competent programming assistance. I am also thankful to Dr.Charles Spierer and Dr.Gabrielle Antille for their help in the early stage of my empirical study.

The typing of this text was a long and arduous task.

All the credit goes to Ms.Cheryl Dotti who, with skill, patience and diligence, typed the different chapters through their several versions and always remained cheerful.

Finally, I shall be failing in my duty if I do not mention the encouragement and support received from my family at every stage of my research.

Geneva, December 1986 Jayalakshmi Krishnakumar

TABLE OF CONTENTS

CHAPTER 1

INTRODUCTION

1.1 General

Any econometric study involves specification and estimation of one or more economic relationships. Let us briefly look at the sequence of steps leading to an estimated equation or an estimated set of equations, starting from the definition of the objective of the study.

In general, the relevant economic variables to be included in the study and the nature of the relationship(s) among them are given by the underlying economic theory relating to the phenomenon under consideration. The choice of the exact functional form to be specified, is relatively more difficult, as the theory may be compatible with more than one functional form. In such cases, the specific function to be retained for the analysis is often determined by : (i) setting certain criteria - economic and/or statistical - to be satisfied by the parameters of the relation; or (ii) the particular context in which the theory is applied, which may provide additional information on the type of the relationship.

Once a particular function (also known as a deterministic model) is specified, the econometrician introduces stochastic elements in it, in order to be able to estimate it. In other words, a stochastic (or econometric) model is formulated. Finally, based on the assumptions concerning the distributional properties of the stochastic elements, econometric theory develops appropriate techniques of estimation using the data available.

Now, let us turn our attention to the type of data available to an econometrician or an economist. In the majority of situations, it is almost impossible for the economic researcher to perform controlled experiments to generate the required data. In such cases, one is simply forced to use the information already available, in the form of observations in real life, in the best possible way.

Generally, economic data can be classified into two broad categories - time-series data and cross-section data. The former are data consisting of measurements of variables during successive time periods or at different dates. The observations are usually successive and equally spaced in time. The second type of data, the cross-section data, represent measurements of variables for different statistical units for a single time period or at a particular date. The units can assume different identities - countries, regions, firms, industries, households, individuals etc.

Any combination of the above two types of data, i.e. cross-section data taken for different time periods, is called pooled data. The cross-sectional units may not necessarily be the same for all time periods. The particular combination which consists of observations on the same units for the same number of successive time periods, is called panel data. Both pooled data and panel data are identified by two indices, one pertaining to sample units and the other pertaining to time periods.

As mentioned earlier, the economic researcher cannot have at his disposal as much information as he/she desires but simply has to make the best use of all the existing information. This is the framework of mind in which we place our work. More specifically, our research is concerned with the utilisation of a combination of time-series and cross-section data, whenever possible, in the estimation of economic models.

Several econometric models have been developed in recent years to enable the use of pooled data for estimation. The error components (EC) model is among the earliest of them and still remains one of the most popular of them to be used in applied work. Until very recently, most panel data models dealt with the estimation of a single relationship. But economic variables are often interdependent and determined simultaneously, and hence a system of simultaneous equations provides a better representation of economic reality than a single equation. Now, how does one estimate a simultaneous equation system using pooled data ? This is the specific question which we try to answer in our work and the solution we propose is a model that combines simultaneous equations and error components.

At this juncture, it should be mentioned that our work was carried out totally independently of the two other studies that have appeared in this topic, namely the papers by Baltagi [9] 1) and Prucha [39]. In fact, the starting point of our work dates back to 1981 when the author submitted her "mémoire de Diplôme" (post-graduate research thesis), parts of which were subsequently published as working papers (cf. Varadharajan [53] , [54] 2)). The reader may also note that our approach to the estimation problem as well as our methodology of estimation are very different from those of Baltagi and Prucha and that our treatment of the model is more complete including more results than these two studies.

1.2 Organization of the Book

This book will be organized as follows. The next Chapter, i.e. Chapter 2, is a review of the literature on panel data models. It is to be treated essentially as a means of getting acquainted with the various possible model specifications existing in this area and obtaining proper references on them for a more detailed analysis.

In Chapter 3, we first present our model of simultaneous equations with error components structure in full detail and in a systematic notation to be retained till the end of the work. Then, we go on to study the reduced form in depth and describe different methods of estimating it, particularly, the covariance estimation, the feasible generalised least squares (fGLS) estimation and the maximum likelihood (ML) estimation.

---

1) Numbers inside square parentheses correspond to those in the list of references given at the end.

2) Varadharajan is the maiden name of the author.

The asymptotic properties and the limiting distributions of all these estimators are derived and compared.

The following two chapters, namely Chapters 4 and 5, are concerned with the direct estimation of structural form parameters. In Chapter 4, we propose generalizations of classical two stage least squares and three stage least squares (which are termed as "generalized" two stage least squares (G2SLS) and "generalized" three stage least squares (G3SLS)), based on the instrumental variables (IV) approach. The G2SLS is a single-equation method whereas the G3SLS is a full-information system method. These methods are made feasible by a prior estimation of the unknown variance components by analysis of variance (AOV) method, with the help of what we call the covariance 2SLS estimation of coefficients. Once again, the asymptotic properties and the limiting distributions of all the proposed estimators are derived and compared.

Chapter 5 is an extensive analysis of the full information maximum likelihood (FIML) estimation of the structural form, starting from the principle of constrained maximum likelihood, suggesting a convenient iterative procedure for solving it, deriving the moments of the limiting distributions of the FIML estimators and finally, comparing them with those obtained in Chapter 4. At the end, the limited information maximum likelihood (LIML) estimation is presented as a special case of FIML.

In Chapter 6, the problem of identification and indirect estimation of structural parameters is discussed. In the case of a single just-identified structural equation, the indirect estimator is compared with the corresponding G2SLS estimator, and in case the whole system is just-identified, interesting relationships among the G2SLS, G3SLS and indirect estimators are established.

Chapter 7 deals with the study of finite sample moments of the estimators developed till then. More specifically, we look at the bias of these estimators in finite samples. Regarding the reduced form estimators, results on the exact bias are obtained. On the other hand, for the structural form estimators, only the bias upto a certain order is calculated by means of an expansion in series of the expression of the estimators.

In Chapter 8, the theory developed in the preceding chapters is applied to a model of residential demand for electricity using data collected during the Residential Energy Consumption Survey (RECS) conducted in the United States for the year 1982-83. Results of estimation of the model using alternative structural estimation methods are compared and commented upon from both statistical and economic points of view. A complete listing of the computer programme, written for this purpose, is appended to this Chapter.

Finally, we end the book with a chapter containing the important results and the major conclusions of our research, and possible areas of extensions.

CHAPTER 2

A SURVEY OF PANEL DATA MODELS

2.1 General

This chapter is an attempt to review and integrate various specifications of econometric models combining cross-section and time series data. It is not our intention to discuss each one of them in depth and neither do we claim the survey to be exhaustive. The models are classified into certain broad groups and within each group different variations are presented bringing out their important characteristics. The estimation of parameters involved is briefly discussed. The reader is invited to consult the articles referred to in each case for detailed treatment of the models concerned.

Most of the literature on panel data models is concerned with the estimation of a single economic relationship. Hence the main part of this chapter will be devoted to single equation panel data models. At the end, some extensions will be presented.

The majority of panel data models can be divided into two broad categories : constant slope variable intercept models and variable coefficient models. The first group (Section 2.2) assumes that the relation under study is the same for all observations apart from possible time and unit effects that enter the equation in an additive manner. Thus the specific effects associated with cross-sectional units and time periods are added to the constant term of the equation but the coefficients of other explanatory variables are assumed to be constant for all observations, hence the name constant slope variable intercept. In this group, a distinction can be made between models that treat the specific effects as fixed (covariance or fixed effects models) and those that treat them as random (random effects models). In the random effects models it can be assumed either that the specific effects and other explanatory variables are independent (error components (EC) models) or that they are correlated. The first case has been studied extensively by econometricians and we will cover the main topics of research concerning this case before proceeding to the second one. Finally, we will briefly look at a random effect model with a lagged endogenous variable and time-varying as well as time-invariant exogenous variables.

The second category of panel data models (Section 2.3) is a generalisation of the first one, in that, not only the constant term but all the coefficients vary over units and/or over time. In general, each coefficient is assumed to be composed of three components - one common to all units varying in time, a second one common to all time periods but varying over units and a third one variable in both dimensions. Again various hypotheses can be made on the type of variation of the coefficient components : fixed coefficient components or random coefficient components; in the second case, either independence between the coefficient components and the explanatory variables (purely random coefficient models) or correlation between the two (transmitted variations model); serial correlation of time components or not; and on the type of data to be processed : stratified or not. All these specifications will be examined in Section 2.3.

We will end this section by presenting a few models that are generalisations of the earlier ones, remaining in the variable coefficients framework : the "mixed" model in which certain coefficients are fixed while others vary; the quantitative effects model with both time-varying but unit-invariant and time-invariant but unit-varying explanatory variables and coefficients; and the general stratified effect components model with random coefficient components using stratified data.

In all these panel data models, prior to the estimation of coefficients of the equation or of the means of coefficients when they are considered random, the variance-covariance matrices of the different random terms have to be estimated. Section 2.4 deals with the various methods of estimation of variance components and mentions some studies making comparisons among these methods.

Up to this point, we were speaking only of models that assume that a balanced set of observations is available i.e. successive observations over time on the same economic units. In practice, in some cases, data on certain units may be missing for certain time periods and data on new units available. In such cases, the sample structure has to be taken into account in the specification of the model and this makes the estimation procedures slightly more complicated. In Section 2.5, we will present a study of an EC model with incomplete panel data.

The last section (Section 2.6) will contain some specifications of systems of several equations using pooled data, namely, a Seemingly Unrelated Regressions (SUR) Model with EC and a Simultaneous Equations Model (SEM) with EC structure. These models are extensions to single-equation panel data model specifications and estimation techniques.

2.2 Constant Slope Variable Intercept Models

2.2.1 Fixed Effects Models

These models, also known as least squares dummy variables models, introduce a dummy variable for each cross-section unit and another one for each time-period (see e.g. Hoch [20], Mundlak [30]). The regression equation can be written as follows :

(2.1) $$y_{it} = \beta' x_{it} + \alpha_{1i} + \alpha_{2t} + \varepsilon_{it}$$

where

(2.2) $$\underset{(1xK)}{\beta'} = [\beta_1 \; \dots \; \beta_K] ; \quad \underset{(1xK)}{x'_{it}} = [x_{1it} \; \dots \; x_{Kit}]$$

and it is assumed that

(2.3) $$E(\varepsilon_{it}) = 0$$

(2.4) $$E(\varepsilon_{it}\, \varepsilon_{i't'}) = \delta_{ii'}\, \delta_{tt'}\, \sigma^2 \; , \; i=1,\dots,N \; ; \; t=1,\dots,T$$

The specific effects denoted by α_{1i} and α_{2t} , $i=1,\dots,N$, $t=1,\dots,T$ are constants to be estimated along with β .

The OLS estimator of β in this model can be easily shown to be the same as the OLS estimator of β in the transformed model[1]:

(2.5) $y_{it} - y_{i.} - y_{.t} + y_{..} = \beta'(x_{it} - x_{i.} - x_{.t} + x_{..}) + \varepsilon_{it}$

This transformation is called the covariance transformation and the estimator of β thus obtained, the covariance estimator. In case OLS is applied directly to the model (2.1), additional restrictions are needed concerning α_{1i} and α_{2t} to be able to identify them. Two kinds of restrictions are possible:

(i) $\sum_i \alpha_{1i} = 0$ and $\sum_t \alpha_{2t} = 0$

or (ii) $\alpha_{1N} = \alpha_{2T} = 0$

2.2.2 Random Effects Models

2.2.2.1 Error Components Models: Classical Estimation Methods

The error component model is one of the earliest models developed to deal with panel data (see e.g. Balestra and Nerlove [6], Wallace and Hussain [55]). The model is as follows:

(2.6) $y_{it} = \beta' x_{it} + u_i + v_t + w_{it}$

where the following assumptions are made on the random error components u_i, v_t and w_{it} :

(2.7) $E(u_i) = E(v_t) = E(w_{it}) = 0$

(2.8) $E(u_i u_j) = \delta_{ij} \sigma_u^2$

(2.9) $E(v_t v_s) = \delta_{ts} \sigma_v^2$

(2.10) $E(w_{it} w_{js}) = \delta_{ij} \delta_{ts} \sigma_w^2$

(2.11) u_i, v_t and w_{it} are pair-wise independent for $i=1,\dots,N$; $t=1,\dots,T$ and are independent of x_{it} .

---

1) A dot in the place of a subscript indicates the average taken over it.

By writing

(2.12) $\varepsilon_{it} = u_i + v_t + w_{it}$

(2.13) $\underset{(NTx1)}{\varepsilon} = [\varepsilon_{11} \cdots \varepsilon_{1T}\ \varepsilon_{21} \cdots \varepsilon_{2T} \cdots \varepsilon_{NT}]'$

it can be easily verified that

(2.14) $E(\varepsilon\varepsilon') = \Omega = \sigma_w^2 I_{NT} + \sigma_u^2 (I_N \otimes \iota_T \iota_T') + \sigma_v^2 (\iota_N \iota_N' \otimes I_T)$

The GLS estimator of ß is thus given by :

(2.15) $\hat{\beta} = (X' \Omega^{-1} X)^{-1} X' \Omega^{-1} y$

where

(2.16)
$$\begin{cases} \underset{(NTx1)}{y} = [y_{11} \cdots y_{1T} \cdots y_{it} \cdots y_{NT}]' \\ \underset{(NTxK)}{X} = [x_{11} \cdots x_{1T} \cdots x_{it} \cdots x_{NT}]' \end{cases}$$

and the model in matrix notation is

(2.17) $y = X\beta + \varepsilon$

Let us add that the spectral decomposition of Ω is (see Nerlove [36]) :

(2.18) $\Omega = \sigma_1^2 \left(\frac{A}{T} - \frac{J_{NT}}{NT} \right) + \sigma_2^2 \left(\frac{B}{N} - \frac{J_{NT}}{NT} \right) + \sigma_3^2 \frac{J_{NT}}{NT} + \sigma_w^2 Q$

where

(2.19) $\sigma_1^2 = \sigma_w^2 + T\sigma_u^2$

(2.20) $\sigma_2^2 = \sigma_w^2 + N\sigma_v^2$

(2.21) $\sigma_3^2 = \sigma_w^2 + T\sigma_u^2 + N\sigma_v^2$

(2.22) $A = I_N \otimes \iota_T \iota_T'$; $B = \iota_N \iota_N' \otimes I_T$; $J_{NT} = \iota_N \iota_N' \otimes \iota_T \iota_T'$

(2.23) $Q = I_{NT} - \frac{A}{T} - \frac{P}{N} + \frac{J_{NT}}{NT}$

and I_N is the identity matrix of order N, ι_N is the unit vector of order N and $\otimes$ denotes the Kronecker product. This decomposition is useful in calculating Ω^{-1} needed for GLS estimation.

In the case of unknown variances, σ_1^2, σ_2^2 and σ_w^2 have to be estimated before applying GLS. The estimation of variance components is considered in a separate section (Section 2.4).

Balestra and Nerlove ([6]) have also considered the maximum likelihood (ML) estimation of the parameters of an EC model. Their model includes lagged endogenous variables among the explanatory variables of the equation and omits time effects. It is the same as (2.6) with $v_t = 0$ and where x'_{it} includes lagged values of y. The stochastic assumptions are (2.7), (2.8) and (2.10) and independence between u_i and w_{it}. The ML estimation yields the following estimates :

(2.24) $$\hat{\beta} = (X'\hat{\Omega}^{-1}X)^{-1} X'\hat{\Omega}^{-1}y$$

(2.25) $$\hat{\rho} = \frac{\sum_{i=1}^{N} \left\{ \left[\sum_{t=1}^{T} \varepsilon_{it} \right]^2 - \sum_{t=1}^{T} \varepsilon_{it}^2 \right\}}{(T-1) \sum_{i=1}^{N} \sum_{t=1}^{T} \varepsilon_{it}^2}$$

(2.26) $$\hat{\sigma}^2 = \frac{\sum_{i=1}^{N} \sum_{t=1}^{T} \varepsilon_{it}^2}{NT}$$

where ε_{it} has to be replaced by $y_{it} - x'_{it}\hat{\beta}$

and

(2.27) $$\rho = \sigma_u^2/\sigma^2 \ ; \ \sigma^2 = \sigma_u^2 + \sigma_w^2$$

(2.28) $$\hat{A} = \hat{\rho}\, \iota_T \iota'_T + (1-\hat{\rho})\, I_T$$

(2.29) $$\hat{\Omega} = \hat{\sigma}^2 (I_N \otimes \hat{A})$$

$\hat{\sigma}_u^2$ and $\hat{\sigma}_w^2$ can be uniquely derived from $\hat{\sigma}^2$ and $\hat{\rho}$ using (2.27). We will come back to this model while discussing dynamic random effects models.

Berzec [10] examines the ML estimation of EC models with the additional assumption of non-zero correlation between unit effects and overall effects (again omitting time effects); i.e., the model is the same as (2.6) with $v_t = 0$ except for the following modification in the assumptions :

(2.30) $E(u_i w_{it}) = \sigma_{uw}$

In this case,

(2.31) $E(\varepsilon\varepsilon') = \sigma^2 (I_N \otimes A)$

with

(2.32) $\sigma^2 = \sigma_u^2 + 2\,\sigma_{uw} + \sigma_w^2$

(2.33) $A = \rho\,\iota_T \iota_T' + (1-\rho)\, I_T$

(2.34) $\rho = (\sigma_u^2 + 2\sigma_{uv}) / \sigma^2$

Note that A is positive definite iff $-(T-1)^{-1} < \rho < 1$.

The Maximum Likelihood estimates are identical to those of (2.24), (2.25) and (2.26) and it can be shown that $-(T-1)^{-1} < \hat{\rho} < 1$ with probability 1 .

The estimate of σ_w^2 is derived as

(2.35) $\hat{\sigma}_w^2 = \hat{\sigma}^2 (1-\hat{\rho})$

But for the EC variance σ_u^2 and covariance σ_{uw} there are an infinite number of solutions. In fact, the system to be solved is :

(2.36) $\sigma_{uw} = \frac{1}{2}(\hat{\sigma}^2\,\hat{\rho} - \sigma_u^2)$

subject to the constraint

(2.37) $\hat{\sigma}^2(1 - \sqrt{1-\hat{\rho}})^2 < \sigma_u^2 < \hat{\sigma}^2(1 + \sqrt{1-\hat{\rho}})^2$

The finite sample properties of EC model estimators are considered by Swamy and Arora [47] and Taylor [50]. Swamy and Arora have shown that the feasible GLS estimator of ß is unbiased and have also derived an expression of its variance-covariance matrix involving conditional expectations over the distribution of variance components. Taylor compares three different estimators (between group, within group and feasible GLS) on the basis of small sample properties and finds that, in general, the feasible GLS estimator dominates the other estimators in efficiency.

2.2.2.2 EC Models : Bayesian Analysis

Up to this point, we were concerned only with classical estimation procedures that use sample information only. There is another approach to the estimation problem that combines sample information and prior information, viz, the Bayesian approach. The Bayesian analysis of EC models is the object of a paper by Swamy and Mehta [48]. The basis is the same model (2.6) with a slightly different notation :

$$(2.38) \quad y_{it} = x'_{it}\,\beta + \alpha_i + \tau_t + u_{it}$$

and the assumptions concerning the errors are :

$$(2.39) \quad E(\alpha) = 0 \;;\; E(\alpha\alpha') = \sigma_\alpha^2\,\Omega_1, \text{ with } \Omega_1 = \rho_1 \iota_N \iota'_N + (1-\rho_1) I_N$$

$$(2.40) \quad E(\tau) = 0 \;;\; E(\tau\tau') = \sigma_\tau^2\,\Omega_2, \text{ with } \Omega_2 = \rho_2 \iota_T \iota'_T + (1-\rho_2) I_T$$

$$(2.41) \quad E(u) = 0 \;;\; E(uu') = \sigma_u^2\,(\Omega_3 \otimes \Omega_4) \text{ with}$$

$$\Omega_3 = \rho_3 \iota_N \iota'_N + (1-\rho_3) I_N$$

$$\Omega_4 = \rho_4 \iota_T \iota'_T + (1-\rho_4) I_T$$

where

$$(2.42) \quad \underset{(NX1)}{\alpha} = [\alpha_1 \; \ldots \; \alpha_N]' \;;\; \underset{(TX1)}{\tau} = [\tau_1 \; \ldots \; \tau_T]' \;;$$

$$\underset{(NTX1)}{u} = [u_{11} \; \ldots \; u_{1T} \; \ldots \; u_{NT}]'$$

Then it follows that

$$(2.43) \quad \Sigma = E(\varepsilon\varepsilon') = \sigma_\alpha^2(\Omega_1 \otimes \iota_T \iota'_T) + \sigma_\tau^2(\iota_N \iota'_N \otimes \Omega_2) + \sigma_u^2(\Omega_3 \otimes \Omega_4)$$

where

$$(2.44) \quad \underset{(NTX1)}{\varepsilon} = [\varepsilon_{11} \; \ldots \; \varepsilon_{1T} \; \ldots \; \varepsilon_{NT}]'$$

with

$$(2.45) \quad \varepsilon_{it} = \alpha_i + \tau_t + u_{it}$$

The equation (2.38), after being written in matrix form, is transformed successively by Q_1, Q_2, Q_3 and Q_4 where

$$(2.46)\quad \begin{cases} Q_1 = \dfrac{\iota_N'}{\sqrt{N}} \otimes \dfrac{\iota_T'}{\sqrt{T}} \quad ; \quad Q_2 = C_2 \otimes \dfrac{\iota_T'}{\sqrt{T}} \\ Q_3 = \dfrac{\iota_N'}{\sqrt{N}} \otimes C_1 \quad ; \quad Q_4 = C_2 \otimes C_1 \end{cases}$$

with C_1, C_2 such that

$O_T = [\iota_T/\sqrt{T}\ C_1']$ is an orthogonal matrix of order T,

$O_N = [\iota_N/\sqrt{N}\ C_2']$ is an orthogonal matrix of order N

and where

$Q = [Q_1'\ Q_2'\ Q_3'\ Q_4']$ is the matrix diagonalizing Σ .

Letting

$$\bar{\varepsilon}_i = Q_i \varepsilon$$

they show that the transformed errors have scalar variance covariance matrices i.e. these matrices are of the form :

$$E(\bar{\varepsilon}_1 \bar{\varepsilon}_1') = \sigma_1^2 \ ; \ E(\bar{\varepsilon}_2 \bar{\varepsilon}_2') = \sigma_2^2 I_{(N-1)}$$

$$E(\bar{\varepsilon}_3 \bar{\varepsilon}_3') = \sigma_3^2 I_{T-1} \ ; \ E(\bar{\varepsilon}_4 \bar{\varepsilon}_4') = \sigma_4^2 I_{(N-1)(T-1)}$$

and

$$E(\bar{\varepsilon}_i \bar{\varepsilon}_j') = 0 \quad \text{for} \quad i \neq j$$

The prior distribution of the unknown parameters is taken as

$$(2.47)\qquad p(\beta, \sigma_2^2, \sigma_3^2, \sigma_4^2) \propto \sigma_2^{-2}\ \sigma_3^{-2}\ \sigma_4^{-2}$$

and σ_1^2 is assumed to be known exactly.

Combining the likelihood function and the prior probability density function via Bayes' theorem, yields the joint posterior distribution of β, σ_2^2, σ_3^2, σ_4^2 given y,X and the prior information. Integrating this posterior distribution from $-\infty$ to ∞ with respect to the constant term gives $p(\delta$[1]$, \sigma_2^2, \sigma_3^2, \sigma_4^2 \mid y, X, (2.47))$ which is a multivariate normal with mean $\tilde{\delta}(\sigma)$[2] and finally integrating out σ_2^2, σ_3^2 and σ_4^2 we get the marginal

1) δ is the subvector of β consisting of the coefficients of all the explanatory variables except the constant term i.e. $\beta = [\beta_o\ \delta']'$.

2) $\sigma = [\sigma_1^2\ \sigma_2^2\ \sigma_3^2\ \sigma_4^2]'$

posterior distribution of δ , $p(\delta|y,X, (2.47))$. In this analysis, they also calculate the information loss due to aggregation of variables across units and give the posterior mean and variance of $\iota'\delta$.

Finally, the Bayesian analysis is repeated including restrictions on variances such as $\sigma_3^2 \geq \sigma_4^2$ or $\sigma_2^2 \geq \sigma_4^2$ or $\sigma_3^2 \geq \sigma_4^2$ & $\sigma_2^2 \geq \sigma_4^2$.

2.2.2.3 EC Models Using Stratified Data

This model is proposed as a generalisation of the EC model (2.6) to enable it to operate on stratified data. For convenience sake, we will use the same stratification notation as in Mazodier and Trognon [29] . The index sets $\bar{N} = \left\{1,2,\ldots,N\right\}$ and $\bar{T} = \left\{1,2,\ldots,T\right\}$ are partitioned into J and S subsets respectively i.e.

(2.48) $\bar{N} = \left\{\bar{N}_1,\ldots,\bar{N}_J\right\}$; $\bar{T} = \left\{\bar{T}_1,\ldots,\bar{T}_S\right\}$

with $\bar{N}_j$ of cardinal N_j , $j=1,\ldots,J$ and $\bar{T}_s$ of cardinal T_s , $s=1,\ldots,S$. Practically, the cross-sectional units are classified into J groups $1,\ldots,J$, the j-th group consisting of N_j units, $j=1,\ldots,J$; and the time periods into S different intervals $1,\ldots,S$, the s-th interval comprising T_s time periods, $s=1,\ldots,S$. Naturally $\sum_j N_j = N$, $\sum_s T_s = T$. Once this is done, the model can be written as :

(2.49) $y_{it} = \beta' x_{it} + u_j + v_s + w_{it}$

for $i \in \bar{N}_j$, $t \in \bar{T}_s$; $i=1,\ldots,N$, $t=1,\ldots,T$.

The assumptions concerning the group effects u_j , the interval effects v_s and the residual effects w_{it} are :

(2.50) $E(u_j) = E(v_s) = E(w_{it}) = 0$

(2.51) $E(u_j\, u_{j'}) = \delta_{jj'}\, \sigma^2_{u_j}$

(2.52) $E(v_s\, v_{s'}) = \delta_{ss'}\, \sigma^2_{v_s}$

(2.53) $E(w_{it}\, w_{i't'}) = \delta_{ii'}\, \delta_{tt'}\, \sigma^2_w$

Thus, it is assumed that the unit effects are common to all units belonging to the same group and their common variance differs from that of other groups. Similar interpretations can be made concerning the time effects and their variability.

Let us denote by y_{it}^{js} the t-th observation on y belonging to the s-th time interval for the i-th unit belonging to the j-th group and similarly define $x_{it}^{js} = [x_{1it}^{js} \ \dots \ x_{Kit}^{js}]'$. Then by writing :

$$
(2.54)\left\{
\begin{aligned}
&\underset{(N_jT_s\times 1)}{y^{js}} = [y_{11}^{js}\dots y_{1T_s}^{js}\dots y_{N_jT_s}^{js}]' \ ; \quad \underset{(N_jT_s\times K)}{X^{js}} = [x_{11}^{js}\dots x_{N_jT_s}^{js}]' \\
&\underset{(NT\times 1)}{y} = [y^{11}\dots y^{1S}\dots y^{JS}]' \ ; \quad \underset{(NT\times K)}{X} = [X^{11}\dots X^{1S}\dots X^{JS}]' \\
&\underset{(J\times 1)}{u} = [u_1 \ \dots \ u_J]' \ ; \quad \underset{(S\times 1)}{v} = [v_1 \ \dots \ v_S]' \\
&\underset{(N_jT_s\times 1)}{w^{js}} = [w_{11}^{js}\dots w_{1T_s}^{js}\dots w_{N_jT_s}^{js}]' ; \quad \underset{(NT\times 1)}{w} = [w^{11}\dots w^{1S}\dots w^{JS}]'
\end{aligned}
\right.
$$

$$
(2.55) \qquad \underset{(N\times J)}{D_1} = \begin{bmatrix} \iota_{N_1} & & \bigcirc \\ & \ddots & \\ \bigcirc & & \iota_{N_J} \end{bmatrix} ; \quad \underset{(T\times S)}{D_2} = \begin{bmatrix} \iota_{T_1} & & \bigcirc \\ & \ddots & \\ \bigcirc & & \iota_{T_S} \end{bmatrix}
$$

the model can be expressed as :

$$(2.56) \qquad y = X\beta + (D_1 \otimes \iota_T)\, u + (\iota_N \otimes D_2)\, v + w$$

with

$$(2.57) \qquad E(\varepsilon\varepsilon') = \Sigma = (D_1 \otimes \iota_T)\Sigma_u(D_1 \otimes \iota_T')+(\iota_N \otimes D_2)\Sigma_v(\iota_N \otimes D_2')+\sigma_w^2 I_{NT}$$

where

$$(2.58) \qquad \varepsilon = (D_1 \otimes \iota_T)\, u + (\iota_N \otimes D_2)\, v + w$$

$$(2.59) \qquad \Sigma_u = \text{diag}\,(\sigma_{u_1}^2 \ \dots \ \sigma_{u_J}^2)$$

$$(2.60) \qquad \Sigma_v = \text{diag}\,(\sigma_{v_1}^2 \ \dots \ \sigma_{v_S}^2)$$

Hence the GLS estimator of β is given by

$$(2.61) \qquad \hat{\beta} = (X'\Sigma^{-1}X)^{-1}\, X'\Sigma^{-1}y$$

2.2.3 Random Effects Models with Non-Zero Correlations between Specific Effects and Exogenous Variables

Mundlak [30] introduces non-zero correlation between the specific effects and the quantitative explanatory variables of an EC model and takes account of such dependence by an auxiliary regression of the effect on the explanatory variables. Let us first rewrite the EC model with no time effect :

(2.62) $y = X \beta + (I_N \otimes \iota_T)\, u + w$

where

(2.63) $\underset{(N\times 1)}{u} = [u_1 \; \ldots \; u_N]'$ and $\underset{(NT\times 1)}{w} = [w_{11} \; \ldots \; w_{1T} \; \ldots \; w_{NT}]'$

The auxiliary regression of u on X is specified as follows :

(2.64) $u_i = x_{it}\, \pi + \xi_{it}$, $t=1,\ldots,T$; $i=1,\ldots,N$

(2.65) $u_i = x_{i.}\, \pi + \xi_{i.}$

It is assumed that $\xi_{i.}$ $(0,\lambda^2)$. By writing $L = I_N \otimes \iota_T$, we have, from (2.65),

(2.66) $Lu = L(L'L)^{-1} L'(X\pi + \xi) = K(X\pi + \xi)$

where

$$\xi = [\xi_{11} \; \ldots \; \xi_{1T} \; \ldots \; \xi_{NT}]' \; ; \; K = L(L'L)^{-1}L'$$

Replacing (2.66) in (2.62) yields

(2.67) $y = X\beta + K(X\pi + \xi) + w$

with

(2.68) $(K\xi + w) \sim (0, (\sigma^2 I_{NT} + T\lambda^2 K) = \Sigma)$

The GLS estimators of β and π are obtained as follows :

(2.69) $$\begin{pmatrix} \hat{\beta} \\ \hat{\pi} \end{pmatrix} = \left[\begin{pmatrix} X' \\ X'K \end{pmatrix} \Sigma^{-1} (X \;\; KX) \right]^{-1} \begin{pmatrix} X' \\ X'K \end{pmatrix} \Sigma^{-1} y$$

After a few simplifications, $\hat{\beta}$ and $\hat{\pi}$ can be expressed as

(2.70) $$\begin{cases} \hat{\beta} = (X'MX)^{-1} X'My \\ \hat{\pi} = (X'KX)^{-1} X'K\, y - (X'MX)^{-1} X'My \end{cases}$$

where $M = I-K$.

It can be noted that $\hat{\beta}$ is also the within group estimator of β in model (2.62). This result is due to Mundlak.

Hausmann and Taylor [18] consider a similar extension to EC models and assume that the individual effects are correlated with certain explanatory variables only and further that the model includes time-invariant observable exogenous variables. In other words, their model is represented by :

$$(2.71) \quad y_{it} = x'_{it}\beta + z'_i\gamma + u_i + w_{it} = x'_{it}\beta + z'_i\gamma + \varepsilon_{it}$$

where $\underset{(1xg)}{z'_i} = [z_{1i} \dots z_{gi}]$ is the vector of observations on the time-invariant exogenous variables and $\varepsilon_{it} = u_i + w_{it}$. Using appropriate notations, the model can be written in matrix form as :

$$(2.72) \quad y = X\,\beta + Z\,\gamma + (I_N \otimes \iota_T)\,u + w = X\,\beta + Z\,\gamma + \varepsilon$$

The stochastic assumptions are :

$$(2.73) \quad E(\varepsilon) = 0\ ,\ E(\varepsilon \mid X,Z) = E(u \mid X,Z) \neq 0$$

$$(2.74) \quad E(w\,w') = \sigma_w^2\, I_{NT}$$

$$(2.75) \quad E(u\,u') = \sigma_u^2\, I_N$$

Further, as mentioned above, the model assumes prior information on which columns of X and Z are asymptotically uncorrelated with u and which are correlated. For fixed T, let

$$(2.76) \quad \begin{cases} \underset{N\to\infty}{\text{plim}} \frac{1}{N} X'_a(I_N \otimes u) = 0\ , & \underset{N\to\infty}{\text{plim}} \frac{1}{N} Z'_a(I_N \otimes u) = 0 \\ \underset{N\to\infty}{\text{plim}} \frac{1}{N} X'_b(I_N \otimes u) = h_x\ , & \underset{N\to\infty}{\text{plim}} \frac{1}{N} Z'_b(I_N \otimes u) = h_z \end{cases}$$

where X'_a contains K_1 variables, X'_b has K_2 variables, Z'_a, g_1 variables and Z'_b, g_2 variables and $h_x,\ h_z \neq 0$.

The within group estimator of β in this model is given by:

$$(2.77) \quad \hat{\beta}_W = (X'Q_vX)^{-1}\,X'Q_vy$$

where

$$(2.78) \quad Q_v = I_{NT} - P_v \quad ; \quad P_v = (I_N \otimes \frac{1}{T}\,\iota_T\,\iota'_T)$$

This estimator is unbiased and consistent as $Q_V X$ is uncorrelated with $Q_V U$ regardless of possible correlation between u_i and x_{it} or z_i .

The between group estimators $\hat{\beta}_B$, $\hat{\gamma}_B$ are obtained by transforming the model by P_V to yield

$$(2.79) \quad y_{i.} = x_{i.} \beta + z_i \gamma + u_i + w_{i.}$$

and performing OLS on the transformed model. These estimators are biased and inconsistent as $E(u_i \mid x_{it}, z_i) \neq 0$.

Thirdly, the GLS estimator is a weighted average of the between group and within group estimators and can be written as follows :

$$(2.80) \quad \begin{pmatrix} \hat{\beta}_{GLS} \\ \hat{\gamma}_{GLS} \end{pmatrix} = \tilde{\Delta} \begin{pmatrix} \hat{\beta}_B \\ \hat{\gamma}_B \end{pmatrix} + (I-\tilde{\Delta}) \begin{pmatrix} \hat{\beta}_W \\ 0 \end{pmatrix}$$

where

$$(2.81) \quad \tilde{\Delta} = (V_B + V_W)^{-1} V_W$$

and V_B and V_W are the covariance matrices of between and within group estimators of β and γ .

Hausmann and Taylor also develop specification tests for testing the hypothesis of non-zero correlation between u and (X,Z). The above three classical estimators of β lead to three different specification tests - GLS versus within, GLS versus between and within versus between.

They note that even if the null hypothesis of no correlation between u and (X,Z) is rejected, it is possible to obtain consistent estimates of both β and γ from within group regression.

Let

$$(2.82) \quad \hat{d} = P_V(y - X \hat{\beta}_W)$$

be the vector of group means estimated from within group residuals. The estimator of γ is given by

$$(2.83) \quad \hat{\gamma}_W = (Z'P_A Z)^{-1} Z'P_A \hat{d}$$

where

$$A = [X_a \quad Z_a]$$

and P_A is the orthogonal projection operator onto its column space.

Finally an instrumental variables estimation of (β,γ) is presented and is as follows. Transform (2.72) by $\Omega^{-\frac{1}{2}}$ to get :

$$(2.84) \qquad \Omega^{-\frac{1}{2}} y = \Omega^{-\frac{1}{2}} X \beta + \Omega^{-\frac{1}{2}} Z \gamma + \Omega^{-\frac{1}{2}} \varepsilon$$

If Ω were known, 2SLS estimates of (β,γ) in (2.84), taking X_a and Z_a as exogenous, would be asymptotically efficient. Also the 2SLS of (β,γ) in (2.84) is equivalent to OLS of (β,γ) in

$$(2.85) \qquad P_A\Omega^{-\frac{1}{2}} y = P_A\Omega^{-\frac{1}{2}} X \beta + P_A\Omega^{-\frac{1}{2}} Z \gamma + P_A\Omega^{-\frac{1}{2}} \varepsilon$$

where P_A is the orthogonal projector operator on to the column space of the instruments $A = [Q_V \quad X_a \quad Z_a]$.

2.2.4 Dynamic Random Effects Models

The reader may recall that, in Section 2.2.2, we had a quick glimpse of a dynamic error component model, namely that of Balestra and Nerlove [6] . Further to this study, Nerlove explored the methodological issues involved in the estimation of dynamic error component models, by means of Monte Carlo experiments (see [34], [35]). In a first series of experiments ([34]), he considered the Balestra-Nerlove model with only the lagged value of the endogenous variable as the explanatory variable, whose coefficient is denoted as α . In a later work ([35]), he introduced an exogenous variable (with coefficient β) in his model and compared different methods of estimation on the basis of small sample properties of their estimates.

A major conclusion of the second set of experiments is that, as far as both the bias and the mean square error are concerned, the two-round estimation procedure (with a first round regression including individual constant terms) yields better results than all the other methods over a wide range of parameter values. The pure GLS also performs well with only a slight bias in the estimation of α . OLS estimates of α are strongly biased upwards but have smaller mean square errors than GLS estimates in certain cases. The instrumental variable estimates behave rather erratically and the performance of the ML estimates is also poor. Another pertinent result is the non-negligible occurrence of boundary solutions in the ML method which, according to Nerlove, is linked to the serial properties of the exogenous variable.

In [52] , Trognon analyses dynamic error component models from the point of view of asymptotic properties of OLS and ML estimators and compares his results with Nerlove's experimental findings. He concludes that his asymptotic results are in perfect agreement with those of Nerlove in the non-explosive pure autoregressive case. He also proves that OLS is consistent in the explosive case. Upon the introduction of an exogenous variable, the bias of the OLS is reduced.

Trognon also concludes that the autocorrelation structure of the exogenous process is a determinant of the existence of boundary solution in the maximum likelihood method. In addition, he derives conditions on parameters for which ML estimation is reduced to OLS estimation.

Now, we turn to the following dynamic random effects model developed by Anderson and Hsiao [2] :

(2.86) $$y_{it} = \beta\, y_{i,t-1} + \delta' z_i + \gamma' x_{it} + v_{it}$$

where z_i represents the $(K_1 X 1)$ vector of time-invariant exogenous variables and x_{it} is the $(K_2 X 1)$ vector of observations on time-varying exogenous variables. The assumptions of the model are :

(2.87) $$v_{it} = \alpha_i + u_{it}$$

(2.88) $$E(\alpha_i) = E(u_{it}) = 0$$

(2.89) $$E(\alpha_i z_i') = E(\alpha_i\, x_{it}') = 0'$$

(2.90) $$E(\alpha_i\, u_{jt}) = 0$$

(2.91) $$E(\alpha_i\, \alpha_j) = \delta_{ij}\, \sigma^2$$

(2.92) $E(u_{it}\ u_{js}) = \delta_{ij}\ \delta_{ts}\ \sigma_u^2 = \delta_{ij}\ \delta_{ts}\ \lambda\ \sigma^2$

In this model the assumption about the initial observations plays a crucial role in interpreting the model and devising consistent estimates. Models of the form (2.86) can be rewritten in two ways :

I. serial correlation model :

(2.93) $$w_{it} = ß\ w_{i,t-1} + u_{it}$$
$$y_{it} = w_{it} + \tilde{\delta}'z_i + \gamma' x_{it} + \eta_i$$

II. state dependence model :

(2.94) $$w_{it} = ß\ w_{i,t-1} + \delta'\ z_i + \gamma' x_{it} + u_{it}$$
$$y_{it} = w_{it} + \eta_i$$

where $\tilde{\delta} = \delta/(1-ß)$, $\eta_i = \alpha_i/(1-ß)$; w_{it}, η_i and u_{it} are unobservable. Equations (2.93) imply :

(2.95) $$y_{it} = ß\ y_{i,t-1} + \delta' z_i + \gamma' x_{it} - ß\ \gamma' x_{i,t-1} + \alpha_i + u_{it}$$

and equations (2.94) imply

(2.96) $$y_{it} = ß\ y_{i,t-1} + \delta' z_i + \gamma' x_{it} + \alpha_i + u_{it}$$

In either case the unobservable variables w_{it} are serially correlated; the impact of a disturbance u_{it} tends to persist for more than one period. In the serial correlation model y_{it} is affected by x_{it} only , not by $x_{i,t-1}$ ... whereas in the state dependence model y_{it} is affected not only by x_{it} but also by $x_{i,t-1}$ ... through $y_{i,t-1}$; the unobservable w_{it} carry the effected x and can be called a state. Inclusion of the time invariant individual effect η_i implies that the aggregate effect of unobservable variables (commonly called "residual") are serially correlated in another way.

With different assumptions about the initial observations, we obtain different models. Concerning the serial correlation model, two assumptions are possible - either w_{io} is fixed or w_{io} is random and stationary with mean zero and variance $\sigma_u^2/(1-ß^2)$. Let us summarize the main results on estimation of the model under these two assumptions.

When w_{io} is fixed, the MLE does not exist as the likelihood function is unbounded as $\sigma_{\eta}^{2} \to 0$. But Anderson and Hsiao go on to find the partial derivatives of the joint likelihood function, setting them equal to zero and suggesting an iterative procedure to solve the resulting system of equations. It is shown that $\hat{\gamma}$ is consistent, the solution for β is consistent when N is fixed and $T \to \infty$ and inconsistent when T is fixed and $N \to \infty$ and the solutions for $w_{io}, \sigma_u^2, \sigma_{\eta}^2$ are inconsistent whether N or $T \to \infty$.

When w_{io} is $N(0, \lambda\sigma^2/1-\beta^2))$, the MLE is consistent when T or N tends to infinity.

Now let us look at the various possibilities in state dependence models :

Case I : y_{io} fixed. In this case a cross-sectional unit may start at some arbitrary position y_{io} and gradually move towards a level determined by $\eta_i + \gamma' \sum_{j=0}^{t-1} x_{i,t-j} \beta^j + \tilde{\delta}' z_i$. The individual effect from η_i does not affect the initial y_{io} but affects all later y_{it}'s .

Case II : y_{io} random and normally distributed with mean μ_{yo} and variance σ_{yo}^2 . Two sub-cases are considered :

IIa . y_{io} independent of η_i

IIb . y_{io} correlated with η_i , the covariance being denoted by $\rho\sigma_{yo}^2$

Case III : w_{io} fixed. The unobserved individual process w_{it} is independent of the individual effect η_i with the starting value arbitrary. Or each of the observed cross-sectional units may start at some arbitrary position y_{io} and gradually move towards a level of $\eta_i + \tilde{\delta}' z_i + \gamma' x_{it} + \beta\gamma' x_{i,t-1} + \beta^2 \gamma' x_{i,t-2} + \ldots$ with η_i affecting y_{io}.

Case IV : w_{io} random. Given that the process $\{w_{it}\}$ and η_i are independent, there are four ways of formulating the initial state w_{io} :

IVa : w_{io} random with common mean μ_w and variance $\sigma_u^2 / (1-\beta^2)$

IVb : w_{io} random with common mean μ_w and variance σ_{wo}^2

IVc : w_{io} random with mean θ_{io} and variance $\sigma_u^2/(1-\beta^2)$

IVd : w_{io} random with mean θ_{io} and variance σ_{wo}^2

IVa and IVb assume that the initial state is a random draw from a distribution with finite mean. IVa assumes that the initial state has the same variance as the later states. IVb allows the initial state to be non-stationary. IVc and IVd assume that the individual states are random draws from different populations with different means.

The consistency results of MLE of the four different models are summarized as follows :

Case		Parameters	N fixed T->∞	T fixed N->∞
I		β,γ	consistent	consistent
y_{io} fixed		λ,σ^2	inconsistent	consistent
II	a)	β,γ	consistent	consistent
y_{io} random		$\mu_{yo},\lambda,\sigma^2,\sigma_{yo}^2$	inconsistent	consistent
	b)	β,γ	consistent	consistent
		$\mu_{yo},\lambda,\sigma^2,\sigma_{yo}^2,\rho$	inconsistent	inconsistent
III		β,γ	consistent	inconsistent
w_{io} fixed		$w_{io},\sigma_u^2,\sigma_\eta^2$	inconsistent	inconsistent
IV	a)	β,γ	consistent	consistent
w_{io} random		$\mu_w,\sigma_u^2,\sigma_\eta^2$	inconsistent	consistent
	b)	β,γ	consistent	consistent
		$\sigma_{wo}^2,\sigma_u^2,\sigma_\eta^2,\mu_w$	inconsistent	consistent
	c)	β,γ	consistent	inconsistent
		$\theta_{io},\sigma_u^2,\sigma_\eta^2$	inconsistent	inconsistent
	d)	β,γ	consistent	inconsistent
		$\theta_{io},\sigma_u^2,\sigma_\eta^2,\sigma_{wo}^2$	inconsistent	inconsistent

Finally a simple consistent four-step estimation method which is independent of the initial condition is described.

In fact, in this article [2] Anderson and Hsiao discuss dynamic models with time-invariant exogenous variables only and those with time-varying exogenous variables only before taking up the general case of models including both types of exogenous variables. We have limited ourselves to the general case only in our presentation.

2.3 Variable Coefficient Models

2.3.1 Fixed Coefficient Component Models

In this model, each coefficient of the regression equation is supposed to be composed of three fixed components as follows :

(2.97) $$\beta_{kit} = \bar{\beta}_k + \delta_{ki} + \gamma_{kt}$$

The model can therefore be written as :

(2.98) $$y_{it} = \sum_k (\bar{\beta}_k + \delta_{ki} + \gamma_{kt})\, x_{kit} + \varepsilon_{it}$$

for $t=1,\ldots,T$, $i=1,\ldots,N$.

The errors ε_{it} are assumed to be independently and identically distributed with zero mean and variance σ^2 .

The BLUE of coefficient components is obviously the OLS estimator. But, as such the model is not identified. Additional restrictions have to be imposed on the coefficient components to overcome the identification problem. Then, they can be estimated by OLS subject to the conditions imposed.

A generalisation of the above model, the stratified fixed coefficient components model, is proposed by Mazodier and Trognon [29] . The stratification is carried out as described in Section 2.2.2.3 . The model is defined by :

(2.99) $$y_{it} = \sum_k (\bar{\beta}_k + \gamma_{ks} + \delta_{kj})\, x_{kit} + \varepsilon_{it}$$

for $i \in \bar{N}_j$, $t \in \bar{T}_s$ and $t=1,\ldots,T_s$, $i=1,\ldots,N_j$, $j=1,\ldots,J$ and $s=1,\ldots,S$. Thus the specific effects are the same for all units belonging to the same group and for all time periods belonging to the same interval. Here again identification problem arises and Mazodier and Trognon have examined it in detail in their paper [29] . The BLUE is once again OLSE subject to identification restrictions.

2.3.2 Random Coefficient Models

2.3.2.1 Purely Random Coefficient Components Model

In this model, the specific effects present as coefficient components in the above model (2.98) are assumed to be random variables. In other words, the coefficients have common means plus random components with zero means to account for temporal and cross-sectional heterogeneity. Thus we have

$$(2.100)\quad y_{it} = \sum_k (\bar{\beta}_k + \gamma_{kt} + \delta_{ki})\, x_{kit} + \varepsilon_{it}$$

with

$$(2.101)\quad E(\gamma_{kt}) = E(\delta_{ki}) = E(\varepsilon_{it}) = 0$$

$$(2.102)\quad E(\delta_{ki}\,\delta_{k'i'}) = \delta_{ii'}\,\delta_{kk'}\,\sigma^2_{\delta_k}$$

$$(2.103)\quad E(\gamma_{kt}\,\gamma_{k't'}) = \delta_{tt'}\,\delta_{kk'}\,\sigma^2_{\gamma_k}$$

$$(2.104)\quad E(\varepsilon_{it}\,\varepsilon_{i't'}) = \delta_{ii'}\,\delta_{tt'}\,\sigma^2$$

$$(2.105)\quad E(\delta_{ki}\,\gamma_{k't}) = E(\delta_{ki}\,\varepsilon_{i't}) = E(\gamma_{kt}\,\varepsilon_{it'}) = 0$$

for $i=1,\dots,N$; $t=1,\dots,T$ and $k=1,\dots,K$.

By introducing the following notations :

$$\underset{(NTx1)}{y} = [y_{11} \cdots y_{1T} \cdots y_{NT}]' \; ; \quad \underset{(Kx1)}{x_{it}} = [x_{1it} \cdots x_{Kit}]'$$

$$\underset{(NTxK)}{X} = [x_{11} \cdots x_{1T} \cdots x_{NT}]' \; ; \quad \underset{(TxK)}{x_i} = [x_{i1} \cdots x_{iT}]'$$

$$\underset{(NTxNK)}{\underset{\sim}{X}} = \begin{bmatrix} X_1 & & 0 \\ & \ddots & \\ 0 & & X_N \end{bmatrix} \; ; \quad \underset{(NTxTK)}{\tilde{X}} = \begin{bmatrix} x_{11} & & 0 \\ & \ddots & \\ 0 & & x_{1T} \\ & \vdots & \\ x_{N1} & & 0 \\ & \ddots & \\ 0 & & x_{NT} \end{bmatrix},$$

$$\underset{(NTx1)}{\varepsilon} = [\varepsilon_{11} \cdots \varepsilon_{1T} \cdots \varepsilon_{NT}]' \; ; \; \underset{(Kx1)}{\bar{\beta}} = [\bar{\beta}_1 \; \cdots \; \bar{\beta}_K]'$$

$$\underset{(KNx1)}{\delta} = [\delta_{11} \cdots \delta_{1K} \cdots \delta_{NK}]' \; ; \; \underset{(KTx1)}{\gamma} = [\gamma_{11} \cdots \gamma_{1K} \cdots \gamma_{TK}]'$$

$$\Delta = \text{diag}(\sigma^2_{\delta_1}, \ldots, \sigma^2_{\delta_K}) \; ; \; \Gamma = \text{diag}(\sigma^2_{\gamma_1}, \ldots, \sigma^2_{\gamma_K})$$

the model can be written in matrix form as:

$$(2.106) \quad y = X\bar{\beta} + \underset{\sim}{X}\delta + \tilde{X}\gamma + \varepsilon = X\bar{\beta} + u \text{ where } u = \underset{\sim}{X}\delta + \tilde{X}\gamma + \varepsilon$$

with the following variance-covariance matrix of errors :

$$(2.107) \quad E(uu') = \Sigma = \underset{\sim}{X}(I_N \otimes \Delta)\underset{\sim}{X}' + \tilde{X}(I_T \otimes \Gamma)\tilde{X}' + \sigma^2 I_{NT}$$

The GLS estimator of $\bar{\beta}$ is given by :

$$(2.108) \quad \hat{\bar{\beta}} = (X'\Sigma^{-1}X)^{-1} X'\Sigma^{-1}y$$

This model has been considered in detail by Hsiao [21]. Two other estimation methods, viz., the minimum norm quadratic unbiased estimation (MINQUE, due to Rao [40] , [41]) and the MLE have been described in this article. Swamy [46] considers a random coefficient model with no time component and with the disturbances having different variances for different units. Swamy and Mehta [49] have also analyzed this model in a slightly different form with the coefficient vector being composed of a constant mean, a random individual component and a random overall component variable both over time and over units. We will just write down their model for future reference :

$$(2.109) \quad y_{it} = \sum_{\ell=1}^{K} x_{\ell it} \beta_{\ell it}$$

where

$$(2.110) \quad \beta_{\ell it} = \bar{\beta}_\ell + \alpha_{\ell i} + \xi_{\ell it}$$

with the following stochastic assumptions :

Denoting :

$$\underset{(Kx1)}{\alpha_i} = [\alpha_{1i} \; \ldots \; \alpha_{Ki}]' ; \quad \underset{(Kx1)}{\xi_{it}} = [\xi_{1it} \; \ldots \xi_{Kit}]' ;$$

$$\underset{(Kx1)}{\bar{\beta}} = [\bar{\beta}_1 \; \ldots \; \bar{\beta}_K]' ,$$

we have :

$$(2.111) \begin{cases} E(\alpha_i) = 0 \quad ; \; E(\alpha_i \;\alpha_j') = \delta_{ij} \;\Delta \\ E(\xi_{it}) = 0 \; ; \; E(\xi_{it} \;\xi_{js}') = \delta_{ij} \;\delta_{ts} \;\Delta_{ii} \\ E(\alpha_i \;\xi_{i't}') = 0 \end{cases}$$

The reader is referred to [49] for a description of the estimation procedures for this model.

2.3.2.2 Transmitted Variations Model

This model, due to Mundlak [31], is a variable coefficient model in which the coefficient vector varies over individuals and is correlated with the explanatory variables. This dependence is taken into account by introducing K auxiliary regressions of coefficients on the exogenous variables. The model is specified as follows :

$$(2.112) \quad y_{it} = \underline{x}_{it}' \;\underline{\beta}_i + \varepsilon_{it}$$

with

$$(2.113) \quad \beta_{ji} = \underline{X}_{i.}' \;\underline{\gamma}_j + w_{ji}$$

where

$$\underline{\gamma}_j' = [\bar{\beta}_j \quad \underline{\pi}_j']$$

and

$$\underline{X}_{i.}' = [x_{1i.} \; \ldots \; x_{Ki.}] \; ; \; x_{ji.} = \frac{1}{T} \sum_t x_{jit}$$

It is assumed that

$$(2.114) \quad \underline{W}_i \sim (0,\Omega) \quad \text{and} \quad E(\underline{W}_i \;\underline{W}_{i'}') = 0 \quad \text{for } i \neq i'$$

where

$$\underline{W}_i' = (w_{1i} \; \ldots \; w_{Ki})$$

and

$$(2.115) \quad \varepsilon_i \sim (0,\sigma_i^2 \;I_T) \text{ and } E(\varepsilon_i \;\varepsilon_{i'}') = 0 \qquad \text{for } i \neq i'$$

where

$$\varepsilon_i' = [\varepsilon_{i1} \ \cdots \ \varepsilon_{iT}]$$

It can be verified that the variance of y has a block diagonal structure. Mundlak points out that, in this case, if the GLSE of $\underline{\gamma} = [\underline{\gamma}_1' \ \cdots \ \underline{\gamma}_K']'$ exists, it is a matrix weighted average of the estimators $\hat{\underline{\gamma}}_{gi}$ where such estimators are obtained by any solution of the GLS normal equations of each cross-sectional unit and that the optimality of the estimator depends on the choice of matrix weights. The reader is referred to [31] for further details.

2.3.2.3 Random Coefficient Models with Stochastically Convergent Parameters

In this model, developed by Rosenberg [45], coefficients of certain independent variables vary only over time and coefficients of the other independent variables vary both across units and over time; moreover, the variation of all the coefficients follows a defined stochastic process. The model is :

$$(2.116) \quad y_{it} = \sum_{j=1}^{K_1} w_{jit} c_{jt} + \sum_{j=1}^{K_2} z_{jit} a_{jit} + u_{it}$$

where

$$(2.117) \quad E(u_{it}) = 0 \ ; \ E(u_{is} u_{jt}) = \delta_{st} \sigma^2(\delta_{ij} R_i + R_G)$$

$$(2.118) \quad \begin{cases} c_{t+1} = c_t + \gamma_t \\ a_{i,t+1} = \bar{a}_t + \Delta_\phi(a_{it} - \bar{a}_t) + \eta_{it} \end{cases}$$

$$(2.119) \quad E(\gamma_t) = 0 \ ; \ E(\gamma_s \gamma_t') = \delta_{st} \sigma^2 Q_c$$

$$(2.120) \quad E(\eta_{it}) = 0 \ ; \ E(\eta_{is} \eta_{jt}') = \delta_{st} \sigma^2(\delta_{ij} Q_a + Q_G)$$

$$(2.121) \quad E(u_{it}\gamma_t') = 0 \ ; \ E(u_{is}\eta_{jt}') = 0 \ ; \ E(\gamma_s \eta_{it}') = \delta_{st} \sigma^2 Q_{ca}$$

with $\underset{(K_1 x 1)}{c_t} = [c_{1t} \ \cdots \ c_{K_1 t}]'$; $\underset{(K_2 x 1)}{a_{it}} = [a_{1it} \ \cdots \ a_{K_2 it}]'$;

$$\underset{(K_2 x K_2)}{\Delta_\phi} = \text{diag}(\phi_i)$$

Assuming that the stochastic terms follow a normal distribution, the unknown parameters are estimated by both MLE and Bayesian analysis. In both the methods, recursive formulae are arrived at, that yield (i) for any set of $R_1,\dots,R_N,R_G,Q_c,Q_{ca},Q_a,Q_G,\phi_1,\dots,\phi_{K_2}$ (let θ denote any such set), numerical values of sample likelihood $\mathcal{L}(\theta|\,y^T)$ 1) and the marginal posterior distribution of θ, $p''(\theta|y^T)$; (ii) ML estimators $\hat{\beta}_{T/T}(\theta)$ 2) and $\hat{\sigma}^2_{ML}(\theta)$ and conditional posterior distributions $p''(\sigma^2|\theta,y^T)$ and $p''(\beta^T|\theta,y^T)$ 1), conditional on that θ. Repeated application of the recursive formulae over θ until convergence gives the respective estimates. Rosenberg also develops approximate formulae of estimation and examines the statistical efficiency and validity of approximations.

2.3.2.4 Random Coefficient First-Order Autoregressive Model

In this model, Lon Mu Liu and Tiao [25] analyze a first-order autoregressive model with pooled data. The equation under study is :

(2.122) $$Z_{it} = \phi_i\, Z_{i(t-1)} + a_{it}$$

for $i=1,\dots,m$, $t=1,2,\dots,n$; where $|\phi_i| < 1$ and the a_{it}'s are i.i.d. $N(0,\sigma_a^2)$. The coefficients ϕ_i, $i=1,\dots,m$ are assumed to be independent drawings from a rescaled beta distribution :

(2.123) $$p(\phi|p,q) = \prod_{i=1}^{m} p(\phi_i|p,q)$$

where

$$\underset{(m\times 1)}{\phi} = [\phi_1,\phi_2 \dots \phi_m]'$$

and

(2.124) $$p(\phi_i|p,q) = \frac{1}{2^{p+q-1}\,B(p,q)}\,(1+\phi_i)^{p-1}\,(1-\phi_i)^{q-1},$$

$$|\phi_i| < 1 \;;\; p,q \geq 1$$

---

1) y^T is the vector $[y_1' \dots y_T']'$ of all observations through period T ; similarly for β^T with $\beta_t = (c'a_1' \dots a_n')_t'$ for $t=1,\dots,T$.

2) $\hat{\beta}_{T|T}$ denotes the estimator of β^T conditional on regression information upto and including period T.

The estimation of ϕ is considered both when p,q,σ_a^2 are known and when they are variable parameters. Estimation is carried out using Bayesian approach. In the first case, combining $p(\phi_i|p,q)$ and the likelihood function of ϕ and σ_a^2 gives the posterior distribution of ϕ_i given p,q,σ_a^2.

In the second case, a prior distribution is assumed for u,v where $u = \frac{p-q}{p+q-1}$, $v=p+q-1$. Combining this prior with the sample Z gives the posterior of (u,v), $p(u,v|Z)$. Then the joint posterior distribution of ϕ is written as

$$(2.125) \quad p(\phi|Z) = \int_1^{\infty}\int_{v^{-1}-1}^{1-v^{-1}} p(\phi|u,v,Z).\ p(u,v|Z)\ du\ dv$$

where $p(\phi|u,v,Z)$ is obtained by combining $p(p,q,\sigma_a^2|Z)$ with $p(\phi_i|p,q,\sigma_a^2,Z)$ and integrating out σ_a^2. $p(p,q,\sigma_a^2|Z)$ is in turn given by $p(p,q)$ (prior of (p,q)) and the likelihood function of (p,q,σ_a^2) while $p(\phi_i|p,q,\sigma_a^2,Z)$ is the posterior distribution of ϕ_i given p,q,σ_a^2 .

2.3.3 The "Mixed" Model

The term "mixed" refers to the case in which, in a regression equation, some coefficients are assumed to be constant over all units and others assumed to vary over units. Both types of coefficients are assumed to be time-invariant (see Mundlak [31]) : Thus for the i-th unit we have :

$$(2.126) \quad y_i = X_i^r\, \beta^r + Z_i\, \beta_i^p + \varepsilon_i \quad , \quad i=1,\ldots,N$$

where

$$\underset{(T\times 1)}{y_i} = [y_{i1} \cdots y_{iT}]' \; ; \qquad \underset{(T\times r)}{X_i^r} = \begin{bmatrix} x_{i11} & \cdots & x_{ir1} \\ \vdots & & \vdots \\ x_{i1T} & \cdots & x_{irT} \end{bmatrix}$$

$$\underset{(T\times p)}{Z_i} = \begin{bmatrix} z_{i11} & \cdots & z_{ip1} \\ \vdots & & \vdots \\ z_{i1T} & \cdots & z_{ipT} \end{bmatrix} \; ; \qquad \underset{(T\times 1)}{\varepsilon_i} = [\varepsilon_{i1} \cdots \varepsilon_{iT}]'$$

$$\underset{(r\times 1)}{\beta^r} = [\beta_1^r \ldots \beta_r^r]' \; ; \qquad \underset{(p\times 1)}{\beta^p} = [\beta_1^p \ldots \beta_p^p]'$$

$$p + r = K$$

Auxiliary regressions are introduced to take account of correlations between the individual coefficients and the explanatory variables - there are p of them :

$$(2.127) \quad \beta_i^p = A_i^p \gamma^p + W_i^p \qquad i=1,\dots,p$$

where $\quad A_i^p = I_p \otimes X_i' \; ; \; \gamma^p = [\gamma_1' \dots \gamma_p']' \; ; \; \gamma_j' = [\beta_j' \; \pi_j']$

It is assumed that

$$(2.128) \quad \varepsilon_i \sim (o, \sigma_i^2 I_T) \text{ and } W_i^p \sim (0, \Omega_p)$$

First β^r is estimated by performing OLS on a transformed version of the transmitted model[1)] and using this, GLS is applied to the same model to estimate γ^p . The reader is invited to consult [31] for a detailed treatment of this model.

2.3.4 Quantitative Effects Model

This model, also termed as generalised EC model, is due to Wansbeek [56]. It is defined as follows :

$$(2.129) \quad y_{it} = \underbrace{\beta' x_{it}}_{A} + \underbrace{\sum_{j=1}^{p} \alpha_{1tj} Z_{1ij}}_{B} + \underbrace{\sum_{j=1}^{q} \alpha_{2ij} Z_{2tj}}_{C} + \underbrace{\varepsilon_{it}}_{D}$$

Four types of relationship are distinguished in this equation:

(i) the invariant component (A)

(ii) a systematic change : a relationship changing over time for all units in the same way (B)

(iii) an individual variation : a relationship constant over time but differing among units (C)

(iv) a relationship represented by a purely random disturbance term varying both over time and units (D).

---

1) Equations (2.126), (2.127) and (2.128) constitute the transmitted model.

In the fixed effects case (i.e. when α_{1tj} and α_{2ij} are assumed to be fixed parameters for i=1,...,N, t=1,...,T), OLS is applied taking into account the special features of this particular model namely perfect multi-collinearity, computational complexity and problems posed by the fact that the number of regressors increase with the size of the sample.

In the random effects case, α_{1tj}'s are assumed to have a common mean $\bar{\alpha}_{1j}$ (and $\bar{\alpha}_{2j}$ for α_{2ij}'s) plus a random component ξ_{1tj} (and ξ_{2ij} for α_{2ij}). Let

(2.130)

$$\underset{(NTx1)}{y} = [y_{11} \cdots y_{1T} \cdots y_{NT}]' ; \quad \underset{(Nxp)}{Z_1} = \begin{bmatrix} z_{111} \cdots & z_{11p} \\ \vdots & \vdots \\ z_{1N1} \cdots & z_{1Np} \end{bmatrix} ; \quad \underset{(Txq)}{Z_2} = \begin{bmatrix} z_{211} \cdots & z_{21q} \\ \vdots & \vdots \\ z_{2T1} \cdots & z_{2Tq} \end{bmatrix}$$

$$\underset{(NTxTp)}{\tilde{Z}_1} = I_T \otimes Z_1 ; \quad \underset{(NTxNq)}{\tilde{Z}_2} = Z_2 \otimes I_N ; \quad \underset{(NTxK)}{X} = [x_{11} \cdots x_{1T} \cdots x_{NT}]'$$

$$\tilde{X} = [X \; Z_1 \; Z_2] ; \quad \underset{(px1)}{\bar{\alpha}_1} = [\bar{\alpha}_{11} \cdots \bar{\alpha}_{1p}]' ; \quad \underset{(qx1)}{\bar{\alpha}_2} = [\bar{\alpha}_{21} \cdots \bar{\alpha}_{2q}]'$$

$$\underset{(Tpx1)}{\alpha_1} = [\alpha_{111} \cdots \alpha_{1T1} \cdots \alpha_{1Tp}]' ; \quad \underset{(Nqx1)}{\alpha_2} = [\alpha_{211} \cdots \alpha_{21N} \cdots \alpha_{2qN}]'$$

$$\xi_1 = \alpha_1 - \iota_T \otimes \bar{\alpha}_1 ; \quad \xi_2 = \alpha_2 - \bar{\alpha}_2 \otimes \iota_N$$

$$\tilde{\beta} = [\beta' \; \bar{\alpha}_1' \; \bar{\alpha}_2']' ; \quad \varepsilon = [\varepsilon_{11} \cdots \varepsilon_{1T} \cdots \varepsilon_{NT}]'$$

Then the model is :

(2.131) $y = \tilde{X}\,\tilde{\beta} + Z_1\xi_1 + Z_2\xi_2 + \varepsilon$

with

(2.132) $E(\xi_1) = E(\xi_2) = E(\varepsilon) = 0$

(2.133) $E(\xi_1\xi_1') = I_T \otimes \Sigma_1 \; ; \; E(\xi_2\xi_2') = \Sigma_2 \otimes I_N$

(2.134) $E(\varepsilon\varepsilon') = \sigma^2 I_{NT}$

and the GLS of $\tilde{\beta}$ is given by

(2.135) $\hat{\tilde{\beta}} = (\tilde{X}'\,\Omega^{-1}\,\tilde{X})^{-1}\,\tilde{X}'\,\Omega^{-1}\,y$

where

$$(2.136) \quad \Omega = \sigma^2 I_{NT} + I_T \otimes \iota_N \Sigma_1 \iota_N' + \iota_T \Sigma_2 \iota_T' \otimes I_N$$

See Wansbeek [56] for further details.

2.3.5 General Stratified Effect Component Models

This model is proposed by Mazodier and Trognon [29] as a generalisation of both constant slope variable intercept models and coefficient components models (excluding any correlation between coefficients and explanatory variables). The model operates on stratified data and hence we will follow the same notations for stratification as in Section 2.2.2.2.

The model is specified as follows :

For $i \in \bar{N}_j$, $t \in \bar{T}_s$, $i' \in \bar{N}_{j'}$, $t' \in \bar{T}_{s'}$

$$(2.137) \quad y_{it} = \sum_{k=1}^{K} (a_k + b'_{kj} + c'_{ks}) x_{kit} + w_{it}$$

and

$$(2.138) \quad \begin{cases} E(w_{it}) = 0 \quad ; \quad E(w_{it} w_{i't'}) = \delta_{ii'} \delta_{tt'} \sigma^2 \\ E(b'_{kj}) = b_{kj} \quad ; \quad E(b'_{kj} b'_{k'j'}) = \delta_{kk'} \delta_{jj'} \sigma^{2(g)}_{k_j} \\ E(c'_{ks}) = c_{ks} \quad ; \quad E(c'_{ks} c'_{k's'}) = \delta_{kk'} \delta_{ss'} \sigma^{2(p)}_{k_s} \\ b'_{kj} , c'_{ks} \text{ and } w_{it} \text{ are uncorrelated two by two.} \end{cases}$$

Thus each coefficient has a three component structure - one component constant for all observations and two specific effects components each composed of two parts in turn : one fixed (the mean value of the component) and a second one assumed to be random. The specific effects are assumed to be equal for observations relating to all the units belonging to the same group and during the same time interval. No correlations are assumed between groups or between time intervals whereas non-zero correlations are assumed among units of the same group or among time periods of the same interval.

Mazodier and Trognon define the model and note the complexity involved in its identification and estimation. In Varadharajan[1)] [54], we have shown that for a certain set of identification conditions analagous to that of dummy variables models, GLS estimation is reduced to OLSE and thus OLSE is the BLUE of parameters involved.

2.3.6 Bayesian Analysis of Some Varying Coefficient Models

Swamy and Mehta [49] apply Bayesian approach to their random coefficient model (2.109) (see Section 2.3.2.1). Let us rewrite the model in matrix form as :

(2.139) $y = X\,\bar{\beta} + D\,\alpha + D_x\,\xi$

where

$$\underset{(NT\times1)}{y} = [y_{11} \cdots y_{1T} \cdots y_{NT}]' \;; \quad \underset{(T\times K)}{X_i} = \begin{bmatrix} x_{1i1} & \cdots & x_{Ki1} \\ \vdots & & \vdots \\ x_{1iT} & \cdots & x_{KiT} \end{bmatrix}$$

$$\underset{(NT\times K)}{X} = [X_1'\, X_2' \ldots X_N']' \;; \quad \underset{(NT\times NK)}{D} = \text{diag}\,(X_1,\ldots,X_N)$$

$$\underset{(NK\times1)}{\alpha} = [\alpha_1' \ldots \alpha_N']' \;; \quad \underset{(NT\times TK)}{D_{x_i}} = \text{diag}\,(x_{i1}',\ldots,x_{iT}'),\; x_{it}'$$

being the t-th row of X_i

$$\underset{(NT\times NTK)}{D_x} = \text{diag}\left(D_{x_1},\ldots,D_{x_N}\right) \;; \quad \underset{(NTK\times1)}{\xi} = [\,\xi_1' \ldots \xi_N'\,]'$$

with

(2.140) $E(y) = X\,\bar{\beta}$

and

(2.141) $E(y-Ey)(y-Ey)' = \Sigma = \text{diag}\,\left[X_1\Delta X_1' + \Sigma_{11},\ldots,X_N\Delta X_N' + \Sigma_{NN}\right],$

Σ_{ii} being equal to $D_{x_i}\,(I_T \otimes \Delta_{ii})\,D_{x_i}'$, $i=1,2,\ldots,N$.

---

1) Varadharajan is the maiden name of the author.

Three cases are considered for estimation.

Case 1. $\bar{\beta}$, Δ, $\Delta_{11},\ldots,\Delta_{NN}$ known and Σ is non singular. In this case the minimum mean square error predictor of β is

$$(2.142)\quad \hat{\beta} = \Sigma_2^{-1} D_x \Sigma^{-1} y + \left[I_{NTK} - \Sigma_2^{-1} D_x' \Sigma^{-1} D_x\right] (\iota_{NT} \otimes \bar{\beta})$$

where

$$(2.143)\quad \Sigma_2 = \mathrm{diag}\left[(\iota_T\iota_T' \otimes \Delta) + (I_T \otimes \Delta_{11}),\ldots,(\iota_T\iota_T' \otimes \Delta) + (I_T \otimes \Delta_{NN})\right]$$

Case 2. $\bar{\beta}$ is unknown, Δ, $\Delta_{11},\ldots,\Delta_{NN}$ known and Σ non-singular. Here, the prior distribution of $\bar{\beta}$ has to be specified. It is assumed that the mean and covariance matrix of the prior distribution of $\bar{\beta}$ are γ and ψ respectively. Combining prior information and sample information, the minimum average risk (linear) estimator of $\bar{\beta}$ is given by :

$$(2.144)\quad \bar{\beta}^* = (X'\Sigma^{-1}X + \psi^{-1})^{-1}(X'\Sigma^{-1}y + \psi^{-1}\gamma) \text{ if } \psi \text{ non-singular}$$

$$= \psi X'(X\psi X'+\Sigma)^{-1}y + \left[I_K - \psi X'(X\psi X'+\Sigma)^{-1}X\right]\gamma \text{ otherwise}$$

Case 3. Neither $\bar{\beta}$ nor Δ, $\Delta_{11},\ldots,\Delta_{NN}$ are known and Σ non-singular. In this case, in addition to the prior distribution of $\bar{\beta}$, we need prior mean and variance matrices of Δ_{ii}, $i=1,\ldots,N$. Using prior distribution of Δ_{ii} and sample information, Δ_{ii} are estimated from time series data of the i-th individual. This leads to the following estimator of Δ :

$$(2.145)\quad \hat{\Delta} = \frac{S}{N-1} - \frac{1}{N} \sum_{i=1}^{N} (X_i' \hat{\Sigma}_{ii}^{-1} X_i)^{-1}$$

where

$$(2.146)\quad S = \sum_{i=1}^{N} b_i b_i' - \frac{1}{N} \sum_{i=1}^{N} b_i \sum_{i=1}^{N} b_i'$$

and

$$(2.147)\quad b_i = (X_i' \hat{\Sigma}_{ii}^{-1} X_i)^{-1} X_i' \hat{\Sigma}_{ii}^{-1} y_i$$

$$(2.148)\quad \hat{\Sigma}_{ii} = D_{x_i} (I_T \otimes \hat{\Delta}_{ii}) D_{x_i}'$$

Finally, the minimum average risk linear estimator of $\bar{\beta}$ is obtained as :

$$(2.149) \quad \hat{\bar{\beta}} = (X'\hat{\Sigma}^{-1} X + \psi^{-1})^{-1} (X'\hat{\Sigma}^{-1} y + \psi^{-1} \gamma)$$

where $\hat{\Sigma}$ is obtained by replacing Δ by $\hat{\Delta}$ and Δ_{ii} by $\hat{\Delta}_{ii}$ in (2.141) and where $\hat{\Delta}_{ii}$ is obtained using the time series data of the i-th unit separately.

Lon Mu Liu and Hanssens [24] apply Bayesian approach to time varying coefficient models of the form :

$$(2.150) \quad y_{it} = x'_{it} \beta_t + \varepsilon_{it}$$

with

$$(2.151) \quad (\beta_t - \beta) = \Phi(\beta_{t-1} - \beta) + a_t$$

$$(2.152) \quad \begin{cases} \varepsilon_{it} \sim N(0,\sigma_\varepsilon^2) \\ a_t \sim N(0,A) \text{ independent of } \varepsilon_{it} \end{cases}$$

for $t=1,\ldots,n$ and $i=1,\ldots,m_t$.

Equation (2.150) is the measurement equation and equation (2.151), the process equation. Φ is the first-order autoregressive parameter matrix, also called transition matrix.

Case 1 . Φ, β, A, σ_ε^2 known

The prior distribution of $y | b,\sigma_\varepsilon^2$ is expressed as :

$$p(y| b,\sigma_\varepsilon^2) \propto (\sigma_\varepsilon^2)^{-m./2} \exp \left\{ - \frac{1}{2\sigma_\varepsilon^2} [s_\varepsilon^2 + (b-\hat{b})'D(b-\hat{b})] \right\}$$

where

$$b' = [\beta'_1 \ldots \beta'_n]' \; ; \; D = \text{diag}\,(X'_1X_1,\ldots,X'_tX_t,\ldots,X'_nX_n)$$

$$\hat{\beta}_t = (X'_tX_t)^{-1}X'_ty_t \; ; \; s_\varepsilon^2 = \sum_{t=1}^{n} (y'_ty_t - \hat{\beta}'_tX'_tX_t) \; ; \; \hat{b}' = [\hat{\beta}_1 \ldots \hat{\beta}_n]'$$

$$m. = \sum_t m_t$$

The joint distribution of β'_ts given Φ, β, A is

$$p(b| \beta,\Phi,A) \propto |\Gamma|^{-\frac{1}{2}} |A|^{-(n-1)/2} \exp \left\{ - \frac{1}{2} [(b-\iota_n \otimes \beta)' \; \Sigma^{-1} \; (b-\iota_n \otimes \beta)] \right\}$$

where

$$\Sigma^{-1} = \begin{bmatrix} \Phi' A^{-1}\Phi + \Gamma^{-1} & -\Phi'\Gamma^{-1} & 0 & 0 & 0 \\ -A^{-1}\Phi & A^{-1}+\Phi' A^{-1} & -\Phi' A^{-1} & 0 & 0 \\ & & & & \\ 0 & 0 & & A^{-1}+\Phi'\bar{A}^{1}\Phi & -\Phi' A^{-1} \\ 0 & 0 & & -A^{-1}\Phi & A^{-1} \end{bmatrix}$$

$P(\Gamma) = (I - \Phi \otimes \Phi)^{-1} P(A)$, P denoting the pack operator.

Combining $p(y|b,\sigma_\varepsilon^2)$ and $p(b|\beta,\Phi,A)$ we obtain the posterior distribution of $b|\Phi,\beta,A,\sigma_\varepsilon^2$ which can be simplified to the following :

$$\exp\left\{-\frac{1}{2\sigma_\varepsilon^2}\left[(b-\tilde{b})'(D+H)(b-\tilde{b})\right]\right\} = \left[N(\tilde{b},\sigma_\varepsilon^2(D+H)^{-1})\right]$$

where

$$\tilde{b} = (D+H)^{-1}(D\hat{b} + H(\iota_N \otimes \beta)) , \quad H=\sigma_\varepsilon^2 \Sigma^{-1} , \quad W=A/\sigma_\varepsilon^2$$

Case 2 . $\beta,\Phi,W,\sigma_\varepsilon^2$ unknown

The prior distributions of the unknowns are assumed as follows:

$$p(\beta) \propto C \; ; \; p(\Phi,W,\sigma_\varepsilon^2) \propto \sigma_\varepsilon^{-2}$$

Combining $p(y|b,\sigma_\varepsilon^2)$ with $p(b|\beta,\Phi,A)$ under prior distribution yields $p(b,\Phi,W|y)$ which can be factorised into $p(b|\Phi,W,y)$ and $p(\Phi,W|y)$. The former is seen to be a $t_{nk}(\bar{b}, \bar{\sigma}_\varepsilon^{2} C, m.-p)$ and given Φ,W ; $\beta_t \sim t(\bar{\beta}_t,\bar{\sigma}_\varepsilon^{2},c_{ii},m.-p)$ where

$$\bar{b} = [\bar{\beta}_1 \ldots \bar{\beta}_n]' = (D+H)^{-1}(D\hat{b} + \bar{\beta} H \iota_n)$$

$$\bar{\beta} = \iota_n' D(D+H)^{-1} H \hat{b} / (\iota_n' D (D+H)^{-1} H \iota_n)$$

$$C = \left\{D+H - (\iota_n' H \iota_n)^{-1} H \iota\iota' H\right\}^{-1}$$

c_{ii} is the i-th diagonal element of C

$$\bar{\sigma}_\varepsilon^{2} = \left[(s_\varepsilon^2 + (\hat{b} - \iota_n \otimes \bar{\beta})' D(D+H)^{-1} H(\iota_n \otimes \bar{\beta})\right] / (m.-p)$$

Integrating $p(\beta_t|\Phi,W,y)$ over $p(\Phi,W|y)$ gives the unconditional posterior distribution of β_t .

Bayesian analysis of two other variable coefficients models, namely that of Rosenberg [45] and that of Lon Mu Liu and Tiao [25] , was already briefly described in Sections 2.3.2.3 and 2.3.2.4 respectively where these models were presented.

2.4 Estimation of Variance Components in Panel Data Models

2.4.1 Analysis of Variance Methods

Let us recall the stochastic structure of a variance components model :

$$\varepsilon_{it} = u_i + v_t + w_{it} \tag{2.153}$$

$$E(u_i) = E(v_t) = E(w_{it}) = 0 \tag{2.154}$$

$$E(u_i u_{i'}) = \delta_{ii'} \sigma_u^2 \tag{2.155}$$

$$E(v_t v_{t'}) = \delta_{tt'} \sigma_v^2 \tag{2.156}$$

$$E(w_{it} w_{i't'}) = \delta_{ii'} \delta_{tt'} \sigma_w^2 \tag{2.157}$$

$$E(u_i v_t) = E(u_i w_{i't}) = E(v_t w_{it'}) = 0 \tag{2.158}$$

The variance components to be estimated are σ_u^2, σ_v^2 and σ_w^2 .

The analysis of variance (AOV) gives the following estimates :

$$\hat{\sigma}_w^2 = \frac{1}{(N-1)(T-1)} \varepsilon' Q \varepsilon \tag{2.159}$$

$$\hat{\sigma}_u^2 = \frac{1}{T} \left[\frac{1}{N-1} \varepsilon' \left(\frac{A}{T} - \frac{J_{NT}}{NT} \right) \varepsilon - \hat{\sigma}_w^2 \right] \tag{2.160}$$

$$\hat{\sigma}_v^2 = \frac{1}{N} \left[\frac{1}{T-1} \varepsilon' \left(\frac{B}{T} - \frac{J_{NT}}{NT} \right) \varepsilon - \hat{\sigma}_w^2 \right] \tag{2.161}$$

where

$$\underset{(NT\times 1)}{\varepsilon} = [\varepsilon_{11} \dots \varepsilon_{1T} \dots \varepsilon_{NT}]'$$

$$A = I_N \otimes \iota_T\iota_T' \; ; \; B = \iota_N\iota_N' \otimes I_T \; ; \; J_{NT} = \iota_N\iota_N' \otimes \iota_T\iota_T'$$

$$Q = I_{NT} - \frac{A}{T} - \frac{B}{N} + \frac{J_{NT}}{NT}$$

See e.g. Amemiya [1], Graybill [16]. In practice, the ε_{it}'s are not observable and hence have to be predicted before applying the above formulae. This is done using the variance components model

(2.162) $y = X\beta + \varepsilon$

There are two ways of predicting the errors :

(i) Apply OLS to (2.162) and compute the residuals :

(2.163) $\hat{\beta}_{OLS} = (X'X)^{-1}X'y$

and

(2.164) $\hat{\varepsilon} = y - X\hat{\beta}_{OLS}$

(ii) Apply least squares with dummy variables on (2.162) and compute the residuals i.e.

(2.165) $\hat{\varepsilon} = y - X(X'QX)^{-1}X'Qy$

Amemiya [1] shows that the AOV estimators based on (i) are less efficient than those based on (ii).

2.4.2 "Fitting of Constants" Method

In this method (cf. Fuller and Battese [14]), the following residual-vectors are first computed :

(2.166) $\hat{w} = Qy - QX(X'QX)^{+}X'Qy$

(2.167) $\hat{u} = (Q+\frac{A}{T}-\frac{J_{NT}}{NT})\left\{y - X\left[X'(Q+\frac{A}{T}-\frac{J_{NT}}{NT})X\right]^{+} X'(Q+\frac{A}{T}-\frac{J_{NT}}{NT})\,y\right\}$

(2.168) $\hat{v} = (Q+\frac{B}{N}-\frac{J_{NT}}{NT})\left\{y - X\left[X'(Q+\frac{B}{N}-\frac{J_{NT}}{NT})X\right]^{+} X'(Q+\frac{B}{N}-\frac{J_{NT}}{NT})\,y\right\}$

where $\underset{(NT\times 1)}{w} = [w_{11} \dots w_{1T} \dots w_{NT}]'$

$\underset{(N\times 1)}{u} = [u_1 \dots u_N]'$; $\underset{(T\times 1)}{v} = [v_1 \dots v_T]'$

and M^{+} denotes the Moore-Penrose generalised inverse of M .

Then the unbiased estimators for variance components are given by :

$$(2.169)\quad \hat{\sigma}_w^2 = \frac{\hat{w}'\hat{w}}{(N-1)(T-1)-K-1}$$

$$(2.170)\quad \hat{\sigma}_u^2 = \frac{\hat{u}'\hat{u} - \left[T(N-1) - K\right]\hat{\sigma}_w^2}{T(N-1)-T\ \mathrm{tr}\left\{\left[X'(Q+\frac{A}{T}-\frac{J_{NT}}{NT})\ X\right]^{+} X'(\frac{A}{T}-\frac{J_{NT}}{NT})\ X\right\}}$$

$$(2.171)\quad \hat{\sigma}_v^2 = \frac{\hat{v}'\hat{v} - \left[N(T-1) - K\right]\hat{\sigma}_w^2}{N(T-1)-N\ \mathrm{tr}\left\{\left[X'(Q+\frac{B}{N}-\frac{J_{NT}}{NT})\ X\right]^{+} X'(\frac{B}{N}-\frac{J_{NT}}{NT})\ X\right\}}$$

2.4.3. "Swamy and Arora" Method :

Estimation of variance components using mean squares residuals of the between groups, between time-periods and within groups regressions (see [47]).

2.4.4 MINQUE

This method due to Rao [40] , [41] applies to a more general model :

$$(2.172)\quad y = X\,\beta + U_1\,\xi_1 + \ldots + U_K\,\xi_K$$

where U_i's are given matrices and ξ_i a vector of uncorrelated random variables with zero mean and dispersion matrix $\sigma_i^2\ I$, i=1,...,K. Further ξ_i and ξ_j are uncorrelated. The variance covariance matrix of the stochastic part of the equation can be written as

$$(2.173)\quad \sigma_1^2\ V_1 + \ldots + \sigma_K^2\ V_K$$

where $V_i = U_i\ U_i'$, i=1,...,K .

The variance components to be estimated are $\sigma_1^2,\ldots,\sigma_K^2$.

Let $\Sigma\, p_i\sigma_i^2$ be a linear function of the variances to be estimated. The quadratic form y'Ay is said to be a MINQUE of

$\sum p_i \sigma_i^2$ if the matrix $A = [a_{ij}]$ is chosen such that $\| A \|$, the Euclidean norm of A, which is the same as the square root of trace A^2 , is a minimum subject to the conditions :

$$(2.174) \begin{cases} \text{(i)} \; A X = 0 \\ \text{(ii)} \; \sum_{i=1}^{n} a_{ii}\sigma_i^2 = \sum_{i=1}^{n} p_i \sigma_i^2 \end{cases}$$

Condition (i) implies that the estimator is invariant of the origin from which the ß parameters are expressed as deviations. Condition (ii) implies that y'Ay is an unbiased estimator of the given linear function of the distinct variances. Let us note that the individual component σ_j^2 is obtained by setting $p_i = 0$ for $i \neq j$ and $p_j = 1$. The reader is referred to [40] and [41] for more details.

2.4.5 Comparison of Alternate Estimation Methods

Maddala and Mount [26] have made a comparative study of alternate estimators of variance components by Monte Carlo experiments and found that all the estimators performed well on the basis of mean-squared errors. But in practice, if there is a large difference between different estimates or if estimates of variance components are negative, then an error in specification is to be suspected.

Baltagi [8] has also compared different procedures for estimation of coefficients which include different estimation methods of variance components. Comparison was made again by means of Monte Carlo experiments. The main results are that all two stage methods performed reasonably well; as long as variance components are not relatively small and close to zero, there is a gain in efficiency in performing feasible GLS rather than LS or LSDV; better estimates of variance components do not necessarily lead to better estimates of coefficients. He also concludes by saying that if different two stage GLS yield widely different estimates, specification is to be tested.

2.5 Estimation of Models using Incomplete Time-Series Cross-Section Data

In this Section, we will present a specification of an EC model for a non-overlapping sample i.e. for a set of data in which the cross-sectional units observed differ from one period to the next (cf. Biorn[12]).

The model is :

(2.175) $y_{it} = f(x_{it} \; ; \; \beta) + e_{it}$

(2.176) $e_{it} = u_i + v_t + w_{it}$

(2.177) $$E(e_{it} e_{ks}) = \begin{cases} \sigma^2 & k=i \; , \; s=t \\ \rho\sigma^2 & k=i \; , \; s \neq t \\ \omega\sigma^2 & k \neq i \; , \; s=t \\ 0 & k \neq i \; , \; s \neq t \end{cases}$$

with

(2.178) $\sigma^2 = \sigma_u^2 + \sigma_v^2 + \sigma_w^2 \quad ; \; \rho = \sigma_u^2/\sigma^2 \quad ; \; \omega = \sigma_v^2/\sigma^2$

The sample is structured as follows : In period 1, N individuals are observed : 1,...,N ; in period 2 only individuals (m+1), (m+2) ... N are observed among the N individuals observed earlier but m additional (new) individuals are included who have index numbers N+1,...,N+m($0 \leq m \leq N$). In the same way, in every period m individuals drop out from the sample of the previous period and m new individuals are added. The sample can be represented as follows :

Period	Individuals observed
1	1,2,...,N
2	m+1,...,N,N+1,...,N+m
3	2m+1,...........,N+2m
.	.
.	.
t	(t-1)m+1,...,(t-1)m+N
T	(T-1)m+1,...,(T-1)m+N

Total number of individuals = (T-1) m+N, say H.

Total number of observations = TN .

Let ε_{it} relate to the i-th of the individuals observed in period t (this individual's number in the population is (t-1)m+i). Let

$$\underset{(N\times 1)}{\varepsilon_t} = [\varepsilon_{1t} \cdots \varepsilon_{Nt}]' , \quad \underset{(H\times 1)}{e_t} = [e_{1t} \cdots e_{Ht}]'$$

Then we can write

$$\varepsilon_t = D_t\, e_t \tag{2.179}$$

where D_t, the sample design (selection) matrix, is given by

$$D_t = [0_{N,(t-1)m} \quad I_N \quad 0_{N,(T-t)m}]$$

We have

$$E(e_t e_s') = \begin{cases} \sigma^2 \{(1-\omega)\, I_H + \omega\, E_H\} & s=t \\ \sigma^2\, \rho\, I_H & s\neq t \end{cases} \tag{2.180}$$

where E_H denotes a (HxH) matrix of ones and

$$D_t\, E_H\, D_t' = E_N \quad \forall\ t \tag{2.181}$$

Thus

$$E(\varepsilon_t \varepsilon_s') = D_t\, E(e_t e_s')\, D_s' = \sigma^2\, \Omega_{ts}$$

$$= \begin{cases} \sigma^2\, ((1-\omega)\, I_N + \omega\, E_N) & s=t \\ \sigma^2\, \rho\, D_t\, D_s' & s\neq t \end{cases} \tag{2.182}$$

Denoting

$$\varepsilon = [\varepsilon_1'\varepsilon_2' \cdots \varepsilon_T']' \ ; \ D = [D_1'D_2' \cdots D_T']'$$

we can write

$$E(\varepsilon\varepsilon') = \sigma^2\, \Omega$$

$$\begin{aligned} &= \sigma^2 \{ I_T \otimes [(1-\omega)I_N + \omega E_N] + \rho\, [DD' - I_T \otimes I_N]\} \\ &= \sigma^2 \{(1-\rho-\omega)\, I_T \otimes I_N + \omega\, (I_T \otimes E_N) + \rho\, DD'\} \end{aligned} \tag{2.183}$$

Estimation of ß, Ω and σ^2 is done by maximum likelihood method which involves an iterative procedure.

The special case in which N=2m is considered in detail and the iterative procedure worked out.

On comparison with similar models using complete cross-section time-series data, the author finds "striking similarities but also notable differences".

2.6. Extensions

2.6.1 SUR with EC

This model was first developed by Avery [3]. The model is specified as follows :

There are M regression equations :

$$(2.184) \quad y_j = X^j \beta_j + \varepsilon_j \qquad j=1,\ldots,M$$

where

$$\underset{(NT\times 1)}{y_j} = [y_{j11} \cdots y_{j1T} \cdots y_{jNT}]' \; ; \quad \underset{(NT\times K_j)}{X^j} = \begin{bmatrix} x^j_{111} & \cdots & x^j_{K_j 11} \\ \vdots & & \vdots \\ x^j_{11T} & \cdots & x^j_{K_j 1T} \\ \vdots & & \vdots \\ x^j_{1NT} & \cdots & x^j_{K_j NT} \end{bmatrix}$$

$$\underset{(K_j\times 1)}{\beta_j} = [\beta_{j1} \cdots \beta_{jK_j}]' \; ; \quad \underset{(NT\times 1)}{\varepsilon_j} = [\varepsilon_{j11} \cdots \varepsilon_{j1T} \cdots \varepsilon_{jNT}]'$$

The errors ε_{jit} are assumed to be of an error components structure :

$$(2.185) \quad \varepsilon_{jit} = u_{ji} + v_{jt} + w_{jit}$$

with the following assumptions on their distribution :

$$(2.186) \quad E(u_{ji}\, u_{\ell i'}) = \delta_{ii'}\, \sigma^2_{uj\ell}$$

$$(2.187) \quad E(v_{jt}\, v_{\ell t'}) = \delta_{tt'}\, \sigma^2_{vj\ell}$$

$$(2.188) \quad E(w_{jit}\, w_{\ell i't'}) = \delta_{ii'}\, \delta_{tt'}\, \sigma^2_{wj\ell}$$

$$(2.189) \quad E(u_{ji}\, v_{\ell t}) = E(v_{jt}\, w_{j'it'}) = E(u_{ji}\, w_{j'i't}) = 0$$

and

$$(2.190) \quad E(u_{ji}) = E(v_{jt}) = E(w_{jit}) = 0$$

From these assumptions it follows that

$$(2.191) \quad E(\varepsilon_j \varepsilon_\ell') = \Sigma_{j\ell} = \sigma^2_{uj\ell}\, A + \sigma^2_{vj\ell}\, B + \sigma^2_{wj\ell}\, I_{NT}$$

where

$$A = I_N \otimes \iota_T \iota_T' \quad \text{and} \quad B = \iota_N \iota_N' \otimes I_T$$

Combining all the M equations, we get :

$$(2.192) \quad y = X \beta + \varepsilon$$

where

$$y = [y_1' \ldots y_M']' \quad , \quad X = \begin{bmatrix} X_1 & & \bigcirc \\ & \ddots & \\ \bigcirc & & X_M \end{bmatrix} ,$$

$$\beta = [\beta_1' \ldots \beta_M']' \quad , \quad \varepsilon = [\varepsilon_1' \ldots \varepsilon_M']'$$

and

$$(2.193) \quad E(\varepsilon\varepsilon') = \Sigma = \Sigma_u \otimes A + \Sigma_v \otimes B + \Sigma_w \otimes I_{NT}$$

where

$$\Sigma_u = [\sigma^2_{uj\ell}] \; ; \; \Sigma_v = [\sigma^2_{vj\ell}] \; ; \; \Sigma_w = [\sigma^2_{wj\ell}] \quad , j,\ell=1,\ldots,M.$$

The GLS estimator of β is given by :

$$(2.194) \quad \hat{\beta} = (X' \Sigma^{-1} X)^{-1} X' \Sigma^{-1} y$$

The variance and covariance components can be estimated using AOV formulae as follows :

$$(2.195) \quad \hat{\sigma}^2_{wj\ell} = \frac{1}{(N-1)(T-1)} \hat{\varepsilon}_j' Q \hat{\varepsilon}_\ell$$

$$(2.196) \quad \hat{\sigma}^2_{uj\ell} = \frac{1}{T} \left[\frac{1}{N-1} \hat{\varepsilon}_j' \left(\frac{A}{T} - \frac{J_{NT}}{NT}\right) \hat{\varepsilon}_\ell - \hat{\sigma}^2_{wj\ell} \right]$$

$$(2.197) \quad \hat{\sigma}^2_{vj\ell} = \frac{1}{N} \left[\frac{1}{T-1} \hat{\varepsilon}_j' \left(\frac{B}{N} - \frac{J_{NT}}{NT}\right) \hat{\varepsilon}_\ell - \hat{\sigma}^2_{wj\ell} \right]$$

Avery [3] used OLS residuals for ε_j , ε_ℓ , whereas in a later article [7] , Baltagi proposed to use Least Squares Dummy Variables (Covariance) residuals. In [38] Prucha shows that both these estimates of variance-covariance components lead to asymptotically efficient feasible GLS estimators of β .

2.6.2 SEM with EC

The second extension, namely Simultaneous Equation Models with Error Components, being the main object of study of our thesis, is presented and discussed at length in the following chapters.

CHAPTER 3

PRESENTATION OF SIMULTANEOUS EQUATIONS MODEL WITH ERROR COMPONENTS STRUCTURE AND ESTIMATION OF THE REDUCED FORM

3.1 The Model

3.1.1 Notations

We consider a complete linear system in M jointly dependent or current endogenous variables and K predetermined variables. Each variable is observed for N cross-sectional units during T successive time periods. This subsection will be entirely devoted to the presentation of a systematic notation which will serve as a basis for all future derivations and calculations.

Let y_{mit} denote t-th observation for the i-th cross-sectional unit on the m-th jointly dependent variable and x_{jit}, the t-th observation for the i-th cross-sectional unit on the j-th predetermined variable.

Since the system is complete by assumption, it consists of M structural equations and the structural form can be written as :

$$(3.1) \qquad \sum_{m'=1}^{M} \gamma^*_{m'm}\, y_{m'it} + \sum_{j=1}^{K} \beta^*_{jm}\, x_{jit} + u_{mit} = 0, \quad m=1,\ldots,M$$

for $t=1,\ldots,T$; $i=1,\ldots,N$;

where

$\gamma^*_{m'm}$ is the coefficient in the m-th structural equation, of the m'-th jointly dependent variable $y_{m'it}$,

β^*_{jm} is the coefficient in the m-th structural equation, of the j-th predetermined variable x_{jit} ,

u_{mit} is the unobserved disturbance at the t-th observation for the i-th unit in the m-th structural equation.

Let

$$(3.2) \quad \begin{cases} y_{it} = [y_{1it}\ y_{2it}\ \cdots\ y_{Mit}]' & (M\times 1) \\ x_{it} = [x_{1it}\ x_{2it}\ \cdots\ x_{Kit}]' & (K\times 1) \end{cases}$$

for $t=1,\ldots,T$; $i=1,\ldots,N$;

and

$$(3.3)\quad \begin{cases} \gamma^*_m = [\gamma^*_{1m}\ \gamma^*_{2m}\ \dots\ \gamma^*_{Mm}]' & (M\times 1) \\ \beta^*_m = [\beta^*_{1m}\ \beta^*_{2m}\ \dots\ \beta^*_{Km}]' & (K\times 1) \end{cases}$$

for $m=1,\dots,M$.

Then the m-th structural equation can be written as

$$(3.4)\quad y'_{it}\ \gamma^*_m + x'_{it}\ \beta^*_m + u_{mit} = 0$$

for $t=1,\dots,T$, $i=1,\dots,N$.

Now, let us introduce the NT x K matrix X and the NT x M matrix Y of, respectively the values taken by the predetermined variables and those taken by the jointly dependent variables ; and also the NT x 1 vector of disturbances of the m-th structural equation, u_m .

$$(3.5)\quad \underset{(NT\times K)}{X} = \begin{bmatrix} x'_{11} \\ \vdots \\ x'_{1T} \\ x'_{21} \\ \vdots \\ x'_{2T} \\ \vdots \\ x'_{NT} \end{bmatrix} ;\quad \underset{(NT\times M)}{Y} = \begin{bmatrix} y'_{11} \\ \vdots \\ y'_{1T} \\ y'_{21} \\ \vdots \\ y'_{2T} \\ \vdots \\ y'_{NT} \end{bmatrix} ;\quad \underset{(NT\times 1)}{u_m} = \begin{bmatrix} u_{m11} \\ \vdots \\ u_{m1T} \\ u_{m21} \\ \vdots \\ u_{m2T} \\ \vdots \\ u_{mNT} \end{bmatrix}$$

Then (3.4) can be written in matrix notation as :

$$(3.6)\quad Y\ \gamma^*_m + X\ \beta^*_m + u_m = 0$$

Using the notation :

$$(3.7)\quad \begin{cases} \underset{(1\times NT)}{y'_m} = [y_{m11}\dots y_{m1T}y_{m21}\dots y_{m2T}\dots y_{mNT}] , & m=1,\dots,M \\ \underset{(1\times NT)}{x'_j} = [x_{j11}\dots x_{j1T}x_{j21}\dots x_{j2T}\dots x_{jNT}] , & j=1,\dots,K \end{cases}$$

we can also write X and Y as :

$$(3.8)\quad X = [x_1\ x_2\ \dots\ x_K] ;\ Y = [y_1\ \dots\ y_M]$$

Finally, let

$$(3.9) \qquad \underset{(M\times M)}{\Gamma} = \left[\gamma_1^* \ \ldots \ \gamma_M^*\right] \quad ; \quad \underset{(K\times M)}{B} = \left[\beta_1^* \ \ldots \ \beta_M^*\right] \quad ;$$

and

$$(3.10) \qquad \underset{(NT\times M)}{U} = \left[u_1 \ \ldots \ u_M\right]$$

With all these notations, the structural form (3.1) can be written compactly as :

$$(3.11) \qquad Y \Gamma + X B + U = 0$$

The matrix Γ is square because there are as many equations as jointly dependent variables. It is assumed non-singular, so that the system can be solved to get the reduced form :

$$(3.12) \qquad Y = - X B \Gamma^{-1} - U \Gamma^{-1}$$

or

$$(3.13) \qquad Y = X \Pi + V$$

where

$$(3.14) \qquad \Pi = - B \Gamma^{-1} \quad ; \quad V = - U \Gamma^{-1}$$

3.1.2 Stochastic Specifications

We propose to account for the cross-sectional and temporal heterogeneity by introducing an error component structure in the above system of equations. That is, we assume that each structural equation error is composed of three components : a time effect (varying with time but constant over units), a unit effect (varying over cross-sectional units but constant over time), and an overall effect (variable both in the time and cross-sectional dimensions).

Formally, we have :

$$(3.15) \qquad u_{mit} = \mu_{mi} + \nu_{mt} + \varepsilon_{mit}$$

for $t=1,\ldots,T$; $i=1,\ldots,N$; $m=1,\ldots,M$.

The expectations of all the three components are assumed to vanish :

$$(3.16) \qquad E(\mu_{mi}) = E(\nu_{mt}) = E(\varepsilon_{mit}) = 0$$

for $t=1,\ldots,T$; $i=1,\ldots,N$; $m=1,\ldots,M$.

It is further assumed that :

$$(3.17)\quad \begin{cases} E(\mu_{mi}\ \mu_{m'i'}) = \delta_{ii'}\ \sigma_{\mu mm'} \\ E(\nu_{mt}\ \nu_{m't'}) = \delta_{tt'}\ \sigma_{\nu mm'} \\ E(\varepsilon_{mit}\ \varepsilon_{m'i't'}) = \delta_{ii'}\ \delta_{tt'}\ \sigma_{\varepsilon mm'} \\ E(\mu_{mi}\ \nu_{m't}) = E(\mu_{mi}\ \varepsilon_{m'i't'}) = E(\nu_{mt}\ \varepsilon_{m'i't'}) = 0 \end{cases}$$

for $t,t'=1,\ldots,T$; $i,i'=1,\ldots,N$; $m,m'=1,\ldots,M$;

where δ denotes the Kronecker delta.

From these assumptions, it follows that

$$(3.18)\quad E(u_{mit}u_{m'i't'}) = \delta_{ii'}\sigma_{\mu mm'} + \delta_{tt'}\sigma_{\nu mm'} + \delta_{ii'}\delta_{tt'}\sigma_{\varepsilon mm'}$$

Let us note that u_m can be written as

$$(3.19)\quad u_m = (I_N \otimes \iota_T)\ \mu^m + (\iota_N \otimes I_T)\ \nu^m + \varepsilon_m$$

where

$$(3.20)\quad \begin{cases} \mu^m = \left[\mu_{mi}\right]\ i=1,\ldots,N\ ;\quad \nu^m = \left[\nu_{mt}\right]\ t=1,\ldots,T\ ; \\ \varepsilon_m = \left[\varepsilon_{mit}\right]\ t=1,\ldots,T\ ,\quad i=1,\ldots,N\ ; \end{cases}$$

I_N is a NxN identity matrix, ι_T is a Tx1 vector of ones and $\otimes$ denotes the Kronecker product.

Now, it is easily verified that

$$(3.21)\quad \begin{cases} E(\mu^m) = 0\ ,\ E(\nu^m) = 0\ ,\ E(\varepsilon_m) = 0\ ,\ m=1,\ldots,M\ ; \\ E(\mu^m\ \mu^{m''}) = \sigma_{\mu mm'}\ I_N \\ E(\nu^m\ \nu^{m''}) = \sigma_{\nu mm'}\ I_T \\ E(\varepsilon_m\ \varepsilon'_{m'}) = \sigma_{\varepsilon mm'}\ I_{NT} \end{cases}$$

and consequently

$$(3.22)\quad E(u_m u'_{m'}) = \Sigma_{mm'} = \sigma_{\mu mm'}\ A + \sigma_{\nu mm'}\ B + \sigma_{\varepsilon mm'}\ I_{NT}$$

where A and B are NT x NT matrices defined as

(3.23) $A = I_N \otimes \iota_T \iota_T'$; $B = \iota_N \iota_N' \otimes I_T$

$\Sigma_{mm'}$ can also be written as (cf. Nerlove [36]) :

(3.24) $$\Sigma_{mm'} = \sigma_{1mm'} M_1 + \sigma_{2mm'} M_2 + \sigma_{3mm'} M_3 + \sigma_{4mm'} M_4$$

where

(3.25) $$\begin{cases} \sigma_{1mm'} = \sigma_{\varepsilon mm'} + T\,\sigma_{\mu mm'} \quad , \\ \sigma_{2mm'} = \sigma_{\varepsilon mm'} + N\,\sigma_{\nu mm'} \quad , \\ \sigma_{3mm'} = \sigma_{\varepsilon mm'} + T\,\sigma_{\mu mm'} + N\,\sigma_{\nu mm'} \qquad \text{and} \\ \sigma_{4mm'} = \sigma_{\varepsilon mm'} \end{cases}$$

are the distinct characteristic roots of $\Sigma_{mm'}$ of multiplicity $m_1 = (N-1)$, $m_2 = (T-1)$, $m_3 = 1$ and $m_4 = (N-1)(T-1)$ respectively and where

(3.26) $$\begin{cases} M_1 = \frac{A}{T} - \frac{J_{NT}}{NT} \quad , \\ M_2 = \frac{B}{N} - \frac{J_{NT}}{NT} \quad , \\ M_3 = \frac{J_{NT}}{NT} \quad , \\ M_4 = Q = I_{NT} - \frac{A}{T} - \frac{B}{N} + \frac{J_{NT}}{NT} \quad , \text{ with } J_{NT} = \iota_{NT}\,\iota_{NT}' \ . \end{cases}$$

We have :

(3.27) $$\begin{cases} M_i M_j = \delta_{ij} M_i \\ M_1 + M_2 + M_3 + M_4 = I \end{cases}$$

and

(3.28) $$\begin{cases} \iota' M_i = 0 \qquad \text{for } i=1,2,4 \\ \iota' M_3 = \iota' \end{cases}$$

Hence the variance-covariance structure of the structural form can be summarized in the following way :

(3.29) $$\Sigma = E(\text{vec } U)(\text{vec } U)' = \Sigma_1 \otimes M_1 + \Sigma_2 \otimes M_2 + \Sigma_3 \otimes M_3 + \Sigma_4 \otimes M_4$$

with

$$
(3.30)\quad
\begin{cases}
\underset{(M\times M)}{\Sigma_1} = \Sigma_\varepsilon + T\,\Sigma_\mu\ , & \underset{(M\times M)}{\Sigma_\varepsilon} = \left[\sigma_{\varepsilon\, mm'}\right]\ , \quad \underset{(M\times M)}{\Sigma_\mu} = \left[\sigma_{\mu\, mm'}\right] \\
 & \qquad m,m'=1,\ldots,M \\
\underset{(M\times M)}{\Sigma_2} = \Sigma_\varepsilon + N\,\Sigma_\nu\ , & \underset{(M\times M)}{\Sigma_\nu} = \left[\sigma_{\nu\, mm'}\right]\ , \quad m,m'=1,\ldots,M \\
\underset{(M\times M)}{\Sigma_3} = \Sigma_\varepsilon + T\,\Sigma_\mu + N\,\Sigma_\nu & \\
\underset{(M\times M)}{\Sigma_4} = \Sigma_\varepsilon &
\end{cases}
$$

Given the particular form of Σ (3.29), the inverse and determinant of Σ are[1)]:

$$(3.31)\qquad \Sigma^{-1} = \Sigma_1^{-1} \otimes M_1 + \Sigma_2^{-1} \otimes M_2 + \Sigma_3^{-1} \otimes M_3 + \Sigma_4^{-1} \otimes M_4$$

$$(3.32)\qquad |\Sigma| = |\Sigma_1|^{m_1}\ |\Sigma_2|^{m_2}\ |\Sigma_3|^{m_3}\ |\Sigma_4|^{m_4}$$

Finally, no correlation is assumed between elements of X and those of U .

---

1) The inverse of Σ is given in Baltagi [7] and can be checked directly. For the determinant, the following proof is offered : Consider the general case

$$A = \sum_{i=1}^{S} A_i \otimes M_i$$

where each M_i is idempotent of rank r_i , $M_i M_j = 0 \quad i \neq j$ and $\sum_{i=1}^{S} M_i = I$. First we reverse the order of the Kronecker products using the commutation matrix P (which is orthogonal). Second, since M_i is idempotent, we can write $M_i = C_i C_i'$ where the r_i columns of C_i are orthonormal vectors. We then construct the orthogonal matrix $C=[C_1 C_2 \ldots C_S]$. Clearly $C \otimes I$ is also orthogonal. Finally, we compute the determinant of A :

$$|A| = |PAP| = |(C' \otimes I)\ PAP\ (C \otimes I)| = \left| \sum_{i=1}^{S} C'M_iC \otimes A_i \right|$$

$$= \begin{vmatrix} I_{r1} \otimes A_1 & & 0 \\ & \ddots & \\ 0 & & I_{rs} \otimes A_s \end{vmatrix} = |A_1|^{r_1} \cdots |A_s|^{r_s}$$

3.2 Estimation of the Reduced Form

3.2.1 Derivation of Stochastic Properties of Reduced Form Errors and Interpretation of the Reduced Form

Before proceeding to the estimation of reduced form parameters, let us first rewrite the reduced form (3.13) as follows :

(3.33) $[y_1 \;\cdots\; y_M] = X\,[\pi_1 \;\cdots\; \pi_M] + [v_1 \;\cdots\; v_M]$

where

(3.34) $$\begin{cases} \underset{(1xK)}{\pi'_m} = [\pi_{1m}\; \pi_{2m} \cdots \pi_{Km}] \;, & m=1,\dots,M \;; \\ \underset{(1xNT)}{v'_m} = [v_{m11} \cdots v_{m1T}\; v_{m21} \cdots v_{m2T} \cdots v_{mNT}] \;, & m=1,\dots,M; \end{cases}$$

and where v_m denotes the m-th column of V , m=1,...,M .

Thus we have M sets of NT equations each :

(3.35) $y_m = X\,\pi_m + v_m$, m=1,...,M .

Let us now examine the errors of (3.35) in detail. We have :

(3.36) $V = -\,U\,\Gamma^{-1}$

If we use $\bar{T}$ to represent $-\Gamma^{-1}$, we can write :

$$[v_1 \;\cdots\; v_M] = [u_1 \;\cdots\; u_M]\,\bar{T}$$
$$= \begin{bmatrix} u'_{11} \\ \vdots \\ u'_{1T} \\ \vdots \\ u'_{NT} \end{bmatrix} [\bar{T}_1 \;\cdots\; \bar{T}_M]$$

where $\bar{T}_m$ denotes the m-th column of $\bar{T}$,m=1,...,M and

$\underset{(1xM)}{u'_{it}} = [u_{1it}\; u_{2it} \cdots u_{Mit}]$ for i=1,...,N ; t=1,...,T .

Thus,

(3.37) $v_m = U\,\bar{T}_m$, m=1,...,M

or

$$(3.38)\quad v_{mit} = u'_{it}\,\bar{T}_m = \bar{T}'_m\,u_{it} = \bar{T}'_m\,\mu_i + \bar{T}'_m\,\nu_t + \bar{T}'_m\,\varepsilon_{it}$$

where

$$(3.39)\quad \underset{(Mx1)}{\mu_i} = \begin{bmatrix}\mu_{1i}\\ \cdot\\ \cdot\\ \cdot\\ \mu_{Mi}\end{bmatrix};\quad \underset{(Mx1)}{\nu_t} = \begin{bmatrix}\nu_{1t}\\ \cdot\\ \cdot\\ \cdot\\ \nu_{Mt}\end{bmatrix};\quad \underset{(Mx1)}{\varepsilon_{it}} = \begin{bmatrix}\varepsilon_{1it}\\ \cdot\\ \cdot\\ \cdot\\ \varepsilon_{Mit}\end{bmatrix}$$

From assumptions (3.17) it follows that

$$(3.40)\quad E(\mu_i\mu'_{i'}) = \delta_{ii'}\Sigma_\mu \quad \text{where } \Sigma_\mu = [\sigma_{\mu mm'}] \quad m,m'=1,\dots,M$$

for $i,i'=1,\dots,N$;

$$(3.41)\quad E(\nu_t\nu'_{t'}) = \delta_{tt'}\Sigma_\nu \quad \text{where } \Sigma_\nu = [\sigma_{\nu mm'}] \quad m,m'=1,\dots,M$$

for $t,t'=1,\dots,T$;

$$(3.42)\quad E(\varepsilon_{it}\varepsilon'_{i't'}) = \delta_{ii'}\,\delta_{tt'}\Sigma_\varepsilon \quad \text{where } \Sigma_\varepsilon = [\sigma_{\varepsilon mm'}] \quad m,m'=1,\dots,M$$

for $i,i'=1,\dots,N$; $t,t'=1,\dots,T$;

$$(3.43)\quad E(\mu_i\nu'_t) = E(\mu_i\varepsilon'_{i't}) = E(\nu_t\varepsilon'_{it'}) = 0$$

for $i,i'=1,\dots,N$; $t,t'=1,\dots,T$.

Let us introduce some more notations which will simplify future calculations :

$$(3.44)\quad \tilde{\mu}_{mi} = \bar{T}'_m\,\mu_i\ ;\ \tilde{\nu}_{mt} = \bar{T}'_m\,\nu_t\ ;\ \tilde{\varepsilon}_{mit} = \bar{T}'_m\,\varepsilon_{it}\ ;$$

$$(3.45)\quad \omega_{\mu mm'} = \bar{T}'_m\Sigma_\mu\bar{T}_{m'}\ ;\ \omega_{\nu mm'} = \bar{T}'_m\Sigma_\nu\bar{T}'_{m'}\ ;\ \omega_{\varepsilon mm'} = \bar{T}'_m\Sigma_\varepsilon\bar{T}'_{m'}$$

Then we can write :

$$(3.46)\quad v_{mit} = \tilde{\mu}_{mi} + \tilde{\nu}_{mt} + \tilde{\varepsilon}_{mit}$$

$$(3.47)\quad E(\tilde{\mu}_{mi}) = E(\tilde{\nu}_{mt}) = E(\tilde{\varepsilon}_{mit}) = 0 \text{ using (3.44) and (3.16)}$$

$$
(3.48)\quad
\begin{cases}
E(\tilde{\mu}_{mi}\,\tilde{\mu}_{m'i'}) = \delta_{ii'}\,\omega_{\mu mm'} & \text{using (3.44),(3.45) \& (3.40)}\\
E(\tilde{\nu}_{mt}\,\tilde{\nu}_{m't'}) = \delta_{tt'}\,\omega_{\nu mm'} & \text{using (3.44),(3.45) \& (3.41)}\\
E(\tilde{\varepsilon}_{mit}\,\tilde{\varepsilon}_{m'i't'}) = \delta_{ii'}\,\delta_{tt'}\,\omega_{\varepsilon mm'} & \text{using (3.44),(3.45) \& (3.42)}\\
E(\tilde{\mu}_{mi}\,\tilde{\nu}_{m't}) = E(\tilde{\mu}_{mi}\,\tilde{\varepsilon}_{m'i't}) = E(\tilde{\nu}_{mt}\,\tilde{\varepsilon}_{m'it'}) = 0 & \text{using (3.44) \& (3.43)}
\end{cases}
$$

and

$$(3.49)\quad E(v_{mit}v_{m'i't'}) = \delta_{ii'}\omega_{\mu mm'} + \delta_{tt'}\omega_{\nu mm'} + \delta_{ii'}\delta_{tt'}\omega_{\varepsilon mm'} \quad \text{using (3.46) and (3.48)}$$

By denoting

$$(3.50)\quad \underset{(N\times 1)}{\tilde{\mu}^m} = [\tilde{\mu}_{mi}]_{i=1,\dots,N}\;;\quad \underset{(T\times 1)}{\tilde{\nu}^m} = [\tilde{\nu}_{mt}]_{t=1,\dots,T}\;;\quad \underset{(NT\times 1)}{\tilde{\varepsilon}_m} = [\tilde{\varepsilon}_{mit}]_{\substack{i=1,\dots,N\\ t=1,\dots,T}}$$

we can write :

$$(3.51)\quad v_m = (I_N \otimes \iota_T)\,\tilde{\mu}^m + (\iota_T \otimes I_N)\,\tilde{\nu}^m + \tilde{\varepsilon}_m$$

which, along with (3.48) and (3.49), allows us to write the stochastic properties of reduced form errors in vector notation as :

$$
(3.52)\quad
\begin{cases}
E(\tilde{\mu}^m) = E(\tilde{\nu}^m) = E(\tilde{\varepsilon}_m) = 0\\
E(\tilde{\mu}^m\,\tilde{\mu}^{m'\prime}) = \omega_{\mu mm'}\,I_N\\
E(\tilde{\nu}^m\,\tilde{\nu}^{m'\prime}) = \omega_{\nu mm'}\,I_T\\
E(\tilde{\varepsilon}_m\,\tilde{\varepsilon}'_{m'}) = \omega_{\varepsilon mm'}\,I_{NT}
\end{cases}
$$

and hence

$$(3.53)\quad E(v_m v'_{m'}) = \Omega_{mm'} = \omega_{\mu mm'}\,A + \omega_{\nu mm'}\,B + \omega_{\varepsilon mm'}\,I_{NT}$$

Two important remarks can be made at this stage by looking at (3.46), (3.47) and (3.48) :

(1) Each reduced form error has an error component structure being composed of three independent components : a unit effect, a time effect and a residual effect, in the same way as any structural equation error.

(2) The first-order and second-order properties of the distribution of reduced form errors are also those of error components. In addition, no correlations exist between the explanatory variables and errors of reduced form equations which is not the case in structural equations.

Thus, the reduced form, represented by

$$(3.54) \qquad y_m = X \pi_m + v_m \quad , \quad m=1,\dots,M$$

is a combination of seemingly unrelated regressions and error components.

3.2.2 Feasible GLS Estimation of Reduced Form Parameters

Let us rewrite the above M reduced form equations one below the other as follows :

$$\begin{bmatrix} y_1 \\ \vdots \\ y_M \end{bmatrix} = \begin{bmatrix} X & 0 & 0 \\ 0 & \ddots & 0 \\ 0 & 0 & X \end{bmatrix} \begin{bmatrix} \pi_1 \\ \vdots \\ \pi_M \end{bmatrix} + \begin{bmatrix} v_1 \\ \vdots \\ v_M \end{bmatrix}$$

and give this system the following compact notation :

$$(3.55) \qquad y = Z \delta + w$$

where

$$(3.56) \qquad Z = I \otimes X \ , \ y = \text{vec } Y \ , \ \delta = \text{vec } \Pi \ \text{ and } w = \text{vec } V$$

The variance-covariance matrix of the error vector of the above system, w, can be written as :

$$(3.57) \qquad \Omega = E(w\, w') = \left[E(v_m v'_{m'})\right] = \left[\Omega_{mm'}\right] \qquad m,m'=1,\dots,M$$

where

(3.58) $\Omega_{mm'} = \omega_{\mu mm'} A + \omega_{\nu mm'} B + \omega_{\varepsilon mm'} I_{NT}$

as derived in (3.53). This matrix can also be written as :

(3.59) $\Omega_{mm'} = \omega_{1mm'} M_1 + \omega_{2mm'} M_2 + \omega_{3mm'} M_3 + \omega_{4mm'} M_4$

where M_1, M_2, M_3 and M_4 are defined in (3.26) (page 51)

and

(3.60) $$\begin{cases} \omega_{1mm'} = \omega_{\varepsilon mm'} + T\,\omega_{\mu mm'} \\ \omega_{2mm'} = \omega_{\varepsilon mm'} + N\,\omega_{\nu mm'} \\ \omega_{3mm'} = \omega_{\varepsilon mm'} + T\,\omega_{\mu mm'} + N\,\omega_{\nu mm'} \\ \omega_{4mm'} = \omega_{\varepsilon mm'} \end{cases}$$

Also $\omega_{1mm'}$, $\omega_{2mm'}$, $\omega_{3mm'}$ and $\omega_{4mm'}$ are the distinct characteristic roots of $\Omega_{mm'}$ of multiplicity (N-1), (T-1), 1 and (N-1)(T-1) respectively.

Replacing $\Omega_{mm'}$ in (3.57) by its expression given in (3.59) we get :

(3.61) $\Omega = \Omega_1 \otimes M_1 + \Omega_2 \otimes M_2 + \Omega_3 \otimes M_3 + \Omega_4 \otimes M_4$

where

(3.62) $\Omega_1 = [\omega_{1mm'}]$, $\Omega_2 = [\omega_{2mm'}]$, $\Omega_3 = [\omega_{3mm'}]$, $\Omega_4 = [\omega_{4mm'}]$

$m = 1,\dots,M$
$m' = 1,\dots,M$

Denoting

(3.63) $\Omega_\mu = [\omega_{\mu mm'}]$, $\Omega_\nu = [\omega_{\nu mm'}]$, $\Omega_\varepsilon = [\omega_{\varepsilon mm'}]$ $m,m'=1,\dots,M$

we have the following relationships :

(3.64) $$\begin{cases} \Omega_1 = T\,\Omega_\mu + \Omega_\varepsilon \\ \Omega_2 = N\,\Omega_\nu + \Omega_\varepsilon \\ \Omega_3 = T\,\Omega_\mu + N\,\Omega_\nu + \Omega_\varepsilon = \Omega_1 + \Omega_2 - \Omega_\varepsilon \\ \Omega_4 = \Omega_\varepsilon \end{cases}$$

Again it can be shown that

$$(3.65)\qquad \Omega^{-1} = \Omega_1^{-1} \otimes M_1 + \Omega_2^{-1} \otimes M_2 + \Omega_3^{-1} \otimes M_3 + \Omega_4^{-1} \otimes M_4$$

Thus the GLS estimator of δ or vec Π is given by

$$(3.66)\qquad \text{vec}\ \hat{\Pi}_{GLS} = (Z'\ \Omega^{-1}\ Z)^{-1}\ Z'\ \Omega^{-1}\ y$$

$$= \left[\sum_i Z'(\Omega_i^{-1} \otimes M_i)\ Z\right]^{-1} \sum_i Z'(\Omega_i^{-1} \otimes M_i)\ y$$

Noting that $Z = I \otimes X$ and simplifying, we obtain :

$$(3.67)\qquad \text{vec}\ \hat{\Pi}_{GLS} = \left[\sum_i \Omega_i^{-1} \otimes X'M_iX\right]^{-1} \left[\sum_i \Omega_i^{-1} \otimes X'M_i\right] y$$

The above expression for vec $\hat{\Pi}_{GLS}$ also proves that in a seemingly unrelated regressions model with error components, the fact that each equation has an identical set of explanatory variables is not a sufficient condition for GLS performed on the whole system to be equivalent to GLS performed on each equation separately. This result is due to Baltagi [7] .

Before proceeding further with the estimation of Ω_1, Ω_2, Ω_3, Ω_4, we will devote a short paragraph to a reformulation of the reduced form equations which separates the constant terms of all equations from the other explanatory variables. We will later see that, in case we assume that the relations of the model contain a constant term in general, the reformulation mentioned above is necessary for the study of the asymptotic properties and the limiting distribution of GLS estimators.

Let us consider the m-th set of reduced form equations

$$y_m = X\ \pi_m + v_m$$

and write it in a slightly different form as

$$(3.68)\qquad y_m = \iota_{NT}\ \pi_{om} + \underline{X}\ \pi_{*m} + v_m$$

where π_m' is partitioned as $[\pi_{om} \;\; \pi_{*m}']$ and X as $[\iota_{NT} \; \underline{X}]$.

Then, combining all the M equations, we get :

$$\begin{bmatrix} y_1 \\ \vdots \\ y_M \end{bmatrix} = \begin{bmatrix} \iota_{NT} & & \\ & \ddots & \\ & & \iota_{NT} \end{bmatrix} \begin{bmatrix} \pi_{o1} \\ \vdots \\ \pi_{oM} \end{bmatrix} + \begin{bmatrix} \underline{X} & & \\ & \ddots & \\ & & \underline{X} \end{bmatrix} \begin{bmatrix} \pi_{*1} \\ \vdots \\ \pi_{*M} \end{bmatrix} + \begin{bmatrix} v_1 \\ \vdots \\ v_M \end{bmatrix}$$

or

$$(3.69) \qquad y = X_o \, \pi_o + X_* \, \pi_* + w$$

where

$$(3.70) \quad \begin{cases} X_o = \begin{bmatrix} \iota_{NT} & & \bigcirc \\ & \ddots & \\ \bigcirc & & \iota_{NT} \end{bmatrix} = (I_M \otimes \iota_{NT}) \; ; \\[2ex] X_* = \begin{bmatrix} \underline{X} & & \bigcirc \\ & \ddots & \\ \bigcirc & & \underline{X} \end{bmatrix} = (I_M \otimes \underline{X}) \; ; \\[2ex] \underset{(Mx1)}{\pi_o'} = [\pi_{o1} \cdots \pi_{oM}] \quad \text{and} \quad \underset{(M(K-1)x1)}{\pi_*'} = [\pi_{*1}' \cdots \pi_{*M}'] \end{cases}$$

Equation (3.69) can also be written as

$$(3.71) \qquad y = \tilde{X} \, \tilde{\pi} + w$$

where

$$(3.72) \qquad \tilde{X} = \begin{bmatrix} X_o & X_* \end{bmatrix} \quad \text{and} \quad \tilde{\pi} = \begin{bmatrix} \pi_o \\ \pi_* \end{bmatrix}$$

Now, let us write the GLS estimator of $\tilde{\pi}$:

$$(3.73) \qquad \hat{\tilde{\pi}}_{GLS} = (\tilde{X}'\Omega^{-1} \tilde{X})^{-1} \tilde{X}'\Omega^{-1} y$$

$$= \left[\begin{pmatrix} X_o' \\ X_*' \end{pmatrix} \Omega^{-1} (X_o \; X_*) \right]^{-1} \begin{pmatrix} X_o' \\ X_*' \end{pmatrix} \Omega^{-1} y$$

$$= \begin{bmatrix} X_o' \Omega^{-1} X_o & X_o' \Omega^{-1} X_* \\ X_*' \Omega^{-1} X_o & X_*' \Omega^{-1} X_* \end{bmatrix}^{-1} \begin{bmatrix} X_o' \Omega^{-1} y \\ X_*' \Omega^{-1} y \end{bmatrix}$$

Substituting $(I_M \otimes \iota_{NT})$ for X_o , $(I_M \otimes \underline{X})$ for X_* and (3.65) for Ω^{-1} and simplifying, we get :

$$(3.74) \qquad \hat{\tilde{\pi}}_{GLS} = \begin{bmatrix} \sum_i \Omega_i^{-1} \otimes \iota' M_i \iota & \sum_i \Omega_i^{-1} \otimes \iota' M_i \underline{X} \\ \sum_i \Omega_i^{-1} \otimes \underline{X}' M_i \iota & \sum_i \Omega_i^{-1} \otimes \underline{X}' M_i \underline{X} \end{bmatrix}^{-1} \begin{bmatrix} \sum_i (\Omega_i^{-1} \otimes \iota' M_i) y \\ \sum_i (\Omega_i^{-1} \otimes \underline{X}' M_i) y \end{bmatrix}$$

Noting that $\iota' M_i = 0$ for i=1,2,4 and $\iota' M_3 = \iota'$, we can write:

$$(3.75) \qquad \hat{\tilde{\pi}}_{GLS} = \begin{bmatrix} \Omega_3^{-1} \cdot NT & \Omega_3^{-1} \otimes \iota' \underline{X} \\ \Omega_3^{-1} \otimes \underline{X}' \iota & \sum_i \Omega_i^{-1} \otimes \underline{X}' M_i \underline{X} \end{bmatrix}^{-1} \begin{bmatrix} (\Omega_3^{-1} \otimes \iota') y \\ \sum_i (\Omega_i^{-1} \otimes \underline{X}' M_i) y \end{bmatrix}$$

Now, let us see how the GLS estimation of vec Π or $\tilde{\pi}$ can be made feasible.

Different feasible Aitken estimators of vec Π could be obtained by constructing different estimators of Ω_1, Ω_2, Ω_3, Ω_4 . Here we present a method leading to an asymptotically efficient estimator of vec Π.

If V, the NT x M matrix of disturbances, is known, the best quadratic unbiased estimators of the variance-covariance matrices Ω_4, Ω_1, Ω_2 and Ω_3 are respectively :

$$(3.76) \quad \begin{cases} \tilde{\Omega}_4 = \dfrac{1}{(N-1)(T-1)} V' Q V \\ \tilde{\Omega}_1 = \dfrac{1}{(N-1)} V' M_1 V \\ \tilde{\Omega}_2 = \dfrac{1}{(T-1)} V' M_2 V \\ \text{and} \quad \tilde{\Omega}_3 = \tilde{\Omega}_1 + \tilde{\Omega}_2 - \tilde{\Omega}_4 \end{cases}$$

These are the analysis of variance (AOV) estimators of the variance components as discussed by Amemiya [1] and Graybill [16] for the single-equation case. Since the v_m(s) are not observable, estimates should be used. The method presented below uses the so-called covariance estimator of the coefficients to predict the errors v_m.

Let us consider the m-th reduced form equation as written in (3.68) :

$$y_m = \iota_{NT}\, \pi_{om} + \underline{X}\, \pi_{*m} + v_m$$

The covariance estimators of π_{om} and π_{*m} are obtained by performing least squares with dummy variables on the above equation. More explicitly, this equation is first transformed by Q . Then π_{*m} is estimated by performing OLS on the transformed equation and π_{om} by the mean of the resulting residuals :

$$(3.77)\quad \begin{cases} \hat{\pi}_{*m(cov)} = (\underline{X}'\, Q\, \underline{X})^{-1}\, \underline{X}'\, Q\, y_m \\ \hat{\pi}_{om(cov)} = \frac{1}{NT}\, \iota'_{NT}\, (y_m - \underline{X}\, \hat{\pi}_{*m(cov)}) \end{cases}$$

Finally, v_m is predicted as :

$$(3.78)\quad \hat{v}_m = y_m - \iota_{NT}\, \hat{\pi}_{om(cov)} - \underline{X}\, \hat{\pi}_{*m(cov)}$$

Repeating the above procedure for all equations, we can form

$$(3.79)\quad \hat{V} = [\hat{v}_1\ \hat{v}_2\ \ldots\ \hat{v}_M]$$

At this point, it may be useful to give the following compact notation for the covariance estimators of the coefficients of all the equations combined :

$$(3.80)\quad \hat{\pi}'_{cov} = [\hat{\pi}_{01(cov)}\ \ldots\ \hat{\pi}_{0M(cov)}\ \hat{\pi}'_{*1(cov)}\ \ldots\ \hat{\pi}'_{*M(cov)}]$$

In case the model does not have a constant term, we can directly estimate π_m as

$$\hat{\pi}_{m(cov)} = (X'\, Q\, X)^{-1}\, X'\, Q\, y_m$$

or estimate Π as :

$$\hat{\Pi}_{cov} = [\,\hat{\pi}_{1(cov)}\ \ldots\ \hat{\pi}_{M(cov)}] = (X'\, Q\, X)^{-1}\, X'\, Q\, Y$$

and predict V as

$$\hat{V} = Y - X\, \hat{\Pi}_{cov}$$

Substituting $\hat{V}$ for V in (3.76) we get $\hat{\Omega}_4$, $\hat{\Omega}_1$, $\hat{\Omega}_2$ and $\hat{\Omega}_3$ respectively. (The consistency of these estimators is proved in Appendix 3.A.3, page 80.) There is no guarantee that the matrix $\hat{\Omega}_3$ in (3.76) is positive definite in a given sample.

The usual procedure is to assume that the data are centered[1], so that $X'M_3X$ is equal to zero. The sum in (3.75) therefore runs only over the indices 1,2 and 4 and $\hat{\Omega}_3$ is not needed. Note also that in the case of only individual effects, there is no such problem.

Substituting $\hat{\Omega}_4$, $\hat{\Omega}_1$, $\hat{\Omega}_2$ and $\hat{\Omega}_3$ for Ω_4, Ω_1, Ω_2 and Ω_3 respectively in (3.67), we obtain the feasible Aitken estimator of vec Π :

$$(3.81)\qquad \text{vec}\,\hat{\Pi}_{fGLS} = \left[\sum_i \hat{\Omega}_i^{-1} \otimes X'M_iX\right]^{-1} \left[\sum_i \hat{\Omega}_i^{-1} \otimes X'M_i\right] y$$

and making the same substitutions in (3.75) we get the feasible Aitken estimator of $\hat{\tilde{\pi}}$:

$$(3.82)\qquad \hat{\tilde{\pi}}_{fGLS} = \begin{bmatrix} \hat{\Omega}_3^{-1}\; NT & \hat{\Omega}_3^{-1} \otimes \iota'\underline{X} \\ \hat{\Omega}_3^{-1} \otimes \underline{X}'\iota & \sum_i \hat{\Omega}_i^{-1} \otimes \underline{X}'M_i\underline{X} \end{bmatrix}^{-1} \begin{bmatrix} (\hat{\Omega}_3^{-1} \otimes \iota')y \\ \sum_i (\hat{\Omega}_i^{-1} \otimes \underline{X}'M_i)y \end{bmatrix}$$

Note that $\hat{\tilde{\pi}}_{fGLS}$ simply represents a rearrangement of the elements of vec $\hat{\Pi}_{fGLS}$ but both the estimators are identical. The consistency of $\hat{\tilde{\pi}}_{fGLS}$ is proved in Appendix 3.A.4.[2]

In order to determine the limiting distribution of the feasible GLS estimator, we will use the above representation of the estimator, in the form of $\tilde{\pi}_{fGLS}$, as plim of $\frac{1}{NT}\sum_i \hat{\Omega}_i^{-1} \otimes X'M_iX$ appearing in the expression of vec $\hat{\Pi}_{fGLS}$ is singular when X contains ι_{NT} as its first column and hence cannot be inversed.

---

1) Note that if the data are centered, the constant term disappears.

2) All consistency proofs hold exactly for the case in which X contains observations on exogenous variables only. When lagged endogenous variables are present, many precautions have to be taken, in particular, the assumptions concerning the probability limits of $\frac{1}{NT}\underline{X}'\underline{X}$ and $\frac{1}{NT}\underline{X}'\iota$ have to be reexamined carefully.

The details of derivation of the limiting distribution are given in Appendix 3.B. The final result is as follows :

$$(3.83) \quad D^{-\frac{1}{2}} (\hat{\tilde{\pi}}_{fGLS} - \tilde{\pi}) \sim N(0,\bar{D})$$

where

$$(3.84) \quad D = \begin{bmatrix} \frac{1}{N} I_M & 0 \\ 0 & \frac{1}{NT} I_{M(K-1)} \end{bmatrix}$$

$$(3.85) \quad \bar{D} = \begin{bmatrix} \Omega_\mu + \Omega_\nu & 0 \\ 0 & \Omega_\varepsilon \otimes R^{-1} \end{bmatrix}$$

In other words,

$$(3.86) \quad \begin{cases} \sqrt{N}(\hat{\pi}_{o,fGLS} - \pi_o) \sim N(0,(\Omega_\mu + \Omega_\nu)) \\ \sqrt{NT}(\hat{\pi}_{*,fGLS} - \pi_*) \sim N(0, \Omega_\varepsilon \otimes R^{-1}) \end{cases}$$

Note that when there is no constant term in the model, $\sqrt{NT}$ (vec $\hat{\Pi}_{fGLS}$ - vec Π) has the limiting distribution $N(0, \Omega_\varepsilon \otimes R^{-1})$ where $R = \text{plim} \frac{1}{NT} X'M_4X$.

By following a similar procedure as in Appendix 3.B, it can be easily verified that $D^{-\frac{1}{2}}(\hat{\tilde{\pi}}_{cov} - \tilde{\pi})$ also has the same limiting distribution as that of $D^{-\frac{1}{2}}(\hat{\tilde{\pi}}_{fGLS} - \tilde{\pi})$ given above in (3.86).

3.2.3 Maximum Likelihood Estimation of the Reduced Form

The log-likelihood function of the unrestricted reduced form, neglecting the term which is constant with respect to the parameters to be estimated, is :

$$\log L(Y/\Pi,\Omega_\mu,\Omega_\nu,\Omega_\varepsilon) = -\frac{1}{2}\sum_{i=1}^{4} m_i \log|\Omega_i|$$

$$-\frac{1}{2}\sum_{i=1}^{4} (\text{vec }(Y-X\Pi))'(\Omega_i^{-1} \otimes M_i)\,\text{vec}(Y-X\Pi)$$

$$(3.87) \qquad = -\frac{1}{2}\sum_{i=1}^{4} m_i \log|\Omega_i|$$

$$-\frac{1}{2}\sum_{i=1}^{4} \text{tr }(Y-X\Pi)'M_i\,(Y-X\Pi)\,\Omega_i^{-1}$$

with

$$(3.88) \quad \begin{cases} \Omega_1 = \Omega_\varepsilon + T\,\Omega_\mu \\ \Omega_2 = \Omega_\varepsilon + N\,\Omega_\nu \\ \Omega_3 = \Omega_\varepsilon + T\,\Omega_\mu + N\,\Omega_\nu \\ \Omega_4 = \Omega_\varepsilon \end{cases}$$

$$m_1 = (N-1) \quad , \quad M_1 = \frac{A}{T} - \frac{J_{NT}}{NT} \quad ,$$

$$m_2 = (T-1) \quad , \quad M_2 = \frac{B}{N} - \frac{J_{NT}}{NT} \quad ,$$

$$m_3 = 1 \quad , \quad M_3 = \frac{J_{NT}}{NT} \quad ,$$

$$m_4 = (N-1)\,(T-1) \; , \; M_4 = Q = I - \frac{A}{T} - \frac{B}{N} + \frac{J_{NT}}{NT} \; .$$

Notice that we parametarise the likelihood function in terms of Ω_μ , Ω_ν and Ω_ε (which are fixed parameters independent of T and N), but we use the Ω_i as a short-hand notation.

The loglikelihood function has to be maximized with respect to Π , Ω_μ , Ω_ν and Ω_ε and subject to the symmetry condition :

$$(3.89) \quad C \text{ vec } \Omega_\mu = 0 \; ; \quad C \text{ vec } \Omega_\nu = 0 \; ; \quad C \text{ vec } \Omega_\varepsilon = 0$$

where C is a known $\frac{M(M-1)}{2} \times M^2$ matrix of full row rank such that C'C is equal to the idempotent matrix $\frac{1}{2}$ (I - P) , P being the commutation matrix. (Note that any matrix F is symmetric iff P vec F = vec F.)

By writing the symmetry condition as in (3.89) above, we avoid extracting the redundant elements of the different covariance matrices. This is an alternative approach to the one based on the elimination matrix as adopted, for instance, by Magnus and Neudecker [28] and Balestra [5] or the one based on the equivalent vech operator (which stacks the columns of a symmetric matrix starting each column at its diagonal element) as in Henderson and Searle [19]. Our procedure is inspired by an article by F.J.H. Don [13] .

Let us form the Lagrangian function corresponding to the above maximisation problem :

$$(3.90) \quad \mathfrak{L} = \log L - \lambda_\mu' C \text{ vec } \Omega_\mu - \lambda_\nu' C \text{ vec } \Omega_\nu - \lambda_\varepsilon' C \text{ vec } \Omega_\varepsilon$$

where λ_μ, λ_ν and λ_ε are appropriate multiplier vectors.

In order to solve this maximisation problem, we first write the first order differential of the Lagrangian :

$$(3.91) \quad d\mathfrak{L} = -\frac{1}{2} \sum_{i=1}^{4} m_i \text{ tr } \Omega_i^{-1} (d\Omega_i) - \sum_{i=1}^{4} \text{tr } V' M_i (dV) \, \Omega_i^{-1}$$

$$+ \frac{1}{2} \sum_{i=1}^{4} \text{tr } V' M_i \, V \, \Omega_i^{-1} \, d\Omega_i \, \Omega_i^{-1}$$

$$-(d\lambda_\mu)' \, C \text{ vec } \Omega_\mu - \lambda_\mu' C \, d \text{ vec } \Omega_\mu - (d\lambda_\nu)' \, C \text{ vec } \Omega_\nu$$

$$- \lambda_\nu' \, C \, d \text{ vec } \Omega_\nu - (d\lambda_\varepsilon)' \, C \text{ vec } \Omega_\varepsilon - \lambda_\varepsilon' \, C \, d \text{ vec } \Omega_\varepsilon$$

writing V for Y - X Π and noting that

$$\text{tr } (dV)' \, M_i \, V \, \Omega_i^{-1} = \text{tr } V' M_i \, dV \, \Omega_i^{-1}$$

In writing the above differential, we have used the following basic results (see [27]) :

For any two matrices A, B of appropriate orders :

$$(3.92)\quad \begin{cases} d(AB) = (dA)\, B + A(dB) \\ d\ \mathrm{tr}\ AB = \mathrm{tr}\ (dA)B + \mathrm{tr}\ A(dB) \\ d\ A^{-1} = -\ A^{-1}\ dA\ A^{-1} \\ d\ \log |A| = \mathrm{tr}\ A^{-1}\ dA \quad , |A| > 0 \end{cases}$$

Substituting

$$(3.93)\quad \begin{cases} d\Omega_1 = d\ \Omega_\varepsilon + T\ d\ \Omega_\mu \\ d\Omega_2 = d\ \Omega_\varepsilon + N\ d\ \Omega_\nu \\ d\Omega_3 = d\ \Omega_\varepsilon + T\ d\ \Omega_\mu + N\ d\ \Omega_\nu \\ d\Omega_4 = d\ \Omega_\varepsilon \\ dV = -\ X\ d\ \Pi \end{cases}$$

in (3.91) and using the relationship that

$$\mathrm{tr}\ A'B = (\mathrm{vec}\ A)'\ \mathrm{vec}\ B$$

we can rearrange (3.91) as :

$$(3.94)\quad \begin{aligned} d\mathcal{L} = & -\frac{1}{2}\sum_{i=1}^{4}(\mathrm{vec}(m_i\Omega_i^{-1}))'d\ \mathrm{vec}\ \Omega_\varepsilon + \frac{1}{2}\sum_{i=1}^{4}(\mathrm{vec}(\Omega_i^{-1}\ V'M_iV\ \Omega_i^{-1}))'d\ \mathrm{vec}\ \Omega_\varepsilon \\ & -\frac{T}{2}\sum_{1,3}(\mathrm{vec}(m_i\Omega_i^{-1}))'d\ \mathrm{vec}\ \Omega_\mu + \frac{T}{2}\sum_{1,3}(\mathrm{vec}(\Omega_i^{-1}\ V'M_iV\ \Omega_i^{-1}))'d\ \mathrm{vec}\ \Omega_\mu \\ & -\frac{N}{2}\sum_{2,3}(\mathrm{vec}(m_i\Omega_i^{-1}))'d\ \mathrm{vec}\ \Omega_\nu + \frac{N}{2}\sum_{2,3}(\mathrm{vec}(\Omega_i^{-1}\ V'M_iV\ \Omega_i^{-1}))'d\ \mathrm{vec}\ \Omega_\nu \\ & + \sum_{i=1}^{4}(\mathrm{vec}(X'M_iV\ \Omega_i^{-1}))'d\ \mathrm{vec}\ \Pi - \lambda_\varepsilon'\ C\ d\ \mathrm{vec}\ \Omega_\varepsilon - \lambda_\mu'C\ d\ \mathrm{vec}\ \Omega_\mu \\ & - \lambda_\nu'\ C\ d\ \mathrm{vec}\ \Omega_\nu - (d\ \lambda_\varepsilon)'C\ \mathrm{vec}\ \Omega_\varepsilon - (d\ \lambda_\mu)'C\ \mathrm{vec}\ \Omega_\mu - (d\ \lambda_\nu)'C\ \mathrm{vec}\ \Omega_\nu \end{aligned}$$

The first-order conditions of maximisation are given by $d\mathcal{L} = 0$ for all $d \text{ vec } \Pi \neq 0$, $d \text{ vec } \Omega_j \neq 0$, $j = \mu,\nu,\varepsilon$ and $d\lambda_j \neq 0$, $j = \mu,\nu,\varepsilon$. Thus we will have the following system of equations (in which we have come back to the expression $Y - X\Pi$ for V) :

$$(3.95) \qquad \text{vec} \sum_{i=1}^{4} X'M_i(Y-X\Pi)\,\Omega_i^{-1} = 0$$

$$(3.96) \qquad \text{vec}\left(-\frac{1}{2}\sum_i m_i\,\Omega_i^{-1} + \frac{1}{2}\sum_i \Omega_i^{-1}(Y-X\Pi)'M_i(Y-X\Pi)\Omega_i^{-1} - C'\lambda_\varepsilon\right) = 0$$

$$(3.97) \qquad \text{vec}\left(-\frac{T}{2}\sum_{1,3} m_i\,\Omega_i^{-1} + \frac{T}{2}\sum_{1,3} \Omega_i^{-1}(Y-X\Pi)'M_i(Y-X\Pi)\Omega_i^{-1} - C'\lambda_\mu\right) = 0$$

$$(3.98) \qquad \text{vec}\left(-\frac{N}{2}\sum_{2,3} m_i\,\Omega_i^{-1} + \frac{N}{2}\sum_{2,3} \Omega_i^{-1}(Y-X\Pi)'M_i(Y-X\Pi)\Omega_i^{-1} - C'\lambda_\nu\right) = 0$$

Premultiplying the last three equations (3.96), (3.97), (3.98) by C we can easily see that $\lambda_\mu = \lambda_\nu = \lambda_\varepsilon = 0$ as the matrices inside the parentheses in each of these three equations are symmetric. Thus we can neglect the symmetry conditions and proceed to the direct maximisation of log L. It should, however, be kept in mind, that the symmetry constraints (3.89) are relevant for the computation of the information matrix.

Upon setting $\lambda_\mu = \lambda_\nu = \lambda_\varepsilon = 0$ in equations (3.96), (3.97) and (3.98), we can rewrite these equations as :

$$(3.99) \qquad \text{vec } \Pi = \left[\sum_{i=1}^{4}(\Omega_i^{-1} \otimes X'M_iX)\right]^{-1} \sum_{i=1}^{4}(\Omega_i^{-1} \otimes X'M_i)\text{ vec } Y$$

$$(3.100) \qquad \Omega_i = \frac{1}{m_i}(Y-X\Pi)'\,M_i(Y-X\Pi) + \frac{1}{m_i}\Omega_i\,W\,\Omega_i \qquad i=1,2$$

$$(3.101) \qquad \Omega_4 = \frac{1}{m_4}(Y-X\Pi)'\,M_4(Y-X\Pi) - \frac{1}{m_4}\Omega_4\,W\,\Omega_4$$

where

$$(3.102)\quad W = \Omega_3^{-1}(Y-X\Pi)'\,M_3(Y-X\Pi)\,\Omega_3^{-1} - m_3\Omega_3^{-1}\ ,\quad \Omega_3=\Omega_1+\Omega_2-\Omega_4$$

Equations (3.100) and (3.101) are straightforward whereas (3.99) is obtained by applying the result :

$$\text{vec } ABC = (C' \otimes A)\ \text{vec } B$$

on (3.95) :

$$\sum_i \text{vec }(X'M_iY\,\Omega_i^{-1} - X'M_iX\,\Pi\,\Omega_i^{-1}) = 0$$

$$\sum_i (\Omega_i^{-1} \otimes X'M_i)\ \text{vec } Y - (\Omega_i^{-1} \otimes X'M_iX)\ \text{vec }\Pi = 0$$

and hence

$$\text{vec }\Pi = (\sum_i \Omega_i^{-1} \otimes X'M_iX)^{-1} \sum_i (\Omega_i^{-1} \otimes X'M_i)\ \text{vec } Y$$

Contrary to the case of only unit effects (see Magnus [27]), in the present case, no explicit expressions for the Ω_i in terms of Π can be obtained from the system (3.99) to (3.102). However, the whole system can be solved iteratively as follows :

Step 1 : Initial conditions : $\Omega_i^{-1} = 0$, i=1,2,3 and $\Omega_4 = I$.

Step 2 : Use (3.99) to compute vec Π .

Step 3 : Compute Ω_i , i=1,2,4 from (3.100) and (3.101) using on the right hand side of these equations the current estimate for Π and the previous ones for Ω_i . (Note that in the first iteration W = 0). Compute also $\Omega_3 = \Omega_1 + \Omega_2 - \Omega_4$ and W from (3.102).

Step 4 : Go back to Step 2 and repeat until convergence.

If the above initial values are chosen, then it is easily verified that :

vec Π = vec $\hat{\Pi}_{cov}$ at the first iteration ;

vec Π = vec $\hat{\Pi}_{fGLS}$ at the second iteration.

In order to find the limiting distribution of the ML estimators, once again if we assume the presence of constant terms in the model, we have to rewrite the first-order conditions using the notation introduced in equations (3.68) to (3.72) of Section 3.2.2 or in short, vec V = vec $(Y - X\Pi)$ has to be replaced by $y - \tilde{X}\tilde{\pi}$ and d vec V by $-\tilde{X}\, d\tilde{\pi}$. Thus, instead of writing tr $V'M_i(dV)\Omega_i^{-1}$ as $-$vec $X'M_iV\,\Omega_i^{-1}$ d vec Π as we did in going from expression (3.91) to (3.94) of $d\mathcal{L}$, now we have to write

$$(3.103)\quad \text{tr } V'M_i dV\Omega_i^{-1} = (\text{vec } V)'(\Omega_i^{-1} \otimes M_i)\, d \text{ vec } V$$

$$= -\,(y - \tilde{X}\tilde{\pi})'(\Omega_i^{-1} \otimes M_i)\,\tilde{X}\, d\,\tilde{\pi}$$

$$= \left[-\, y'(\Omega_i^{-1} \otimes M_i)\tilde{X} + \tilde{\pi}'\tilde{X}'(\Omega_i^{-1} \otimes M_i)\tilde{X}\right] d\tilde{\pi}$$

Hence, equation (3.95) will be replaced by

$$\sum_{i=1}^{4} -\, y'(\Omega_i^{-1} \otimes M_i)\,\tilde{X} + \tilde{\pi}'\tilde{X}'(\Omega_i^{-1} \otimes M_i)\,\tilde{X} = 0$$

and equation (3.99) becomes :

$$(3.104)\quad \tilde{\pi} = \left[\tilde{X}'\,(\sum_i \Omega_i^{-1} \otimes M_i)\,\tilde{X}\right]^{-1} \tilde{X}'(\sum_i \Omega_i^{-1} \otimes M_i)\, y$$

The solution of the maximum likelihood problem for the parameters $\tilde{\pi}$, Ω_μ, Ω_ν, Ω_ε is given by iteratively solving (3.104), (3.100), (3.101) and (3.102) using the same iteration procedure as before. In the first iteration, one has to be careful while calculating $\tilde{\pi}$ using (3.104) as for the initial conditions of the iteration, $\tilde{X}'\,(\sum_i \Omega_i^{-1} \otimes M_i)\,\tilde{X}$ is singular. Thus the formula has to be modified slightly for the first step :

$$(3.105)\quad \pi_*^{(1)} = \left[X_*'(\sum_i \Omega_i^{-1} \otimes M_i)\, X_*\right]^{-1} X_*'(\sum_i \Omega_i^{-1} \otimes M_i)\, y$$

$$(3.106)\quad \pi_o^{(1)} = \frac{1}{NT}\begin{bmatrix} \iota_{NT}' & & \bigcirc \\ & \ddots & \\ \bigcirc & & \iota_{NT}' \end{bmatrix} \begin{bmatrix} y_1 - \underline{X}\,\pi_{*1}^{(1)} \\ \vdots \\ y_M - \underline{X}\,\pi_{*M}^{(1)} \end{bmatrix}$$

where initial conditions $\Omega_i^{-1} = 0$ for $i=1,2,3$ and $\Omega_4 = I$ have to be substituted. Note that once the initial conditions are replaced in (3.105), (3.106), $\tilde{\pi}^{(1)} = [\pi_o^{(1)\prime}\pi_*^{(1)\prime}]'$ is exactly the covariance estimator $\hat{\hat{\pi}}_{cov}$ defined in (3.80).

The moments of the limiting distribution of ML estimators can be calculated using the inverse of the bordered information-matrix. The different steps leading to the bordered information matrix, namely, the computations of the second-order differential of the loglikelihood function, its expectation, the information matrix and its limit and inverse, are presented in detail in Appendix 3.C. Here we will state only the final result.

Before that, let us mention that we derive the limiting distribution of the coefficient parameters in the following form :

$$\sqrt{N}\ (\hat{\pi}_{o,ML} - \pi_o)$$
$$\sqrt{NT}\ (\hat{\pi}_{*,ML} - \pi_*)$$

and that, as far as the covariance parameters are concerned, the limiting distributions of

$$\sqrt{NT}\ (\text{vec}\ \hat{\Omega}_{\varepsilon,ML} - \text{vec}\ \Omega_\varepsilon)$$
$$\sqrt{N}\ (\text{vec}\ \hat{\Omega}_{\mu,ML} - \text{vec}\ \Omega_\mu)$$
$$\sqrt{T}\ (\text{vec}\ \hat{\Omega}_{\nu,ML} - \text{vec}\ \Omega_\nu)$$

are derived, as already done by Amemiya [1] for the single-equation error components model.

The results are as follows :

$$(3.107)\quad \begin{cases} \sqrt{N}^{1)}\ (\hat{\pi}_{o,ML} - \pi_o) \sim N\Big(0\ ,\ (\Omega_\mu + \Omega_\nu)\ \Big) \\ \sqrt{NT}\ (\hat{\pi}_{*,ML} - \pi_*) \sim N\Big(0\ ,\ \Omega_\varepsilon \otimes R^{-1}\Big) \end{cases}$$

---

(1) Note that we show in Appendix 3.C that $\sqrt{T}\ (\hat{\pi}_{o,ML} - \pi_o)$ has the same limiting distribution as $\sqrt{N}\ (\hat{\pi}_{o,ML} - \pi_o)$. (See footnote 2) on page 115.)

$$(3.108)\quad \begin{cases} \sqrt{NT}\ \mathrm{vec}(\hat{\Omega}_{\varepsilon,ML} - \Omega_\varepsilon) \sim N\left(0\ ,\ \frac{1}{2}(I+P)(2\,\Omega_\varepsilon \otimes \Omega_\varepsilon\ )\ \frac{1}{2}(I+P)\right) \\ \sqrt{N}\ \ \mathrm{vec}(\hat{\Omega}_{\mu,ML} - \Omega_\mu) \sim N\left(0\ ,\ \frac{1}{2}(I+P)(2\,\Omega_\mu \otimes \Omega_\mu\ )\ \frac{1}{2}(I+P)\right) \\ \sqrt{T}\ \ \mathrm{vec}(\hat{\Omega}_{\nu,ML} - \Omega_\nu) \sim N\left(0\ ,\ \frac{1}{2}(I+P)(2\,\Omega_\nu \otimes \Omega_\nu\ )\ \frac{1}{2}(I+P)\right) \end{cases}$$

By combining the final results of Section 3.2.2 and the ones above, we establish the asymptotic equivalence of $\hat{\hat{\pi}}_{ML}$, $\hat{\tilde{\pi}}_{fGLS}$ and $\hat{\hat{\pi}}_{cov}$, in the sense that all the three have the same limiting distribution.

APPENDIX 3.A : Proof of the Consistency of the Feasible GLS Estimator of Reduced Form Coefficients[1)]

3.A.1 Basic Assumptions and Some Preliminary Results

We make the following assumptions :

(3.A.1) $\operatorname*{plim}_{\substack{N\to\infty \\ T\to\infty}} \frac{1}{NT} \underline{X}'\underline{X} = \underline{R}$, a finite non-singular matrix,

(3.A.2) $\operatorname*{plim}_{N\to\infty} \frac{1}{NT} \underline{X}'\underline{X} = \underline{R}^{(1)}$, a finite non-singular matrix,

(3.A.3) $\operatorname*{plim}_{T\to\infty} \frac{1}{NT} \underline{X}'\underline{X} = \underline{R}^{(2)}$, a finite non-singular matrix,

(3.A.4) $\operatorname*{plim}_{\substack{N\to\infty \\ T\to\infty}} \frac{1}{NT} \underline{X}'\iota = r$, a finite vector,

(3.A.5) $\operatorname*{plim}_{N\to\infty} \frac{1}{NT} \underline{X}'\iota = r^{(1)}$, a finite vector,

(3.A.6) $\operatorname*{plim}_{T\to\infty} \frac{1}{NT} \underline{X}'\iota = r^{(2)}$, a finite vector,

(3.A.7)
- $\tilde{\mu}$, $\tilde{\nu}$ and $\tilde{\varepsilon}$ are pair-wise independent vectors,
- $\tilde{\varepsilon}_{mit}$ and $\tilde{\varepsilon}_{m'i't'}$ are <u>independent</u> for $i \neq i'$ or $t \neq t'$ or both,
- $\tilde{\mu}_{mi}$ and $\tilde{\mu}_{m'i'}$ are <u>independent</u> for $i \neq i'$,
- $\tilde{\nu}_{mt}$ and $\tilde{\nu}_{m't'}$ are <u>independent</u> for $t \neq t'$.

We will now derive certain probability limits that will be useful in future.

---

1) See footnote 2) on page 62.

Plim[1)] of $\frac{1}{NT}\,\underline{X}'M_1\underline{X}$

From (3.26) we have

$$M_1 = \frac{A}{T} - \frac{J_{NT}}{NT} = \frac{I_N \otimes \iota_T \iota_T'}{T} - \frac{\iota_{NT}\,\iota_{NT}'}{NT}$$

and we know that

$$M_1^2 = M_1$$

Let us write :

$$\underline{X} = \begin{bmatrix} x_{211} & \cdots & x_{K11} \\ \vdots & & \vdots \\ x_{21T} & \cdots & x_{K1T} \\ \vdots & & \vdots \\ x_{2N1} & \cdots & x_{KN1} \\ \vdots & & \vdots \\ x_{2NT} & \cdots & x_{KNT} \end{bmatrix} = \begin{bmatrix} \underline{X}_1' \\ \cdot \\ \cdot \\ \cdot \\ \underline{X}_N' \end{bmatrix} \quad \text{where } \underline{X}_i' = \begin{bmatrix} x_{2i1} & \cdots & x_{Ki1} \\ \vdots & & \vdots \\ x_{2iT} & \cdots & x_{KiT} \end{bmatrix}$$

$i=1,\ldots,N.$

Thus

$$M_1X = \begin{bmatrix} \iota_T & \frac{\iota_T'\underline{X}_1'}{T} \\ \vdots & \\ \iota_T & \frac{\iota_T'\underline{X}_N'}{T} \end{bmatrix} - \begin{bmatrix} \iota_T & \frac{\iota_T'(\underline{X}_1'+\ldots+\underline{X}_N')}{NT} \\ \vdots & \\ \iota_T & \frac{\iota_T'(\underline{X}_1'+\ldots+\underline{X}_N')}{NT} \end{bmatrix}$$

$$= \begin{bmatrix} \iota_T\,(\underline{X}_{1.}' - \underline{X}'_{..}) \\ \vdots \\ \iota_T\,(\underline{X}_{N.}' - \underline{X}'_{..}) \end{bmatrix}$$

where

$$\underline{X}_{i.}' = \frac{1}{T}\,\iota_T'\underline{X}_i' = \left[x_{ki.}\right] \qquad k = 2,\ldots,K \quad , \quad i=1,\ldots,N$$

$$\underline{X}'_{..} = \frac{1}{NT}\,\iota_T'(\underline{X}_1' + \ldots + \underline{X}_N') = \left[x_{k..}\right] \qquad k=2,\ldots,K$$

with

$$x_{ki.} = \frac{1}{T}\sum_t x_{kit} \qquad k=1,\ldots,K \;;$$

$$x_{k..} = \frac{1}{NT}\sum_i \sum_t x_{kit} \qquad k=1,\ldots,K \;.$$

1) N -> ∞ or T -> ∞ or both.

Therefore

$$\underline{X}'\,M_1\,\underline{X} = \underline{X}'\,M_1\,M_1\,\underline{X} = (M_1\underline{X})'\,M_1\,\underline{X}$$

$$= T\sum_i (\underline{X}_{i.} - \underline{X}_{..})(\underline{X}'_{i.} - \underline{X}'_{..})$$

A typical element (j,k) of the above matrix is

$$T\sum_i (x_{ji.} - x_{j..})(x_{ki.} - x_{k..})$$

$$= \frac{1}{T}\sum_i\sum_t\sum_{t'} x_{jit}\,x_{kit'} - \frac{1}{NT}\sum_i\sum_t\sum_{i'}\sum_{t'} x_{jit}\,x_{ki't'}$$

and a typical element of $\frac{1}{NT}$ of $\underline{X}'\,M_1\,\underline{X}$ is, therefore,

$$\frac{1}{NT^2}\sum_i\sum_t\sum_{t'} x_{jit}\,x_{kit'} - \frac{1}{N^2T^2}\sum_i\sum_t\sum_{i'}\sum_{t'} x_{jit}\,x_{ki't'}$$

Now, as plim[1] $\frac{1}{NT}\,\underline{X}'\,\underline{X}$ = a constant matrix (assumption (3.A.1) or (3.A.2) or (3.A.3)), we can write, by considering a typical element of this matrix, that :

(3.A.8) $$\text{plim}^{1)}\ \frac{1}{NT}\sum_i\sum_t x_{jit}\,x_{kit} = \text{a constant} \qquad \forall j,k$$

From this it follows that :

$$\text{plim}^{1)}\ \frac{1}{NT^2}\sum_i\sum_t\sum_{t'} x_{jit}\,x_{kit'} = \text{a constant}$$

$$\text{plim}^{1)}\ \frac{1}{N^2T^2}\sum_i\sum_t\sum_{i'}\sum_{t'} x_{jit}\,x_{kit'} = \text{a constant}$$

and hence

(3.A.9) $$\text{plim}^{1)}\ \frac{1}{NT}\,\underline{X}'\,M_1\,\underline{X} = R^{(1)},$$ a finite matrix which we assume to be non-singular.

<u>Plim[1] of $\frac{1}{NT}\,\underline{X}'M_2\underline{X}$:</u>

By noting that

$$M_2 = \frac{B}{N} - \frac{J_{NT}}{NT} = \frac{\iota_N\,\iota_N' \otimes I_T}{N} - \frac{\iota_{NT}\,\iota_{NT}'}{NT}$$

---

1) N-> ∞ or T-> ∞ or both.

and that

$$M_2^2 = M_2$$

we can proceed in a similar way as in the case of $\underline{X}'M_1\underline{X}$ and finally get to the following expression of a typical element of $\frac{1}{NT}\,\underline{X}'M_2\underline{X}$:

$$\frac{1}{N^2T}\sum_i\sum_{i'}\sum_t x_{jit}\,x_{ki't} - \frac{1}{N^2T^2}\sum_i\sum_t\sum_{i'}\sum_{t'} x_{jit}\,x_{ki't'}$$

and by the same reasoning as the one used in the previous case we conclude that both the terms above tend to a constant as N and T tend to ∞ and thus

(3.A.10) $\text{plim}^{1)}\ \frac{1}{NT}\,\underline{X}'M_2\underline{X} = R^{(2)}$, a finite matrix which we assume to be non-singular.

<u>Plim[1] of $\frac{1}{NT}\,\underline{X}'M_3\underline{X}$:</u>

$$\frac{1}{NT}\,\underline{X}'M_3\underline{X} = \frac{1}{NT}\ \underline{X}'\frac{l_{NT}\,l_{NT}'}{NT}\ \underline{X}$$, a typical element

of which is

$$\frac{1}{N^2T^2}\sum_i\sum_t\sum_{i'}\sum_{t'} x_{jit}\,x_{ki't'}$$

Once again from (3.A.8) it follows that

$$\text{plim}\ \frac{1}{N^2T^2}\sum_i\sum_t\sum_{i'}\sum_{t'} x_{jit}\,x_{ki't'} = \text{a constant}$$

and hence

(3.A.11) $\text{plim}\ \frac{1}{NT}\,\underline{X}'M_3\underline{X} = R^{(3)}$, a finite matrix which we assume to be non-singular.

<u>Plim[1] of $\frac{1}{NT}\,\underline{X}'M_4\underline{X}$:</u>

$$\text{plim}\ \frac{1}{NT}\,\underline{X}'M_4\underline{X} = \text{plim}\ \frac{1}{NT}\,\underline{X}'Q\,\underline{X}$$

$$= \frac{1}{NT}\left[\sum_i\sum_t \underline{X}_{it}\underline{X}_{it}' - T\sum_i \underline{X}_{i.}\,\underline{X}_{i.}' - N\,\bar{X}\bar{X}' + NT\,X_{..}X_{..}'\right]$$

1) N-> ∞ or T->∞ or both.

where $\underline{X}'_{it} = [x_{2it} \cdots x_{Kit}]$

$$\bar{X} = \frac{\underline{X}_1 + \dots \underline{X}_N}{N}$$

A typical element of the above matrix is

$$\frac{1}{NT}\sum_i\sum_t x_{jit}x_{kit} - \frac{1}{T}\sum_i\sum_t\sum_{t'} x_{jit}x_{kit'} - \frac{1}{N}\sum_i\sum_{i'}\sum_t x_{jit}x_{ki't}$$

$$+ \frac{1}{NT}\sum_i\sum_t\sum_{i'}\sum_{t'} x_{jit}x_{ki't'}$$

Again using statement (3.A.8) we can say that

(3.A.12) $\text{plim}^{1)} \frac{1}{NT}\underline{X}'Q\,\underline{X} = R$, a finite matrix which we will assume to be positive definite.

Plim of $\frac{1}{NT}\underline{X}'\varepsilon_m$:

A typical element of $\frac{1}{NT}\underline{X}'\varepsilon_m$ is

$$\frac{1}{NT}\sum_i\sum_t x_{kit}\,\varepsilon_{mit}$$

As $\quad E\ \frac{1}{NT}\sum_i\sum_t x_{kit}\,\varepsilon_{mit} = 0$

and Var $\frac{1}{NT}\sum_i\sum_t x_{kit}\,\varepsilon_{mit} = \frac{1}{(NT)^2}\sum_i\sum_t x^2_{kit}\,\sigma_{\varepsilon mm}$

$\rightarrow 0$ as $N\rightarrow\infty$ and $T\rightarrow\infty$

we can conclude (using Result R-1 stated in the beginning of proof of Lemma L-1, page 93) that

(3.A.13) $$\underset{\substack{N\rightarrow\infty\\T\rightarrow\infty}}{\text{plim}} \frac{1}{NT}\sum_i\sum_t x_{kit}\,\varepsilon_{mit} = \underset{\substack{N\rightarrow\infty\\T\rightarrow\infty}}{\text{plim}} \frac{1}{NT}\underline{X}'\varepsilon_m = 0$$

Plim of $\frac{1}{NT}\underline{X}'M_1\,u_m$:

$$\frac{1}{N}\underline{X}'M_1u_m = \frac{1}{N}\underline{X}'M_1\left((I_N\otimes\iota_T)\,\mu^m + (\iota_N\otimes I_T)\,\nu^m + \varepsilon_m\right)$$

1) $N\rightarrow\infty$ or $T\rightarrow\infty$ or both.

$$(3.A.14) \qquad = \frac{1}{N} \underline{X}'(I_N \otimes \iota_T)\, \mu^m + \frac{1}{N^2} \underline{X}'(\iota_N\, \iota_N' \otimes \iota_T)\, \nu^m + \frac{1}{N} \underline{X}' M_1 \varepsilon_m$$

A typical element of the first vector in (3.A.14) is

$$\frac{1}{N} \sum_i \sum_t x_{kit}\, \mu_{mi}$$

whose expectation is zero and whose variance

$$\frac{1}{N^2} \sum_i (\sum_t x_{kit})^2\, \sigma_{\mu mm}$$

$$\frac{1}{N^2} \sum_i \sum_t \sum_{t'} x_{kit}\, x_{kit'}\, \sigma_{\mu mm}$$

tends to zero as N tends to infinity (using assumption (3.A.2)). Thus the first vector in (3.A.14) tends to zero in probability as $N\to\infty$. The second vector has the following typical element :

$$\frac{1}{N^2} \sum_i \sum_{i'} \sum_t x_{kit}\, \mu_{mi'}$$

whose expectation is again zero and whose variance

$$\frac{1}{N^4} N(\sum_i \sum_t x_{kit})^2\, \sigma_{\mu mm}$$

$$= \frac{1}{N^3} \sum_i \sum_t \sum_j \sum_s x_{kit}\, x_{kjs}\, \sigma_{\mu mm}$$

has also a zero limit. Thus the second vector of (3.A.14) also tends in probability to zero as $N\to\infty$. Finally, we can write a typical element of the third vector in (3.A.14) as

$$\frac{1}{N} \left[\frac{1}{T} \sum_i \sum_t \sum_{t'} x_{kit}\, \varepsilon_{mit'} - \frac{1}{NT} \sum_i \sum_t \sum_{i'} \sum_{t'} x_{kit}\, \varepsilon_{mi't'} \right]$$

By calculating the expectation and variance of the above terms it can be verified that the third vector also tends to zero in probability as $N\to\infty$. Thus we can conclude that

$$(3.A.15) \quad \operatorname*{plim}_{N\to\infty} \frac{1}{N} \underline{X}'\, M_1\, u_m = 0 \quad , \; m=1,\ldots,M .$$

In the same way, it can be shown that

$$(3.A.16) \quad \operatorname*{plim}_{T\to\infty} \frac{1}{T} \underline{X}' M_2 u_m = 0 \quad , m=1,\dots,M$$

$$(3.A.17) \quad \operatorname*{plim}_{\substack{N\to\infty \\ T\to\infty}} \frac{1}{NT} \underline{X}' M_3 u_m = 0 \quad , m=1,\dots,M$$

and

$$(3.A.18) \quad \operatorname*{plim}_{\substack{N\to\infty \\ T\to\infty}} \frac{1}{NT} \underline{X}' M_4 u_m = \operatorname*{plim}_{\substack{N\to\infty \\ T\to\infty}} \underline{X}' Q u_m = 0 \quad , m=1,\dots,M$$

Proofs of (3.A.16), (3.A.17), (3.A.18), are omitted as they are very similar to the one leading to (3.A.15).

Finally, (3.A.15), (3.A.16), (3.A.17) and (3.A.18) imply

$$(3.A.19) \quad \begin{cases} \operatorname*{plim}_{N\to\infty} \frac{1}{N} \underline{X}'M_1 v_m = \operatorname*{plim}_{N\to\infty} \frac{1}{N} \underline{X}'M_1 U\bar{T}_m = 0 \text{ using } (3.37) \\ \operatorname*{plim}_{T\to\infty} \frac{1}{N} \underline{X}'M_2 v_m = \operatorname*{plim}_{T\to\infty} \frac{1}{N} \underline{X}'M_2 U\bar{T}_m = 0 \quad \text{"} \\ \operatorname*{plim}_{\substack{N\to\infty \\ T\to\infty}} \frac{1}{NT} \underline{X}'M_3 v_m = \operatorname*{plim}_{\substack{N\to\infty \\ T\to\infty}} \frac{1}{NT} \underline{X}'M_3 U\bar{T}_m = 0 \quad \text{"} \\ \operatorname*{plim}_{\substack{N\to\infty \\ T\to\infty}} \frac{1}{NT} \underline{X}'Q\, v_m = \operatorname*{plim}_{\substack{N\to\infty \\ T\to\infty}} \frac{1}{NT} \underline{X}'M_4 U\bar{T}_m = 0 \quad \text{"} \end{cases}$$

Now, the expression of $\hat{\tilde{\pi}}_{fGLS}$ contains inverses of $\hat{\Omega}_3$, $\hat{\Omega}_2$, $\hat{\Omega}_1$ and $\hat{\Omega}_\varepsilon$ whose typical elements can be denoted respectively as $\hat{\omega}_{3mm'}$, $\hat{\omega}_{2mm'}$, $\hat{\omega}_{1mm'}$, and $\hat{\omega}_{\varepsilon mm'}$. These estimators involve, in turn, predictors of reduced form errors, $\hat{v}_m$, m=1,...,M calculated using covariance estimators of reduced form coefficient vectors $\hat{\pi}_{m(cov)}$, m=1,...,M. First, consistency of $\hat{\pi}_{m(cov)}$ will be proved and then probability limits of $\hat{\omega}_{3mm'}$, $\hat{\omega}_{2mm'}$, $\hat{\omega}_{1mm'}$ and $\hat{\omega}_{\varepsilon mm'}$ will be considered and finally, consistency of $\hat{\tilde{\pi}}_{fGLS}$ will be proved.

3.A.2 Consistency of $\hat{\pi}_{m(cov)}$:

In this Appendix only, we will omit the index "(cov)" to avoid repeating cumbersome notations. The m-th reduced form equation is

(3.A.20) $y_m = \iota_{NT}\, \pi_{om} + \underline{X}\, \pi_{*m} + v_m$

and the covariance estimates of π_{*m} , π_{om} are respectively

(3.A.21) $$\begin{cases} \hat{\pi}_{*m} = (\underline{X}'Q\,\underline{X})^{-1}\,\underline{X}'\,Q\,y_m \\ \hat{\pi}_{om} = \frac{1}{NT}\,\iota'_{NT}(y_m - \underline{X}\,\hat{\pi}_{*m}) \end{cases}$$

and $\hat{\pi}'_m = \left[\hat{\pi}_{om}\ \hat{\pi}'_{*m}\right]$

By substituting (3.A.20) in (3.A.21) and taking plim[1], we get

$$\text{plim}\ \hat{\pi}_{*m} = \pi_{*m} + \text{plim}\ (\underline{X}'\,Q\,\underline{X})^{-1}\,\underline{X}'\,Q\,v_m$$

From (3.A.12) and (3.A.19) we have :

$$\text{plim}\ (\frac{1}{NT}\,\underline{X}'\,Q\,\underline{X})^{-1} \text{ is a finite non-singular matrix}$$

$$\text{plim}\ \frac{1}{NT}\,\underline{X}'Q\,v_m = 0$$

leading to

(3.A.22) $\text{plim}\ \hat{\pi}_{*m} = \pi_{*m}$

As regards $\hat{\pi}_{om}$,

$$\text{plim}\ \hat{\pi}_{om} = \text{plim}\ \frac{1}{NT}\,\iota'_{NT}\,(y_m - \underline{X}\,\hat{\pi}_{*m})$$

$$= \text{plim}\ \frac{1}{NT}\,\iota'_{NT}\,(\iota_{NT}\,\pi_{om} + v_m)$$

$$= \text{plim}\ (\pi_{om} + \frac{1}{NT}\,\iota'_{NT}\,v_m)$$

(3.A.23) $$= \pi_{om}$$

as $\text{plim}\ \frac{1}{NT}\,\iota'_{NT}\,v_m = 0$, the expectation of v_{mit} .

---

1) Unless otherwise explicitly written, plim will be taken as both N and T tend to infinity.

Hence

(3.A.24) $\operatorname{plim} \hat{\pi}_m = \pi_m \quad , \quad m=1,\ldots,M$.

3.A.3 Consistency of AOV Estimators of Eigenvalues (and Variance Components) of $\Omega_{mm'}$

Let us consider the estimates of eigenvalues of $\Omega_{mm'}$, one by one.

I.
$$\hat{\omega}_{\varepsilon mm'} = \frac{1}{(N-1)\ (T-1)} \hat{v}_m' Q \hat{v}_{m'}$$
$$= \frac{1}{(N-1)\ (T-1)} (y - X\hat{\pi}_m)' Q(y - X\hat{\pi}_{m'})$$
$$= \frac{1}{(N-1)\ (T-1)} (X(\pi_m - \hat{\pi}_m) + v_m)'Q(X(\pi_{m'} - \hat{\pi}_{m'}) + v_{m'})$$
$$= \frac{1}{(N-1)\ (T-1)} (\pi_m - \hat{\pi}_m)'X'QX(\pi_{m'} - \hat{\pi}_{m'}) + v_m'QX(\pi_{m'} - \hat{\pi}_{m'}) + (\pi_m - \hat{\pi}_m)'X'Q\, v_{m'} + v_m' Q\, v_{m'}$$

Now,

$$\operatorname{plim} \frac{1}{(N-1)(T-1)} (\pi_m - \hat{\pi}_m)'X'Q\, X(\pi_{m'} - \hat{\pi}_{m'}) = 0$$
using (3.A.24), (3.A.12) and the result $\iota'Q = 0$.

$$\operatorname{plim} \frac{1}{(N-1)(T-1)} (\pi_m - \hat{\pi}_m)'X'Q\, v_{m'} = 0$$
using (3.A.24), (3.A.19) and the result $\iota'Q = 0$.

$$\operatorname{plim} \frac{1}{(N-1)(T-1)} v_m' Q\, X\, (\pi_{m'} - \hat{\pi}_{m'}) = 0$$
using (3.A.24), (3.A.19) and the result $\iota'Q = 0$.

Therefore,

(3.A.25) $$\operatorname{plim} \hat{\omega}_{\varepsilon mm'} = \operatorname{plim} \frac{1}{(N-1)(T-1)} v_m' Q\, v_{m'}$$

Before continuing with these calculations, let us introduce the following lemma whose proof is given at the end of Appendix 3.A (3.A.5, page 93) :

<u>Lemma L-1</u> :

Let $\xi = [\xi_t]$ and $\eta = [\eta_t]$ t=1,...,T be two Tx1 vectors of random variables such that for t,τ=1,...,T

(L-1-1) $E(\xi_t) = E(\eta_t) = 0$

(L-1-2) $\begin{cases} E(\xi_t \xi_\tau) = \delta_{t\tau}\sigma_1^2 \\ E(\eta_t \eta_\tau) = \delta_{t\tau}\sigma_2^2 \\ E(\xi_t \eta_\tau) = \delta_{t\tau}\sigma_{12} \end{cases}$

(L-1-3) $E(\xi_t^4) = \mu_4^1$; $E(\eta_t^4) = \mu_4^2$

(L-1-4) $E(\xi_t^2 \eta_\tau^2) = \delta_{t\tau}\sigma_{12}^{(2)}$

(L-1-5) ξ_t and ξ_τ are independent for $t \neq \tau$, the same for η_t and η_τ

(L-1-6) ξ_t and η_τ are independent for $t \neq \tau$

Then

(L-1-7) $\text{plim}^{1)}\ \bar{\xi} = \text{plim}\ \frac{1}{T}\sum_t \xi_t = 0$

(L-1-8) $\text{plim}\ \bar{\eta} = \text{plim}\ \frac{1}{T}\sum_t \eta_t = 0$

(L-1-9) $\text{plim}\ \frac{1}{T}\sum_t \xi_t^2 = \sigma_1^2$

(L-1-10) $\text{plim}\ \frac{1}{T}\sum_t \eta_t^2 = \sigma_2^2$

(L-1-11) $\text{plim}\ \frac{1}{T}\sum_t \xi_t \eta_t = \sigma_{12}$

(L-1-12) $\text{plim}\ s_1^2 = \text{plim}\ \frac{1}{T}\sum_t (\xi_t - \bar{\xi})^2 = \sigma_1^2$

---

1) Throughout Lemma L-1, plim is taken as $T \to \infty$.

(L-1-13) $\text{plim } s_2^2 = \text{plim } \frac{1}{T} \sum_t (\eta_t - \bar{\eta})^2 = \sigma_2^2$

(L-1-14) $\text{plim } s_{12} = \text{plim } \frac{1}{T} \sum_t (\xi_t - \bar{\xi})(\eta_t - \bar{\eta}) = \sigma_{12}$

(L-1-15) $\text{plim } \bar{\xi}\bar{\eta} = 0$

(L-1-16) $\text{plim } \bar{\xi}^2 = 0$

(L-1-17) $\text{plim } \bar{\eta}^2 = 0$

Now, let us come back to $\hat{\omega}_{\varepsilon mm'}$. From (3.A.25) (page 80), we have :

$$\text{plim } \hat{\omega}_{\varepsilon mm'} = \text{plim } \frac{1}{(N-1)\,(T-1)} v_m' Q\, v_{m'}$$

From the expression of Q in (3.26) (page 51) and that of v_m in (3.51) (page 55) it can be easily verified that :

(3.A.26) $Q\, v_m = Q\, \tilde{\varepsilon}_m$

And since $Q^2 = Q$ we can write :

$$\text{plim } \hat{\omega}_{\varepsilon mm'} = \text{plim } \frac{1}{(N-1)(T-1)} v_m' Q v_{m'} = \text{plim } \frac{1}{(N-1)(T-1)} \tilde{\varepsilon}_m' Q \tilde{\varepsilon}_{m'}$$

(3.A.27) $= \text{plim} \frac{1}{(N-1)(T-1)} \left(\tilde{\varepsilon}_m' \tilde{\varepsilon}_{m'} - \tilde{\varepsilon}_m' \frac{A}{T} \tilde{\varepsilon}_{m'} - \tilde{\varepsilon}_m' \frac{B}{N} \tilde{\varepsilon}_{m'} + \tilde{\varepsilon}_m' \frac{J_{NT}}{NT} \tilde{\varepsilon}_{m'} \right)$

Let us examine the four terms of (3.A.27) one by one.

First,

$$\text{plim } \frac{1}{(N-1)(T-1)} \tilde{\varepsilon}_m' \tilde{\varepsilon}_{m'}$$

$$= \text{plim } \frac{1}{(N-1)(T-1)} \sum_i \sum_t \tilde{\varepsilon}_{mit} \tilde{\varepsilon}_{m'it}$$

(3.A.28) $= \omega_{\varepsilon mm'}$

using (L-1-11) of Lemma L-1, under assumptions (3.A.7), (3.17) and other general assumptions concerning the existence of higher order moments, corresponding to those of Lemma L-1. Here, it may be noted that assumptions relating to higher-order moments of disturbances can be made without loss of generality.

Next,

$$\operatorname{plim} \frac{1}{(N-1)(T-1)} \tilde{\varepsilon}'_m \frac{A}{T} \tilde{\varepsilon}_{m'} = \operatorname{plim} \frac{1}{(N-1)(T-1)} \frac{1}{T} \sum_i \tilde{\varepsilon}'_{mi} 1_T 1'_T \tilde{\varepsilon}_{m'i}$$

$$\left(\text{where } \underset{(T\times 1)}{\tilde{\varepsilon}_{mi}} = \left[\tilde{\varepsilon}_{mit}\right] \; t=1,\dots,T\right)$$

$$= \operatorname{plim} \frac{1}{(N-1)(T-1)} T \sum_i \tilde{\varepsilon}_{mi.} \tilde{\varepsilon}_{m'i}$$

where $\tilde{\varepsilon}_{mi.} = \frac{1}{T} \sum_t \tilde{\varepsilon}_{mit}$

with $E(\tilde{\varepsilon}_{mi.}) = 0$

$$E(\tilde{\varepsilon}_{mi.} \tilde{\varepsilon}_{m'i'.}) = \delta_{ii'} \frac{\omega_{\varepsilon mm'}}{T}$$

and independence two by two of $\tilde{\varepsilon}_{mi.}$, $\tilde{\varepsilon}_{m'i'.}$ $i \neq i'$

for i , $i'=1,\dots,N$.

Assuming that

$E(\tilde{\varepsilon}^2_{mi.} \tilde{\varepsilon}^2_{m'i.})$ exists and is finite

we can apply (L-1-11) of Lemma L-1 to get

$$\operatorname*{plim}_{N\to\infty} \frac{1}{N-1} \sum_i \tilde{\varepsilon}_{mi.} \tilde{\varepsilon}_{m'i.} = \frac{1}{T} \omega_{\varepsilon mm'}$$

and then

$$\operatorname*{plim}_{\substack{N\to\infty \\ T\to\infty}} \frac{1}{(N-1)(T-1)} \tilde{\varepsilon}'_m \frac{A}{T} \tilde{\varepsilon}_{m'} = \operatorname*{plim}_{\substack{N\to\infty \\ T\to\infty}} \frac{T}{T-1} \frac{1}{N-1} \sum_i \tilde{\varepsilon}_{mi.} \tilde{\varepsilon}_{m'i.}$$

$$= \operatorname*{plim}_{T\to\infty} \frac{T}{T-1} \frac{1}{T} \omega_{\varepsilon mm'}$$

$$= 0 \tag{3.A.29}$$

Similarly, it can be proved that

$$(3.A.30)\quad \underset{\substack{N\to\infty \\ T\to\infty}}{\text{plim}} \frac{1}{(N-1)(T-1)} \tilde{\varepsilon}_m' \frac{B}{N} \tilde{\varepsilon}_{m'} = 0$$

Now, the last term of (3.A.27) :

$$\text{plim} \frac{1}{(N-1)(T-1)} \tilde{\varepsilon}_m' \frac{J_{NT}}{NT} \tilde{\varepsilon}_{m'}$$

$$= \text{plim} \frac{1}{(N-1)(T-1)} \frac{1}{NT} (\sum_i \sum_t \tilde{\varepsilon}_{mit})(\sum_j \sum_s \tilde{\varepsilon}_{m'js})$$

$$= \text{plim} \frac{1}{(N-1)(T-1)} NT \tilde{\varepsilon}_{m..} \tilde{\varepsilon}_{m'..}$$

where $\tilde{\varepsilon}_{m..} = \frac{1}{NT} \sum_i \sum_t \tilde{\varepsilon}_{mit} \quad m=1,\ldots,M$

$$= \text{plim}\ \tilde{\varepsilon}_{m..} \tilde{\varepsilon}_{m'..}$$

$= 0$ according to (L-1-15) of Lemma L-1.

Thus

$$(3.A.31)\quad \text{plim} \frac{1}{(N-1)(T-1)} \tilde{\varepsilon}_m' \frac{J_{NT}}{NT} \tilde{\varepsilon}_{m'} = 0$$

Combining (3.A.27), (3.A.28), (3.A.29), (3.A.30), (3.A.31) we get

$$(3.A.32)\quad \underset{\substack{N\to\infty \\ T\to\infty}}{\text{plim}} \hat{\omega}_{\varepsilon mm'} = \underset{\substack{N\to\infty \\ T\to\infty}}{\text{plim}} \frac{1}{NT} v_m' Q v_{m'} = \omega_{\varepsilon mm'}$$

II. $\hat{\omega}_{1mm'} = \frac{1}{(N-1)} \hat{v}_m' M_1 \hat{v}_{m'}$

$$= \frac{1}{(N-1)} (X(\pi_m - \hat{\pi}_m) + v_m)' M_1 (X(\pi_{m'} - \hat{\pi}_{m'}) + v_{m'})$$

Using

$$\underset{N\to\infty}{\text{plim}} \frac{1}{N} \underline{X}' M_1 \underline{X} = R^{(1)} \qquad \text{(cf. (3.A.9))}$$

$$\underset{N\to\infty}{\text{plim}} \frac{1}{N} \underline{X}' M_1 v_m = 0 \qquad \text{(cf. (3.A.19))}$$

and proceeding in the same manner as done on page 80 for $\hat{\omega}_{\varepsilon mm'}$, it can be easily seen that

$$\text{(3.A.33)} \quad \underset{N\to\infty}{\text{plim}} \ \hat{\omega}_{1mm'} = \underset{N\to\infty}{\text{plim}} \ \frac{1}{N-1} v_m' M_1 v_{m'}$$

Substituting the expression given in (3.51) (page 55) for v_m in (3.A.33) and noting that

$$M_1 (\iota_N \otimes I_T) = 0$$

we can write

$$\text{(3.A.34)} \quad \underset{N\to\infty}{\text{plim}} \ \hat{\omega}_{1mm'} = \underset{N\to\infty}{\text{plim}} \ \frac{1}{N-1} (\tilde{\mu}^m)'(I_N \otimes \iota_T')M_1(I_N \otimes \iota_T) \tilde{\mu}^{m'}$$

$$+ \underset{N\to\infty}{\text{plim}} \ \frac{1}{N-1} (\tilde{\mu}^m)'(I_N \otimes \iota_T') M_1 \tilde{\varepsilon}_{m'}$$

$$+ \underset{N\to\infty}{\text{plim}} \ \frac{1}{N-1} \tilde{\varepsilon}_m' M_1 (I_N \otimes \iota_T') \tilde{\mu}^{m'}$$

$$+ \underset{N\to\infty}{\text{plim}} \ \frac{1}{N-1} \tilde{\varepsilon}_m' M_1 \tilde{\varepsilon}_{m'}$$

Let us look at these four terms one by one.

The first term of (3.A.34) :

$$\underset{N\to\infty}{\text{plim}} \ \frac{1}{N-1} (\tilde{\mu}^m)'(I_N \otimes \iota_T') M_1 (I_N \otimes \iota_T) \tilde{\mu}^{m'}$$

$$= \underset{N\to\infty}{\text{plim}} \ \frac{1}{N-1} T \sum_i (\tilde{\mu}_{mi} - \tilde{\mu}_{m.})(\tilde{\mu}_{m'i} - \tilde{\mu}_{m'.})$$

where $\tilde{\mu}_{m.} = \frac{1}{N} \sum_i \tilde{\mu}_{mi} \qquad m=1,\ldots,M$.

From results (3.47), (3.48) we have

$$E(\tilde{\mu}_{mi}) = 0 \qquad i=1,\ldots,N$$

$$E(\tilde{\mu}_{mi} \tilde{\mu}_{m'j}) = \delta_{ij} \ \omega_{\mu mm'}$$

To these we add independence between $\tilde{\mu}_{mi}$ and $\tilde{\mu}_{m'j}$ $i \neq j$ and other assumptions concerning higher order moments needed to be able to apply Lemma L-1. Then we can conclude that

$$\operatorname*{plim}_{N \to \infty} \frac{1}{N-1} \sum_i (\tilde{\mu}_{mi} - \tilde{\mu}_{m.})(\tilde{\mu}_{m'i} - \tilde{\mu}_{m'.}) = \omega_{\mu mm'} \quad \text{using (L-1-14)}$$

and hence

$$(3.A.35) \quad \operatorname*{plim}_{N \to \infty} \frac{1}{N-1} (\tilde{\mu}^m)'(I_N \otimes \iota_T') M_1 (I_N \otimes \iota_T) \tilde{\mu}^{m'} = T \, \omega_{\mu mm'}$$

The second term of (3.A.34) :

$$\operatorname*{plim}_{N \to \infty} \frac{1}{N-1} (\tilde{\mu}^m)' (I_N \otimes \iota_T') M_1 \tilde{\varepsilon}_{m'}$$

$$= \operatorname*{plim}_{N \to \infty} \frac{1}{N-1} \sum_i \sum_t (\tilde{\mu}_{mi} - \tilde{\mu}_{m.})(\tilde{\varepsilon}_{m'it} - \tilde{\varepsilon}_{m'..})$$

$$\text{where } \tilde{\varepsilon}_{m'..} = \frac{1}{NT} \sum_i \sum_t \tilde{\varepsilon}_{m'it} \ , \quad m'=1,\ldots,M$$

$$= \operatorname*{plim}_{N \to \infty} \frac{1}{N-1} \sum_i (\tilde{\mu}_{mi} - \tilde{\mu}_{m.}) \sum_t (\tilde{\varepsilon}_{m'it} - \tilde{\varepsilon}_{m'..})$$

$$= \operatorname*{plim}_{N \to \infty} \frac{1}{N-1} T \sum_i (\tilde{\mu}_{mi} - \tilde{\mu}_{m.})(\tilde{\varepsilon}_{m'i.} - \tilde{\varepsilon}_{m'..})$$

with $\tilde{\varepsilon}_{m'i.} = \frac{1}{T} \sum_t \tilde{\varepsilon}_{m'it} \qquad m'=1,\ldots,M$.

From assumptions (3.47), (3.48), we derive :

$$E(\tilde{\varepsilon}_{m'i.}) = E(\tilde{\mu}_{mi}) = 0$$

$$E(\tilde{\mu}_{mi} \tilde{\varepsilon}_{m'i.}) = 0$$

independence between $\tilde{\mu}_{mi}$ and $\tilde{\varepsilon}_{m'i.}$

and using (L-1-14) of Lemma L-1 we conclude that

$$\operatorname*{plim}_{N \to \infty} \frac{1}{N-1} \sum_i (\tilde{\mu}_{mi} - \tilde{\mu}_{m.}) (\tilde{\varepsilon}_{m'i.} - \tilde{\varepsilon}_{m'..}) = 0$$

Thus

$$(3.A.36)\quad \underset{N\to\infty}{\text{plim}}\ (\tilde{\mu}^m)'(I_N \otimes \iota_T')\ M_1\ \tilde{\varepsilon}_{m'} = 0$$

(3.A.37) The third term of (3.A.34) can be similarly shown to be equal to zero.

The fourth term of (3.A.35) :

$$\underset{N\to\infty}{\text{plim}}\ \frac{1}{N-1}\tilde{\varepsilon}_m'\ M_1\ \tilde{\varepsilon}_{m'}$$

$$= \underset{N\to\infty}{\text{plim}}\ \frac{1}{N-1}\ T\sum_i (\tilde{\varepsilon}_{mi.} - \tilde{\varepsilon}_{m..})(\tilde{\varepsilon}_{m'i.} - \tilde{\varepsilon}_{m'..})$$

By remarking that

$$E(\tilde{\varepsilon}_{mi.}) = 0 \qquad m=1,\dots,M$$

$$E(\tilde{\varepsilon}_{mi.}\ \tilde{\varepsilon}_{m'i.}) = \frac{1}{T}\ \omega_{\varepsilon mm'}$$

we can conclude, applying (L-1-15) of Lemma L-1 (making appropriate assumptions about higher order moments) that :

$$(3.A.38)\quad \underset{N\to\infty}{\text{plim}}\ \frac{1}{N-1}\ \tilde{\varepsilon}_m'\ M_1\ \tilde{\varepsilon}_{m'} = \omega_{\varepsilon mm'}$$

Thus, putting together (3.A.33), (3.A.34), (3.A.35), (3.A.36), (3.A.37) and (3.A.38), we get :

$$(3.A.39)\quad \underset{N\to\infty}{\text{plim}}\ \hat{\omega}_{1mm'} = \underset{N\to\infty}{\text{plim}}\ \frac{1}{N-1}\ v_m' M_1 v_{m'} = T\ \omega_{\mu mm'} + \omega_{\varepsilon mm'} = \omega_{1mm'}$$

III. In an exactly analogous way to II, it can be shown that

$$(3.A.40)\quad \underset{T\to\infty}{\text{plim}}\ \hat{\omega}_{2mm'} = \underset{T\to\infty}{\text{plim}}\ \frac{1}{T-1}\ v_m' M_2 v_{m'} = N\ \omega_{\nu mm'} + \omega_{\varepsilon mm'} = \omega_{2mm'}$$

From I., II. and III. the following can be derived :

a) If we estimate $\omega_{\mu mm'}$ by

$$(3.A.41)\quad \hat{\omega}_{\mu mm'} = \frac{1}{T}\ (\hat{\omega}_{1mm'} - \hat{\omega}_{\varepsilon mm'})$$

then $\operatorname*{plim}_{\substack{N\to\infty \\ T\to\infty}} \hat{\omega}_{\mu mm'} = \operatorname*{plim}_{\substack{N\to\infty \\ T\to\infty}} \frac{1}{T} (\hat{\omega}_{1mm'} - \hat{\omega}_{\varepsilon mm'})$

$$= \operatorname*{plim}_{T\to\infty} \frac{1}{T} (T\, \omega_{\mu mm'} + \omega_{\varepsilon mm'})$$

$$= \operatorname{plim} \omega_{\mu mm'}$$

(3.A.42) $$= \omega_{\mu mm'}$$

b) If we estimate $\omega_{\nu mm'}$ by

(3.A.43) $$\hat{\omega}_{\nu mm'} = \frac{1}{N} (\hat{\omega}_{2mm'} - \hat{\omega}_{\varepsilon mm'})$$

then

(3.A.44) $$\operatorname{plim} \hat{\omega}_{\nu mm'} = \omega_{\nu mm'}$$

the reasoning being similar to that of a).

c) $$\hat{\omega}_{3mm'} = \hat{\omega}_{1mm'} + \hat{\omega}_{2mm'} - \hat{\omega}_{\varepsilon mm'}$$

$$= T\, \hat{\omega}_{\mu mm'} + \hat{\omega}_{\nu mm'} + N\, \hat{\omega}_{\nu mm'} + \hat{\omega}_{\varepsilon mm'} - \hat{\omega}_{\varepsilon mm'}$$

(3.A.45) $$= T\, \hat{\omega}_{\mu mm'} + N\, \hat{\omega}_{\nu mm'} + \hat{\omega}_{\varepsilon mm'}$$

Assuming that

(3.A.46) $$\lim_{\substack{N\to\infty \\ T\to\infty}} \frac{N}{T} = 1$$

we can write

$$\operatorname*{plim}_{\substack{N\to\infty \\ T\to\infty}} \frac{(\hat{\omega}_{3mm'})}{T} = \operatorname*{plim}_{\substack{N\to\infty \\ T\to\infty}} (\hat{\omega}_{\mu mm'} + \frac{N}{T} \hat{\omega}_{\nu mm'} + \frac{1}{T} \hat{\omega}_{\varepsilon mm'})$$

(3.A.47) $$= \omega_{\mu mm'} + \omega_{\nu mm'}$$ (see footnote 1) below)

---

1) We can alternatively divide $\hat{\omega}_{3mm'}$ by N and arrive at the same limit using assumption (3.A.46), i.e. we have $\operatorname*{plim}_{\substack{N\to\infty \\ T\to\infty}} (\frac{1}{N} \hat{\omega}_{3mm'}) = \omega_{\mu mm'} + \omega_{\nu mm'}$.

d) Denoting

(3.A.48) $\hat{\Omega}_\mu = \left[\hat{\omega}_{\mu\, mm'}\right], \quad \hat{\Omega}_\varepsilon = \left[\hat{\omega}_{\varepsilon\, mm'}\right], \quad \hat{\Omega}_1 = \left[\hat{\omega}_{1mm'}\right] \qquad \begin{matrix} m=1,..,M \\ m'=1,..,M \end{matrix}$

we can write

(3.A.49) $\hat{\Omega}_1 = T\,\hat{\Omega}_\mu + \hat{\Omega}_\varepsilon$

Then

$$\underset{\substack{N\to\infty \\ T\to\infty}}{\text{plim}}\ \frac{1}{T}\hat{\Omega}_1 = \lim_{T\to\infty}\frac{1}{T}(T\ \Omega_\mu + \Omega_\varepsilon) \qquad \text{from (3.A.39)}$$

$$= \lim_{T\to\infty}\Omega_\mu$$

Hence

$$\underset{\substack{N\to\infty \\ T\to}}{\text{plim}}\left(\frac{1}{T}\hat{\Omega}_1\right)^{-1} = \lim_{T\to\infty}\Omega_\mu^{-1}$$

or

$$\underset{\substack{N\to\infty \\ T\to\infty}}{\text{plim}}\ \frac{1}{T}\left(\frac{1}{T}\hat{\Omega}_1\right)^{-1} = \lim_{T\to\infty}\frac{1}{T}\Omega_\mu^{-1}$$

or

(3.A.50) $\underset{\substack{N\to\infty \\ T\to\infty}}{\text{plim}}\ \hat{\Omega}_1^{-1} = 0$

e) Similarly, it can be shown that

(3.A.51) $\text{plim}\ \hat{\Omega}_2^{-1} = 0$

writing

(3.A.52) $\hat{\Omega}_2 = \left[\hat{\omega}_{2mm'}\right], \quad \hat{\Omega}_\nu = \left[\hat{\omega}_{\nu\, mm'}\right], \quad m,m'=1,...,M$

f)

(3.A.53) $\hat{\Omega}_3 = T\,\hat{\Omega}_\mu + N\,\hat{\Omega}_\nu + \hat{\Omega}_\varepsilon$ from (3.A.45)

where

(3.A.54) $\hat{\Omega}_3 = \left[\hat{\omega}_{3mm'}\right], \quad m,m'=1,...,M$

Therefore,

$$\text{plim } \hat{\Omega}_3^{-1} = \text{plim } \frac{1}{NT}\left(\frac{1}{NT}\hat{\Omega}_3\right)^{-1}$$

$$= \text{plim } \frac{1}{NT}\left(\frac{1}{N}\hat{\Omega}_\mu + \frac{1}{T}\hat{\Omega}_\nu + \frac{1}{NT}\hat{\Omega}_\varepsilon\right)^{-1}$$

$$= \text{plim } \frac{1}{N}\left(\frac{T}{N}\hat{\Omega}_\mu + \hat{\Omega}_\nu + \frac{1}{N}\hat{\Omega}_\varepsilon\right)^{-1}$$

(3.A.55) $= 0$ using (3.A.46)

Also

(3.A.56) $\text{plim } \left(\frac{\hat{\Omega}_3}{T}\right)^{-1} = (\Omega_\mu + \Omega_\nu)^{-1}$ from (3.A.46) (see footnote 1) below)

g) From (3.A.32)

(3.A.57) $\text{plim } \hat{\Omega}_\varepsilon^{-1} = \Omega_\varepsilon^{-1}$

Now that we have the probability limits of $\hat{\Omega}_1^{-1}$, $\hat{\Omega}_2^{-1}$, $\hat{\Omega}_3^{-1}$ and $\hat{\Omega}_\varepsilon^{-1}$ appearing in the expression of vec $\hat{\Pi}_{fGLS}$ or that of $\hat{\tilde{\pi}}_{fGLS}$, we can proceed to the last step.

3.A.4. Consistency of the Feasible GLS Estimator of Π :

We will consider the feasible GLS estimator $\hat{\tilde{\pi}}_{fGLS}$ as it is a more general representation which brings out the additional points to be given attention to while considering a model with constant terms. But as $\hat{\tilde{\pi}}_{fGLS}$ is just vec $\hat{\Pi}_{fGLS}$ rearranged, if the former is consistent so is the latter.

From (3.82) we have :

$$(3.A.58)\quad \hat{\tilde{\pi}}_{fGLS} = \begin{bmatrix} \hat{\Omega}_3^{-1} NT & \hat{\Omega}_3^{-1} \otimes \iota'\underline{X} \\ \hat{\Omega}_3^{-1} \otimes \underline{X}'\iota & \sum_i \hat{\Omega}_i^{-1} \otimes \underline{X}'M_i\underline{X} \end{bmatrix}^{-1} \begin{bmatrix} (\hat{\Omega}_3^{-1} \otimes \iota')y \\ \sum_i (\hat{\Omega}_i^{-1} \otimes \underline{X}'M_i)y \end{bmatrix}$$

---

1) We also have (see footnote 1) of page 88)

$$\underset{\substack{N\to\infty \\ T\to\infty}}{\text{plim}} \left(\frac{1}{N}\hat{\Omega}_3\right)^{-1} = (\Omega_\mu + \Omega_\nu)^{-1}$$

Substituting $y = [X_o \; X_*]\tilde{\pi} + w$ in (3.A.58) we get

$$(3.A.59) \quad \hat{\tilde{\pi}}_{fGLS} = \begin{bmatrix} \hat{\Omega}_3^{-1} NT & \hat{\Omega}_3^{-1} \otimes \iota'\underline{X} \\ \hat{\Omega}_3^{-1} \otimes \underline{X}'\iota & \sum_i \hat{\Omega}_i^{-1} \otimes \underline{X}'M_i\underline{X} \end{bmatrix}^{-1} \left\{ \begin{bmatrix} (\hat{\Omega}_3^{-1} \otimes \iota')[X_o X_*]\tilde{\pi} \\ \sum_i (\hat{\Omega}_i^{-1} \otimes \underline{X}'M_i)[X_o X_*]\tilde{\pi} \end{bmatrix} + \begin{bmatrix} (\hat{\Omega}_3^{-1} \otimes \iota') w \\ \sum_i (\hat{\Omega}_i^{-1} \otimes \underline{X}'M_i) w \end{bmatrix} \right\}$$

Let us consider the first partitioned vector inside the big parentheses postmultiplying the inverse in (3.A.59) :

$$\begin{bmatrix} (\hat{\Omega}_3^{-1} \otimes \iota') X_o & (\hat{\Omega}_3^{-1} \otimes \iota') X_* \\ \sum_i (\hat{\Omega}_i^{-1} \otimes \underline{X}'M_i) X_o & \sum_i (\hat{\Omega}_i^{-1} \otimes \underline{X}'M_i) X_* \end{bmatrix} \tilde{\pi}$$

$$= \begin{bmatrix} (\hat{\Omega}_3^{-1} \otimes \iota')(I \otimes \iota) & (\hat{\Omega}_3^{-1} \otimes \iota')(I \otimes \underline{X}) \\ \sum_i (\hat{\Omega}_i^{-1} \otimes \underline{X}'M_i)(I \otimes \iota) & \sum_i (\hat{\Omega}_i^{-1} \otimes \underline{X}'M_i)(I \otimes \underline{X}) \end{bmatrix} \tilde{\pi}$$

using $X_o = I \otimes \iota$; $X_* = I \otimes \underline{X}$

$$= \begin{bmatrix} \hat{\Omega}_3^{-1} NT & \hat{\Omega}_3^{-1} \otimes \iota'\underline{X} \\ \sum_i \hat{\Omega}_i^{-1} \otimes \underline{X}'M_i & \sum_i \hat{\Omega}_i^{-1} \otimes \underline{X}'M_i\underline{X} \end{bmatrix}$$

by multiplying out the terms with Kronecker products

$$(3.A.60) \quad = \left[\begin{array}{c|c} \hat{\Omega}_3^{-1} NT & \hat{\Omega}_3^{-1} \otimes \iota'\underline{X} \\ \hat{\Omega}_3^{-1} \otimes \underline{X}'\iota & \sum_i \hat{\Omega}_i^{-1} \otimes \underline{X}'M_i\underline{X} \end{array}\right]$$

using $M_i\iota = 0, i=1,2,4$

$M_3\iota = \iota$

Note that the above matrix is exactly the same matrix as the one appearing inside the inverse in (3.A.59).

Thus we can write :

$$(3.A.61) \quad \hat{\tilde{\pi}}_{fGLS} = \tilde{\pi} + \begin{bmatrix} \hat{\Omega}_3^{-1} NT & \hat{\Omega}_3^{-1} \otimes \iota'\underline{X} \\ \hat{\Omega}_3^{-1} \otimes \underline{X}'\iota & \sum_i \hat{\Omega}_i^{-1} \otimes \underline{X}'M_i \underline{X} \end{bmatrix}^{-1} \begin{bmatrix} (\hat{\Omega}_3^{-1} \otimes \iota') \\ \sum_i (\hat{\Omega}_i^{-1} \otimes \underline{X}'M_i) \end{bmatrix} w$$

Hence

plim $\hat{\tilde{\pi}}_{fGLS} = \tilde{\pi}$ + plim of the second term on the R.H.S. of (3.A.61)

In this second term, let us premultiply both the matrix inside the inverse and the matrix premultiplying w by the following matrix

$$D = \begin{bmatrix} \frac{1}{N} I_M & 0 \\ 0 & \frac{1}{NT} I_{M(K-1)} \end{bmatrix}$$

Doing so does not change its value and yields :

$$\text{plim}\ \hat{\tilde{\pi}}_{fGLS} = \tilde{\pi} + \text{plim} \begin{bmatrix} \frac{1}{N}\hat{\Omega}_3^{-1}\ NT & \frac{1}{N}\hat{\Omega}_3^{-1} \otimes \iota'\underline{X} \\ \frac{1}{NT}\hat{\Omega}_3^{-1} \otimes \underline{X}'\iota & \frac{1}{NT}\sum_i \hat{\Omega}_i^{-1} \otimes \underline{X}'M_i\underline{X} \end{bmatrix}^{-1} \begin{bmatrix} \frac{1}{N}(\hat{\Omega}_3^{-1} \otimes \iota')\, w \\ \frac{1}{NT}\sum_i (\hat{\Omega}_i^{-1} \otimes X'M_i)\, w \end{bmatrix}$$

$$(3.A.62) \quad = \tilde{\pi} + \text{plim} \begin{bmatrix} \left(\frac{1}{T}\hat{\Omega}_3\right)^{-1} & \left(\frac{1}{T}\hat{\Omega}_3\right)^{-1} \otimes \frac{1}{NT}\iota'\underline{X} \\ \hat{\Omega}_3^{-1} \otimes \frac{1}{NT}\underline{X}'\iota & \sum_i \hat{\Omega}_i^{-1} \otimes \frac{1}{NT}\underline{X}'M_i\underline{X} \end{bmatrix}^{-1} \begin{bmatrix} \left(\left(\frac{1}{T}\hat{\Omega}_3\right)^{-1} \otimes \frac{1}{NT}\iota'\right) w \\ \sum_i (\hat{\Omega}_i^{-1} \otimes \frac{1}{NT}\underline{X}'M_i)\, w \end{bmatrix}$$

From (3.A.56), (3.A.4), (3.A.51), (3.A.52), (3.A.55), (3.A.12) we have respectively :

$$\text{plim}\ (\tfrac{1}{T}\hat{\Omega}_3)^{-1} = (\Omega_\mu + \Omega_\nu)^{-1}$$

$$\text{plim}\ \frac{1}{NT}\iota'\underline{X} = r'$$

$$\text{plim}\ \hat{\Omega}_i^{-1} = 0 \quad \text{for } i=1,2,3$$

$$\text{plim}\ \frac{1}{NT}\underline{X}'M_4\underline{X} = \text{plim}\ \frac{1}{NT}\underline{X}'Q\,\underline{X} = R$$

Using these, we can write :

$$\text{plim}\left[(\tfrac{1}{T}\hat{\Omega}_3)^{-1} \otimes \frac{1}{NT}\iota'\right] w = \text{plim}\left[(\Omega_\mu + \Omega_\nu)^{-1} \otimes \frac{1}{NT}\iota'\right] \text{vec } V$$

$$= \text{plim}\left[\sum_{\ell=1}^{M} (\Omega_\mu + \Omega_\nu)^{-1}_{k\ell}\ \frac{1}{NT}\iota' v_\ell\right], k=1,\ldots,M$$

$$(3.A.63) \qquad = 0$$

as $\text{plim}\ \frac{1}{NT}\iota' v_m = \text{plim}\ \frac{1}{N}\iota'_N \tilde{\mu}^m + \frac{1}{T}\iota'_T \tilde{\nu}^m + \frac{1}{NT}\iota'\tilde{\varepsilon}_m$

$= 0$ as each of the three terms tend to their respective means that are assumed to be zero,

and

$$\text{plim } (\hat{\Omega}_4^{-1} \otimes \frac{1}{NT} \underline{X}'M_4) \, w = \text{plim } \left(\Omega_\varepsilon^{-1} \otimes \frac{1}{NT} \underline{X}'Q\right) \text{vec } V$$

$$= \text{plim} \left[\sum_\ell (\Omega_\varepsilon)^{-1}_{k\ell} \frac{1}{NT} \underline{X}'Q \, v_\ell\right] \quad , k=1,\ldots,M$$

(3.A.64)
$$= 0 \text{ as plim } \frac{1}{NT} \underline{X}'Q \, v_\ell = 0 \text{ from (3.A.19)}$$

Substituting the above results in (3.A.62) we obtain :

$$\text{plim } \hat{\tilde{\pi}}_{fGLS} = \tilde{\pi} + \text{plim} \begin{bmatrix} (\Omega_\mu + \Omega_\nu)^{-1} & (\Omega_\mu + \Omega_\nu)^{-1} \otimes r' \\ 0 & \Omega_\varepsilon^{-1} \otimes R \end{bmatrix} . \, 0$$

(3.A.65)
$$= \tilde{\pi}$$

We have thus proved the consistency of $\hat{\tilde{\pi}}_{fGLS}$ and hence equivalently that of vec $\hat{\Pi}_{fGLS}$.

3.A.5 Proof of Lemma L-1 :

In the proof the following well-known result is used :

Result R-1 :

If $\hat{\theta}$ is an estimator of θ such that

$E(\hat{\theta}) \to \theta$

$V(\hat{\theta}) \to 0$

as T goes to infinity, then

$\text{plim } \hat{\theta} = \theta$.

(L-1-7): plim $\bar{\xi}$

(L-1-18) $E(\bar{\xi}) = E(\frac{1}{T} \sum_t \xi_t) = 0$

$$V(\bar{\xi}) = E(\bar{\xi}^2) = E(\frac{1}{T^2} \sum_t \sum_\tau \xi_t \, \xi_\tau) = \frac{1}{T^2} \sum_t \sum_\tau E(\xi_t \xi_\tau)$$

$$= \frac{1}{T^2} T \sigma_1^2 \quad \text{as for } t \neq \tau \, , \, E(\xi_t \xi_\tau) = 0$$

(L-1-19)
$$= \frac{1}{T} \sigma_1^2$$

Thus $\lim_{T\to\infty} E(\bar{\xi}) = 0$ and $\lim_{T\to\infty} V(\bar{\xi}) = 0$

and hence $\text{plim}\ \bar{\xi} = 0$ (using Result R-1)

(L-1-8) can be proved in a similar way.

(L-1-9): $\text{plim}\ \frac{1}{T} \sum_t \xi_t^2$

(L-1-20) $$E(\frac{1}{T} \sum_t \xi_t^2) = \frac{1}{T} \sum_t E(\xi_t^2) = \frac{1}{T}\, T\, \sigma_1^2 = \sigma_1^2$$

$$V(\frac{1}{T} \sum_t \xi_t^2) = E(\frac{1}{T} \sum_t \xi_t^2)^2 - \sigma_1^4$$

$$E(\frac{1}{T} \sum_t \xi_t^2)^2 = \frac{1}{T^2} E(\sum_t \sum_\tau \xi_t^2\, \xi_\tau^2)$$

$$= \frac{1}{T^2} \sum_t \sum_\tau E(\xi_t^2\, \xi_\tau^2)$$

$$= \frac{1}{T^2}\, T\, E(\xi_t^4) + \frac{1}{T^2} \sum_t \sum_{\tau \neq t} E(\xi_t^2)\, E(\xi_\tau^2)$$

using assumption (L-1-15)

$$= \frac{1}{T} \mu_4^1 + \frac{1}{T^2}\, T(T-1)\, \sigma_1^4$$

Therefore

$$V(\frac{1}{T} \sum_t \xi_t^2) = \frac{1}{T} \mu_4^1 + \frac{1}{T^2}\, T(T-1)\, \sigma_1^4 - \sigma_1^4$$

(L-1-21) $$= \frac{1}{T} \mu_4^1 - \frac{1}{T} \sigma_1^4$$

Thus $E(\frac{1}{T} \sum_t \xi_t^2) \to \sigma_1^2$,

$V(\frac{1}{T} \sum_t \xi_t^2) \to 0$.

Hence $\text{plim}\ \frac{1}{T} \sum_t \xi_t^2 = \sigma_1^2$ (using Result R-1, page 93)

(L-1-10) can be proved similarly to (L-1-9).

(L-1-11): $\text{plim}\ \frac{1}{T} \sum_t \xi_t\, \eta_t$

(L-1-22) $$E\left(\frac{1}{T}\sum_t \xi_t \eta_t\right) = \frac{1}{T} E\left(\sum_t \xi_t \eta_t\right) = \frac{1}{T} T \sigma_{12} = \sigma_{12}$$

$$E\left(\frac{1}{T}\sum_t \xi_t \eta_t\right)^2 = \frac{1}{T^2} E\left(\sum_t \xi_t \eta_t\right)\left(\sum_\tau \xi_\tau \eta_\tau\right)$$

$$= \frac{1}{T^2} E\left(\sum_t \sum_\tau \xi_t \eta_t \xi_\tau \eta_\tau\right)$$

$$= \frac{1}{T^2} \left(T\, E(\xi_t^2 \eta_t^2) + \sum_t \sum_{\tau \neq t} E(\xi_t \eta_t \xi_\tau \eta_\tau)\right)$$

$$= \frac{1}{T} \sigma_{12}^{(2)} + \frac{1}{T^2} \sum_t \sum_{\tau \neq t} E(\xi_t \eta_t)\, E(\xi_\tau \eta_\tau)$$

as $\xi_t \eta_t$ is independent of $\xi_\tau \eta_\tau$

(L-1-23) $$= \frac{1}{T} \sigma_{12}^{(2)} + \frac{T(T-1)}{T^2} \sigma_{12}^2$$

Hence $$V\left(\frac{1}{T}\sum_t \xi_t \eta_t\right) = \frac{1}{T} \sigma_{12}^{(2)} + \frac{T(T-1)}{T^2} \sigma_{12}^2 - \sigma_{12}^2$$

(L-1-24) $$= \frac{1}{T} \sigma_{12}^{(2)} - \frac{1}{T} \sigma_{12}^2$$

Thus, by verifying that the expectation goes to σ_{12} and the variance to 0 as T goes to infinity, we have

$$\text{plim}\ \frac{1}{T}\sum_t \xi_t \eta_t = \sigma_{12} \qquad \text{(using Result R-1, page 93)}$$

<u>(L-1-12)</u>: plim s_1^2

$$s_1^2 = \frac{1}{T}\sum_t (\xi_t - \bar{\xi})^2 = \frac{1}{T}\left(\sum_t \xi_t^2 - T\bar{\xi}^2\right)$$

Thus we can write :

(L-1-25) $$E(s_1^2) = E\left(\frac{1}{T}\sum_t \xi_t^2 - \bar{\xi}^2\right)$$

$$= \sigma_1^2 - \frac{\sigma_1^2}{T} \qquad \text{using (L-1-19), (L-1-20)}$$

$$= \frac{T-1}{T} \sigma_1^2$$

and $$V(s_1^2) = V\left(\frac{1}{T}\sum_t \xi_t^2\right) + V(\bar{\xi}^2) - 2\,\text{Cov}\left(\frac{1}{T}\sum_t \xi_t^2, \bar{\xi}^2\right)$$

We have: $V\left(\frac{1}{T}\sum_t \xi_t^2\right) = \frac{1}{T}\mu_4 - \frac{1}{T}\sigma_1^4$ from (L-1-21) ;

$$V(\bar{\xi}^2) = E(\bar{\xi}^4) - (E(\bar{\xi}^2))^2$$

$$= E(\bar{\xi}^4) - (\frac{1}{T} \sigma_1^2)^2 = E(\bar{\xi}^4) - \frac{\sigma_1^4}{T^2}$$

with $E(\bar{\xi}^4) = E(\frac{1}{T} \sum_t \xi_t)^4$

$$= E(\frac{1}{T^4} \sum_t \sum_s \sum_r \sum_q \xi_t \xi_s \xi_r \xi_q)$$

$$= \frac{1}{T^4} (T\, E(\xi_t^4) + 3\, T(T-1)\, E(\xi_t^2 \xi_s^2)) \quad t \neq s\ ;$$

for $t \neq r$, $t \neq s$, $t \neq q$ $E(\xi_t \xi_s \xi_r \xi_q) = 0$

using independence between ξ_t and ξ_s, $t \neq s$

$$= \frac{1}{T^4} (T\, \mu_4^1 + 3\, T(T-1)\, \sigma_1^4) \quad ,$$

i.e. $V(\bar{\xi}^2) = \frac{1}{T^3} \mu_4^1 + \frac{3}{T^3} (T-1)\, \sigma_1^4 - \frac{1}{T^2} \sigma_1^4$

(L-1-26) $$= \frac{1}{T^3} (\mu_4^1 - 3\, \sigma_1^4) + \frac{2}{T^2} \sigma_1^4 \quad ;$$

$$\text{Cov}\,(\frac{1}{T} \sum_t \xi_t^2 , \bar{\xi}^2) = E(\frac{1}{T} \sum_t \xi_t^2\, \bar{\xi}^2) - E(\frac{1}{T} \sum_t \xi_t^2)\, E(\bar{\xi}^2)$$

$$= E(\frac{1}{T}\, \bar{\xi}^2 \sum_t \xi_t^2) - \sigma_1^2 \frac{\sigma_1^2}{T}$$

using (L-1-19), (L-1-20)

with $E(\frac{1}{T}\, \bar{\xi}^2 \sum_t \xi_t^2) = E(\frac{1}{T^3} (\sum_t \xi_t)^2 (\sum_\tau \xi_\tau^2))$

$$= \frac{1}{T^3} E(\sum_t \sum_\tau \sum_s \xi_t \xi_\tau \xi_s^2)$$

$$= \frac{1}{T^3} (T\, \mu_4^1 + T(T-1)\, \sigma_1^4) \; ; \text{ for } t \neq \tau , t \neq s,$$

$$E(\xi_t \xi_\tau \xi_s^2) = 0$$

i.e. $\text{Cov}\,(\frac{1}{T} \sum_t \xi_t^2 , \bar{\xi}^2) = \frac{1}{T^2} \mu_4^1 + \frac{T-1}{T^2} \sigma_1^4 - \frac{1}{T} \sigma_1^4$

$$= \frac{1}{T^2} \mu_4^1 - \frac{1}{T^2} \sigma_1^4$$

Thus

(L-1-27) $V(s_1^2) = \frac{1}{T}\mu_4^1 - \frac{1}{T}\sigma_1^4 + \frac{1}{T^3}\mu_4^1 - \frac{1}{T^3}3\sigma_1^4 + \frac{2}{T^2}\sigma_1^4$

$$- 2\frac{1}{T^2}\mu_4^1 + 2\frac{1}{T^2}\sigma_1^4$$

It is then evident that $\underset{T\to\infty}{\text{plim}}\ s_1^2 = \sigma_1^2$ as

$$\lim_{T\to\infty} E(s_1^2) = \sigma_1^2$$

and $\lim_{T\to\infty} V(s_1^2) = 0$.

<u>(L-1-13)</u> is proved similarly to (L-1-12).

<u>(L-1-14)</u>: plim s_{12}

$$s_{12} = \frac{1}{T}\sum_t (\xi_t - \bar{\xi})(\eta_t - \bar{\eta})$$

$$= \frac{1}{T}\sum_t \xi_t\eta_t - \bar{\xi}\bar{\eta}$$

Thus, $E(s_{12}) = \frac{1}{T}E(\sum_t \xi_t\eta_t) - E(\bar{\xi}\bar{\eta})$

$$= \frac{1}{T}T\sigma_{12} - E(\bar{\xi}\bar{\eta})$$

Now, $E(\bar{\xi}\bar{\eta}) = \frac{1}{T^2}E(\sum_t \xi_t)(\sum_\tau \eta_\tau)$

$$= \frac{1}{T^2}E(\sum_t\sum_\tau \xi_t\eta_\tau)$$

$$= \frac{1}{T^2}(T\sigma_{12}) \quad \text{as for } t\neq\tau,\ E(\xi_\tau\eta_t) = 0$$

Thus

(L-1-28) $E(s_{12}) = \sigma_{12} - \frac{1}{T}\sigma_{12} = \frac{T-1}{T}\sigma_{12}$

Next, $V(s_{12}) = V(\frac{1}{T}\sum_t \xi_t\eta_t) + V(\bar{\xi}\bar{\eta}) - 2\,\text{Cov}(\frac{1}{T}\sum_t \xi_t\eta_t, \bar{\xi}\bar{\eta})$

We have: $V(\frac{1}{T}\sum_t \xi_t\eta_t) = \frac{1}{T}\sigma_{12}^{(2)} - \frac{1}{T}\sigma_{12}^2$ from (L-1-24) ;

$$V(\bar{\xi}\bar{\eta}) = E(\bar{\xi}^2\bar{\eta}^2) - E^2(\bar{\xi}\bar{\eta})$$

with $$E(\bar{\xi}^2 \bar{\eta}^2) = \frac{1}{T^4} E(\sum_t \xi_t)^2 (\sum_t \eta_t)^2$$

$$= \frac{1}{T^4} E(\sum_t \sum_s \xi_t \xi_s)(\sum_r \sum_q \eta_r \eta_q)$$

$$= \frac{1}{T^4} E(\sum_t \sum_s \sum_r \sum_q \xi_t \xi_s \eta_r \eta_q)$$

$$= \frac{1}{T^4} (T \sigma_{12}^{(2)} + T(T-1) \sigma_1^2 \sigma_2^2 + 2\, T(T-1) \sigma_{12}^2)$$

since for $t \neq s$, $t \neq r$, $t \neq q$, $E(\xi_t \xi_s \eta_r \eta_q) = 0$

and $$E(\bar{\xi}\, \bar{\eta}) = \frac{1}{T^2} E(\sum_t \xi_t)(\sum_\tau \eta_\tau)$$

$$= \frac{1}{T^2} E(\sum_t \sum_\tau \xi_t \eta_\tau)$$

(L-1-29) $$= \frac{1}{T^2} T \sigma_{12} = \frac{\sigma_{12}}{T}$$

i.e. $$V(\bar{\xi}\, \bar{\eta}) = \frac{1}{T^3} \sigma_{12}^{(2)} + \frac{T-1}{T^3} \sigma_1^2 \sigma_2^2 + 2\frac{(T-1)}{T^3} \sigma_{12}^2 - \frac{\sigma_{12}^2}{T^2}$$

(L-1-30) $$= \frac{1}{T^3} \sigma_{12}^{(2)} + \frac{T-1}{T^3} \sigma_1^2 \sigma_2^2 + \frac{1}{T^2} \sigma_{12}^2 - \frac{2}{T^3} \sigma_{12}^2 \quad ;$$

$$\text{Cov}\,(\frac{1}{T} \sum_t \xi_t \eta_t, \bar{\xi}\, \bar{\eta}) = E(\frac{1}{T} \sum_t \xi_t \eta_t\, \bar{\xi}\, \bar{\eta}) - E(\frac{1}{T} \sum_t \xi_t \eta_t)\, E(\bar{\xi}\, \bar{\eta})$$

$$= E(\frac{1}{T} \sum_t \xi_t \eta_t\, \bar{\xi}\, \bar{\eta}) - \sigma_{12} \frac{\sigma_{12}}{T}$$

with $$E(\frac{1}{T} \sum_t \xi_t \eta_t\, \bar{\xi}\, \bar{\eta}) = \frac{1}{T^3} E(\sum_t \sum_s \sum_r \xi_t \eta_t \xi_s \eta_r)$$

$$= \frac{1}{T^3} (T \sigma_{12}^{(2)} + T(T-1) \sigma_{12}^2)$$

i.e.

$$\text{Cov}(\frac{1}{T} \sum_t \xi_t \eta_t\, , \bar{\xi}\, \bar{\eta}) = \frac{1}{T^2} \sigma_{12}^{(2)} + \frac{T-1}{T^2} \sigma_{12}^2 - \frac{1}{T} \sigma_{12}^2$$

$$= \frac{1}{T^2} \sigma_{12}^{(2)} - \frac{1}{T^2} \sigma_{12}^2$$

Finally, therefore,

$$V(s_{12}) = \frac{1}{T}\sigma_{12}^{(2)} - \frac{1}{T}\sigma_{12}^{2} + \frac{1}{T^3}\sigma_{12}^{(2)} + \frac{T-1}{T^3}\sigma_1^2\sigma_2^2 + \frac{1}{T^2}\sigma_1^2$$

$$\text{(L-1-31)} \qquad - \frac{2}{T^3}\sigma_{12}^2 - \frac{2}{T^2}\sigma_{12}^{(2)} + \frac{2}{T^2}\sigma_{12}^2$$

Hence $E(s_{12}) \to \sigma_{12}$ as $T \to \infty$

$V(s_{12}) \to 0$ as $T \to \infty$

and

$\text{plim } s_{12} = \sigma_{12}$ (using Result R-1, page 93)

(L-1-15): plim $\bar{\xi}\ \bar{\eta}$

From (L-1-29) and (L-1-30) it can be seen that

$$\lim E(\bar{\xi}\ \bar{\eta}) = \lim \frac{\sigma_{12}}{T} = 0$$

$$\lim V(\bar{\xi}\ \bar{\eta}) = (\lim \frac{1}{T^3}\sigma_{12}^{(2)} + \frac{T-1}{T^3}\sigma_1^2\sigma_2^2 + \frac{1}{T^2}\sigma_{12}^2 - \frac{2}{T^3}\sigma_{12}^2)$$

$$= 0$$

Hence $\text{plim } (\bar{\xi}\ \bar{\eta}) = 0$ (using Result R-1, page 93)

(L-1-16) :plim $\bar{\xi}^2$

From (L-1-19) and (L-1-20) we have :

$$E(\bar{\xi}^2) = \frac{1}{T}\sigma_1^2 \to 0 \text{ as } T \to \infty$$

$$V(\bar{\xi}^2) = \frac{1}{T^3}(\mu_4^1 - 3\mu_1^4) + \frac{2}{T^2}\sigma_1^4 \to 0 \text{ as } T \to \infty$$

Hence $\text{plim } \bar{\xi}^2 = 0$ (using Result R-1, page 93)

(L-1-17) can be proved similarly to (L-1-16).

APPENDIX 3.B Limiting Distribution of the Feasible GLS Estimator of the Reduced Form

We will study the feasible GLS estimator $\hat{\hat{\pi}}_{fGLS}$ as plim of $\frac{1}{NT}\sum_i \hat{\Omega}_i^{-1} \otimes X'M_iX$ appearing in the expression of vec $\hat{\Pi}_{fGLS}$ is singular when X contains ι_{NT} as its first column and hence the matrix (in limit) cannot be inversed. Thus in order to determine the limiting distribution of the feasible GLS estimator, a more careful examination is required which necessitates separation of the constant term and thus reformulation of the estimator in the form of $\hat{\hat{\pi}}_{fGLS}$.

Let us recall that the expression of $\hat{\tilde{\pi}}_{fGLS}$ is given in (3.82). By substituting (3.71) for y in (3.82) and simplifying (see (3.A.61), Appendix 3.A.4, page 72 for calculations), we obtain :

$$(3.B.1)\quad \hat{\hat{\pi}}_{fGLS} - \tilde{\pi} = \begin{pmatrix}\hat{\pi}_o\\ \hat{\pi}_*\end{pmatrix} - \begin{pmatrix}\pi_o\\ \pi_*\end{pmatrix} = \begin{bmatrix}\hat{\Omega}_3^{-1}\cdot NT & \hat{\Omega}_3^{-1}\otimes \iota'\underline{X}\\ \hat{\Omega}_3^{-1}\otimes \underline{X}'\iota & \sum_i \hat{\Omega}_i^{-1}\otimes \underline{X}'M_i\underline{X}\end{bmatrix}^{-1}\begin{bmatrix}\hat{\Omega}_3^{-1}\otimes \iota'\\ \sum_i(\hat{\Omega}_i^{-1}\otimes \underline{X}'M_i)\end{bmatrix} w$$

Now, if we consider $\sqrt{NT}\,(\hat{\hat{\pi}}_{fGLS} - \tilde{\pi})$ as is usually done, the plim of $\frac{1}{NT}$ of the matrix inside the inverse in (3.B.1) becomes singular. In order to avoid this problem, we will derive the limiting distribution of $\begin{pmatrix}\sqrt{N}\,(\hat{\pi}_o - \pi_o)\\ \sqrt{NT}\,(\hat{\pi}_* - \pi_*)\end{pmatrix}$[1] .

---

1) We can alternatively take $\begin{pmatrix}\sqrt{T}\,(\hat{\pi}_o - \pi_o)\\ \sqrt{NT}\,(\hat{\pi}_* - \pi_*)\end{pmatrix}$ but the result is the same in both cases if we assume $\lim\limits_{\substack{N\to\infty\\ T\to\infty}} \frac{N}{T} = \lim\limits_{\substack{N\to\infty\\ T\to\infty}} \frac{T}{N} = 1.$

Let us first define the following matrices :

(3.B.2) $$D = \begin{bmatrix} \frac{1}{N} I_M & 0 \\ 0 & \frac{1}{NT} I_{M(K-1)} \end{bmatrix}$$

(3.B.3) $$D_1 = \begin{bmatrix} \hat{\Omega}_3^{-1}\, NT & \hat{\Omega}_3^{-1} \otimes \iota'\underline{X} \\ \hat{\Omega}_3^{-1} \otimes \underline{X}'\iota & \sum_i \hat{\Omega}_i^{-1} \otimes \underline{X}'M_i\underline{X} \end{bmatrix}$$

(3.B.4) $$D_2 = \begin{bmatrix} \hat{\Omega}_3^{-1} \otimes \iota' \\ \sum_i \hat{\Omega}_i^{-1} \otimes \underline{X}'M_i \end{bmatrix}$$

Now, we can proceed to calculate the limiting distribution of $\begin{pmatrix} \sqrt{N} & (\hat{\pi}_o - \pi_o) \\ \sqrt{NT} & (\hat{\pi}_* - \pi_*) \end{pmatrix} = D^{-\frac{1}{2}} \left(\hat{\tilde{\pi}}_{fGLS} - \tilde{\pi}\right)$.

(3.B.5) $$\begin{aligned} D^{-\frac{1}{2}}\left(\hat{\tilde{\pi}}_{fGLS} - \tilde{\pi}\right) &= D^{-\frac{1}{2}} D_1^{-1} D_2\, w \\ &= D^{-\frac{1}{2}} D_1^{-1} D^{-\frac{1}{2}} D^{\frac{1}{2}} D_2\, w \\ &= \left[D^{\frac{1}{2}} D_1 D^{\frac{1}{2}}\right]^{-1} D^{\frac{1}{2}} D_2\, w \end{aligned}$$

$$\text{plim}^{1)} D^{\frac{1}{2}} D_1 D^{\frac{1}{2}} = \text{plim} \begin{bmatrix} \frac{1}{N}\hat{\Omega}_3^{-1}\, NT & \frac{1}{N\sqrt{T}}\hat{\Omega}_3^{-1} \otimes \iota'\underline{X} \\ \frac{1}{N\sqrt{T}}\hat{\Omega}_3^{-1} \otimes \underline{X}'\iota & \frac{1}{NT}\sum_i \hat{\Omega}_i^{-1} \otimes \underline{X}'M_i\underline{X} \end{bmatrix}$$

$$= \text{plim} \begin{bmatrix} \left(\frac{\hat{\Omega}_3}{T}\right)^{-1} & \frac{1}{\sqrt{T}}\left(\frac{\hat{\Omega}_3}{T}\right)^{-1} \otimes \frac{1}{NT}\iota'\underline{X} \\ \frac{1}{\sqrt{T}}\left(\frac{1}{T}\hat{\Omega}_3\right)^{-1} \otimes \frac{1}{NT}\underline{X}'\iota & \sum_i \hat{\Omega}_i^{-1} \otimes \frac{1}{NT}\underline{X}'M_i\underline{X} \end{bmatrix}$$

$$= \begin{bmatrix} (\Omega_\mu + \Omega_\nu)^{-1} & 0 \\ 0 & \Omega_\varepsilon^{-1} \otimes R \end{bmatrix}$$

using (3.A.4), (3.A.12), (3.A.56) and (3.A.57) of Appendix 3.A.

---

1) Unless otherwise explicitly stated, plim is always taken as both N and T tend to infinity.

Hence

$$(3.B.6)\quad \text{plim}\,(D^{\frac{1}{2}} D_1 D^{\frac{1}{2}})^{-1} = \begin{bmatrix} (\Omega_\mu + \Omega_\nu) & 0 \\ 0 & \Omega_\varepsilon \otimes R^{-1} \end{bmatrix} \equiv \bar{D}$$

Therefore, the limiting distribution of $\left[D^{\frac{1}{2}} D_1 D^{\frac{1}{2}}\right]^{-1} D^{\frac{1}{2}} D_2 w$ is the same as the limiting distribution of $\bar{D} D^{\frac{1}{2}} D_2 w$.

Now,

$$(3.B.7)\quad D^{\frac{1}{2}} D_2 w = \begin{bmatrix} \frac{1}{\sqrt{N}}(\hat{\Omega}_3^{-1} \otimes \iota') \\ \frac{1}{\sqrt{NT}} \sum_i \hat{\Omega}_i^{-1} \otimes \underline{X}'M_i \end{bmatrix} w = \begin{bmatrix} \left(\sqrt{N}\left(\frac{\hat{\Omega}_3}{T}\right)^{-1} \otimes \frac{1}{NT}\iota'\right) w \\ \sum_i (\hat{\Omega}_i^{-1} \otimes \frac{1}{\sqrt{NT}} \underline{X}'M_i) w \end{bmatrix}$$

Let us first look at the first block :

$$\left(\sqrt{N}\left(\frac{\hat{\Omega}_3}{T}\right)^{-1} \otimes \frac{1}{NT}\iota'\right) w$$

$$= \sqrt{N}\left(\frac{\hat{\Omega}_3}{T}\right)^{-1} (I_M \otimes \frac{1}{NT}\iota')\, w$$

Now,

$$\sqrt{N}(I \otimes \frac{1}{NT}\iota')w = \sqrt{N}\begin{bmatrix} \frac{1}{NT}\iota' & & \bigcirc \\ & \ddots & \\ \bigcirc & & \frac{1}{NT}\iota' \end{bmatrix} \begin{bmatrix} v_1 \\ \vdots \\ v_M \end{bmatrix}$$

$$= \sqrt{N}\begin{bmatrix} \frac{1}{NT} & \iota' v_1 \\ & \vdots \\ \frac{1}{NT} & \iota' v_M \end{bmatrix}$$

$$= \sqrt{N}\begin{bmatrix} \frac{1}{NT} & \sum_i \sum_t v_{1it} \\ & \vdots \\ \frac{1}{NT} & \sum_i \sum_t v_{Mit} \end{bmatrix}$$

$$= \sqrt{N}\begin{bmatrix} \frac{1}{N}\sum_i \tilde{\mu}_{1i} + \frac{1}{T}\sum_t \tilde{\nu}_{1t} + \frac{1}{NT}\sum_i\sum_t \tilde{\varepsilon}_{1it} \\ \vdots \\ \frac{1}{N}\sum_i \tilde{\mu}_{Mi} + \frac{1}{T}\sum_t \tilde{\nu}_{Mt} + \frac{1}{NT}\sum_i\sum_t \tilde{\varepsilon}_{Mit} \end{bmatrix}$$

Thus, $\sqrt{N}(I \otimes \frac{1}{NT}\iota')w =$

$$\sqrt{N}\begin{bmatrix} \frac{1}{N}\sum_i \tilde{\mu}_{1i} \\ \vdots \\ \frac{1}{N}\sum_i \tilde{\mu}_{Mi} \end{bmatrix} + \sqrt{N}\begin{bmatrix} \frac{1}{T}\sum_t \tilde{\nu}_{1t} \\ \vdots \\ \frac{1}{T}\sum_t \tilde{\nu}_{Mt} \end{bmatrix} + \sqrt{N}\begin{bmatrix} \frac{1}{NT}\sum_i\sum_t \tilde{\varepsilon}_{1it} \\ \vdots \\ \frac{1}{NT}\sum_i\sum_t \tilde{\varepsilon}_{Mit} \end{bmatrix}$$

$$(3.B.8) = \sqrt{N}\,\frac{1}{N}\sum_i \begin{bmatrix} \tilde{\mu}_{1i} \\ \vdots \\ \tilde{\mu}_{Mi} \end{bmatrix} + \sqrt{N}\,\frac{1}{T}\sum_t \begin{bmatrix} \tilde{\nu}_{1t} \\ \vdots \\ \tilde{\nu}_{Mt} \end{bmatrix} + \sqrt{N}\,\frac{1}{NT}\sum_i\sum_t \begin{bmatrix} \tilde{\varepsilon}_{1it} \\ \vdots \\ \tilde{\varepsilon}_{Mit} \end{bmatrix}$$

Therefore,

$$\left(\sqrt{N}\left(\frac{\hat{\Omega}_3}{T}\right)^{-1} \otimes \frac{1}{NT}\iota'\right) w$$

$$= \sqrt{N}\left(\frac{\hat{\Omega}_3}{T}\right)^{-1}(I \otimes \frac{1}{NT}\iota')\, w$$

$$= \sqrt{N}\left(\frac{\hat{\Omega}_3}{T}\right)^{-1}\left[\frac{1}{N}\sum_i \tilde{\mu}_i + \frac{1}{T}\sum_t \tilde{\nu}_t + \frac{1}{NT}\sum_i\sum_t \tilde{\varepsilon}_{it}\right] \text{ using (3.B.8)}$$

$$(3.B.9) = \sum_i \frac{1}{\sqrt{N}}\left(\frac{\hat{\Omega}_3}{T}\right)^{-1}\tilde{\mu}_i + \sum_t \frac{\sqrt{N}}{T}\left(\frac{\hat{\Omega}_3}{T}\right)^{-1}\tilde{\nu}_t + \sum_i\sum_t \frac{\sqrt{N}}{NT}\left(\frac{\hat{\Omega}_3}{T}\right)^{-1}\tilde{\varepsilon}_{it}$$

Let us leave aside the first block for the moment and examine the second one of (3.B.7) :

$$\frac{1}{\sqrt{NT}}\sum_i (\hat{\Omega}_i^{-1} \otimes \underline{X}'M_i)\, w$$

$$= \frac{1}{\sqrt{NT}}\sum_i (\hat{\Omega}_i^{-1} \otimes I_M)(I_M \otimes \underline{X}'M_i)\, w$$

Now,

$$(I \otimes \underline{X}'M_i)\, w = \begin{bmatrix} \underline{X}'M_i v_1 \\ \vdots \\ \underline{X}'M_i v_M \end{bmatrix} = \left[\underline{X}'M_i v_m\right] \qquad m=1,\dots,M$$

$$(3.B.10) \qquad = \left[\underline{X}'M_i(I \otimes \iota)\tilde{\mu}^m + \underline{X}'M_i(\iota \otimes I)\tilde{\nu}^m + \underline{X}\, M_i\tilde{\varepsilon}_m\right]$$

$$m=1,\dots,M$$

Denoting, for i=1,2,3,4,

$$
(3.B.11)\quad \left\{
\begin{array}{l}
\underline{X}'M_i(I \otimes \iota) = A_i = [A_{i1} \cdots A_{iN}] \\
\text{where } \underset{((K-1)\times 1)}{A_{i\ell}} \text{ is the } \ell\text{-th column of } A_i \\
\underline{X}'M_i(\iota \otimes I) = B_i = [B_{i1} \cdots B_{iT}] \\
\text{where } \underset{((K-1)\times 1)}{B_{it}} \text{ is the t-th column of } B_i \\
\underline{X}'M_i = C_i = [C_{i1} \cdots C_{iNT}] \\
\text{where } \underset{((K-1)\times 1)}{C_{i\ell t}} \text{ is the } \ell t\text{-th column of } C_i
\end{array}
\right.
$$

we can write (3.B.10) as :

$$
(I \otimes \underline{X}'M_i)w = [A_i \tilde{\mu}^m + B_i \tilde{\nu}^m + C_i \tilde{\varepsilon}_m] \qquad m=1,\ldots,M
$$

$$
= \left[\sum_\ell A_{i\ell}\tilde{\mu}_{m\ell} + \sum_t B_{it}\tilde{\nu}_{mt} + \sum_\ell \sum_t C_{i\ell t}\tilde{\varepsilon}_{m\ell t}\right] \qquad m=1,\ldots,M
$$

$$
(3.B.12)\qquad = \sum_\ell (I \otimes A_{i\ell})\tilde{\mu}_\ell + \sum_t (I \otimes B_{it})\tilde{\nu}_t + \sum_\ell \sum_t (I \otimes C_{i\ell t})\tilde{\varepsilon}_{\ell t}
$$

$$
\text{where } \tilde{\mu}_\ell = [\tilde{\mu}_{m\ell}],\ \tilde{\nu}_t = [\tilde{\nu}_{mt}],\ \tilde{\varepsilon}_{\ell t} = [\tilde{\varepsilon}_{m\ell t}] \qquad m=1,\ldots,M .
$$

Hence

$$
\frac{1}{\sqrt{NT}} \sum_i (\hat{\Omega}_i^{-1} \otimes \underline{X}'M_i)\, w
$$

$$
= \frac{1}{\sqrt{NT}} \sum_i (\hat{\Omega}_i^{-1} \otimes I)\left[\sum_\ell (I \otimes A_{i\ell})\tilde{\mu}_\ell + \sum_t (I \otimes B_{it})\tilde{\nu}_t + \sum_\ell\sum_t (I \otimes C_{i\ell t})\tilde{\varepsilon}_{\ell t}\right]
$$

using (3.B.12)

$$
= \frac{1}{\sqrt{NT}} \sum_\ell \sum_i (\hat{\Omega}_i^{-1} \otimes A_{i\ell})\tilde{\mu}_\ell + \sum_t \sum_i (\hat{\Omega}_i^{-1} \otimes B_{it})\tilde{\nu}_t + \sum_\ell \sum_t \sum_i (\hat{\Omega}_i^{-1} \otimes C_{i\ell t})\tilde{\varepsilon}_{\ell t}
$$

$$
(3.B.13) = \sum_\ell \frac{1}{\sqrt{NT}} \sum_i \left(\hat{\Omega}_i^{-1} \otimes A_{i\ell}\right) \tilde{\mu}_\ell + \sum_t \frac{1}{\sqrt{NT}} \sum_i \left(\hat{\Omega}_i^{-1} \otimes B_{it}\right)\tilde{\nu}_t +
$$

$$
\sum_\ell \sum_t \left(\frac{1}{\sqrt{NT}} \sum_i \hat{\Omega}_i^{-1} \otimes C_{i\ell t}\right)\tilde{\varepsilon}_{\ell t}
$$

Now, let us combine (3.B.9) and (3.B.13) and write

$$(3.B.14)\qquad D^{\frac{1}{2}}D_2 w = \sum_{\ell}\begin{bmatrix}\frac{1}{\sqrt{N}}\left(\frac{\hat{\Omega}_3}{T}\right)^{-1}\\ \frac{1}{\sqrt{NT}}\sum_i \hat{\Omega}_i^{-1}\otimes A_{i\ell}\end{bmatrix}\tilde{\mu}_{\ell} + \sum_t \begin{bmatrix}\frac{\sqrt{N}}{T}\left(\frac{\hat{\Omega}_3}{T}\right)^{-1}\\ \frac{1}{\sqrt{NT}}\sum_i \hat{\Omega}_i^{-1}\otimes B_i\end{bmatrix}\tilde{\nu}_t +$$

$$\sum_{\ell}\sum_t\begin{bmatrix}\frac{\sqrt{N}}{NT}\left(\frac{\hat{\Omega}_3}{T}\right)^{-1}\\ \frac{1}{\sqrt{NT}}\sum_i \hat{\Omega}_i^{-1}\otimes C_{i\ell t}\end{bmatrix}\tilde{\varepsilon}_{\ell t}$$

Thus $D^{\frac{1}{2}}D_2 w$ is composed of three independent components. Let us calculate the limiting distribution of each component, one by one.

Before that, let us note that, from the definition of M_i and those of A_i, B_i, C_i (given in (3.B.11)), we have the following results :

$$(3.B.15)\qquad \begin{cases} A_i = 0\ , & i=2,4 \\ B_i = 0\ , & i=1,4 \end{cases}$$

Keeping them in mind, let us proceed further.

a) <u>Limiting Distribution of the first component of (3.B.14)</u> :

$$\sum_{\ell}\begin{bmatrix}\frac{1}{\sqrt{N}}\left(\frac{\hat{\Omega}_3}{T}\right)^{-1}\\ \frac{1}{\sqrt{NT}}\sum_i \hat{\Omega}_i^{-1}\otimes A_{i\ell}\end{bmatrix}\tilde{\mu}_{\ell}$$

$$= \sum_{\ell}\begin{bmatrix}\frac{1}{\sqrt{N}}\left(\frac{\hat{\Omega}_3}{T}\right)^{-1}\\ \frac{1}{\sqrt{NT}}\sum_{i=1,3}\left(\frac{\hat{\Omega}_i}{T}\right)^{-1}\otimes\left(\frac{A_{i\ell}}{T}\right)\end{bmatrix}\tilde{\mu}_{\ell}\qquad \text{as } A_i = 0, i=2,4$$

The limiting distribution of the above vector is the same as that of

$$\sum_{\ell}\operatorname{plim}\begin{bmatrix}\frac{1}{\sqrt{N}}\left(\frac{\hat{\Omega}_3}{T}\right)^{-1}\\ \frac{1}{\sqrt{NT}}\sum_{i=1,3}\left(\frac{\hat{\Omega}_i^{-1}}{T}\right)\otimes\frac{A_{i\ell}}{T}\end{bmatrix}\tilde{\mu}_{\ell}$$

$$(3.B.16)\qquad = \sum_{\ell}\begin{bmatrix}\frac{1}{\sqrt{N}}(\Omega_{\mu}+\Omega_{\nu})^{-1}\\ \frac{1}{\sqrt{NT}}\left((\Omega_{\mu}^{-1}\otimes\bar{A}_{1\ell}) + (\Omega_{\mu}+\Omega_{\nu})^{-1}\otimes\bar{A}_{3\ell}\right)\end{bmatrix}\tilde{\mu}_{\ell}$$

where we have used the following results :

$$\text{plim}\left(\frac{\hat{\Omega}_1^{-1}}{T}\right) = \Omega_\mu^{-1} \;,\; \text{plim}\left(\frac{\hat{\Omega}_3}{T}\right)^{-1} = (\Omega_\mu + \Omega_\nu)^{-1} \;;$$

$$\text{plim}\,\frac{1}{T}A_{1\ell} = \text{plim}\,\frac{1}{T}\left(\ell\text{-th column of } \underline{X}'M_1(I\otimes\iota) = \underline{X}'\left(I\otimes\iota - \frac{\iota_N\iota_N'\otimes\iota_T}{N}\right)\right)$$

$$= \text{plim}\,\frac{1}{T}\begin{bmatrix} x_{2\ell}'\,\iota_T \\ \vdots \\ x_{K\ell}'\,\iota_T \end{bmatrix} - \frac{1}{NT}\begin{bmatrix} \sum_i x_{2i}'\,\iota_T \\ \vdots \\ \sum_i x_{Ki}'\,\iota_T \end{bmatrix}$$

where $x_{k\ell}' = \begin{bmatrix} x_{k\ell 1} & \cdots & x_{k\ell T} \end{bmatrix}$ $,k=2,\ldots,K$

(3.B.17) $= \text{constant say } \bar{A}_{1\ell}$;

$$\text{plim}\,\frac{1}{T}\,A_{3\ell} = \text{plim}\,\frac{1}{T}\left(\ell\text{-th column of } \underline{X}'M_3(I\otimes\iota) = \underline{X}'\left(\frac{\iota_N\iota_N'\otimes\iota_T}{N}\right)\right)$$

(3.B.18) $= \text{constant say } \bar{A}_{3\ell}$.

As $\tilde{\mu}_\ell$(s) are i.i.d. with mean zero and variance-covariance matrix Ω_μ , by applying the Central Limit Theorem, we can conclude that (3.B.16) has a limiting normal distribution with mean zero and variance-covariance matrix equal to the limit of the following matrix :

$$\sum_\ell \begin{bmatrix} \frac{1}{\sqrt{N}}(\Omega_\mu + \Omega_\nu)^{-1} \\ \frac{1}{\sqrt{NT}}(\Omega_\mu^{-1}\otimes\bar{A}_{1\ell} + (\Omega_\mu+\Omega_\nu)^{-1}\otimes\bar{A}_{3\ell}) \end{bmatrix} \Omega_\mu \left[\frac{1}{\sqrt{N}}(\Omega_\mu+\Omega_\nu)^{-1} \quad \frac{1}{\sqrt{NT}}\left(\Omega_\mu^{-1}\otimes\bar{A}_{1\ell}' + (\Omega_\mu+\Omega_\nu)^{-1}\otimes\bar{A}_{3\ell}'\right)\right]$$

$$= \sum_\ell \begin{bmatrix} \frac{1}{N}(\Omega_\mu+\Omega_\nu)^{-1}\Omega_\mu(\Omega_\mu+\Omega_\nu)^{-1} & \frac{1}{N\sqrt{T}}(\Omega_\mu+\Omega_\nu)^{-1}\Omega_\mu(\Omega_\mu^{-1}\otimes\bar{A}_{1\ell}' + (\Omega_\mu+\Omega_\nu)^{-1}\otimes\bar{A}_{3\ell}') \\ \frac{1}{N\sqrt{T}}\left(\Omega_\mu^{-1}\otimes\bar{A}_{1\ell} + (\Omega_\mu+\Omega_\nu)^{-1}\otimes\bar{A}_{3\ell}\right)\Omega_\mu(\Omega_\mu+\Omega_\nu)^{-1} & \frac{1}{NT}(\Omega_\mu^{-1}\otimes\bar{A}_{1\ell} + (\Omega_\mu+\Omega_\nu)^{-1}\otimes\bar{A}_{3\ell})\,\Omega_\mu\left(\Omega_\mu^{-1}\otimes\bar{A}_{1\ell}' + (\Omega_\mu+\Omega_\nu)^{-1}\otimes\bar{A}_{3\ell}'\right) \end{bmatrix}$$

which is

(3.B.19) $$\begin{bmatrix} (\Omega_\mu + \Omega_\nu)^{-1}\,\Omega_\mu(\Omega_\mu + \Omega_\nu)^{-1} & 0 \\ 0 & 0 \end{bmatrix}$$

as

(3.B.20) $\bar{A}_{i\ell}\,\bar{A}_{j\ell}' = 0$ for $i\neq j$

(3.B.21) $$\sum_\ell \text{plim}\,\frac{1}{NT}\,\bar{A}_{1\ell}\,\bar{A}_{1\ell}' = \text{plim}\,\frac{1}{NT}\,\frac{1}{T^2}\,\underline{X}'M_1(I\otimes\iota)(I\otimes\iota')M_1\underline{X}$$

$$= \text{plim}\,\frac{1}{NT}\,\frac{1}{T}\,\underline{X}'M_1\underline{X} = 0$$

(3.B.22) $$\sum_{\ell} \operatorname{plim} \frac{1}{NT} \bar{A}_{3\ell} \bar{A}'_{3\ell} = \operatorname{plim} \frac{1}{NT} \frac{1}{T^2} \underline{X}' M_3 (I \otimes \iota)(I \otimes \iota') M_3 \underline{X}$$

$$= \operatorname{plim} \frac{1}{NT} \frac{1}{T} \underline{X}' M_3 \underline{X} = 0$$

b) Limiting distribution of the second component of (3.B.14) :

$$\sum_t \begin{bmatrix} \frac{\sqrt{N}}{T} \left(\frac{\hat{\Omega}_3}{T}\right)^{-1} \\ \frac{1}{\sqrt{NT}} \sum_i \hat{\Omega}_i^{-1} \otimes B_{it} \end{bmatrix} \tilde{\nu}_t$$

$$= \sum_t \begin{bmatrix} \frac{\sqrt{N}}{T} \left(\frac{\hat{\Omega}_3}{T}\right)^{-1} \\ \frac{1}{\sqrt{NT}} \sum_{i=2,3} \hat{\Omega}_i^{-1} \otimes B_{it} \end{bmatrix} \tilde{\nu}_t \qquad \text{as } B_i = 0 \text{ , } i=1,4$$

The limiting distribution of the above vector is the same as that of

$$\sum_t \operatorname{plim} \begin{bmatrix} \frac{\sqrt{N}}{T} \left(\frac{\hat{\Omega}_3}{T}\right)^{-1} \\ \frac{1}{\sqrt{NT}} \sum_{i=2,3} \left(\frac{\hat{\Omega}_i}{N}\right)^{-1} \otimes \frac{B_{it}}{N} \end{bmatrix} \tilde{\nu}_t$$

(3.B.23) $$= \begin{bmatrix} \frac{\sqrt{N}}{T} (\Omega_\mu + \Omega_\nu)^{-1} \\ \frac{1}{\sqrt{NT}} \left((\Omega_\nu^{-1} \otimes \bar{B}_{2t}) + (\Omega_\mu + \Omega_\nu)^{-1} \otimes \bar{B}_{3t} \right) \end{bmatrix} \tilde{\nu}_t$$

as $\operatorname{plim} \left(\frac{\hat{\Omega}_2}{N}\right)^{-1} = \Omega_\nu^{-1}$;

$\operatorname{plim} \left(\frac{\hat{\Omega}_3}{N}\right)^{-1} = (\Omega_\mu + \Omega_\nu)^{-1}$;

$\operatorname{plim} \frac{1}{N} B_{2t} = \operatorname{plim} \frac{1}{N}$ (t-th column of $\underline{X}' M_2 (\iota \otimes I_T)$)

$= \operatorname{plim} \frac{1}{N}$ (t-th column of $\underline{X}' \left((\iota_N \otimes I) - (\iota_N \otimes \iota_T \iota_T') \frac{1}{T} \right)$)

$$= \operatorname{plim} \frac{1}{N} \begin{bmatrix} \sum_\ell x_{2\ell t} \\ \vdots \\ \sum_\ell x_{K\ell t} \end{bmatrix} - \frac{1}{N} \begin{bmatrix} \frac{1}{T} \sum_\ell \sum_t x_{2\ell t} \\ \vdots \\ \frac{1}{T} \sum_\ell \sum_t x_{K\ell t} \end{bmatrix}$$

(3.B.24) $= \text{constant, say } \bar{B}_{2t}$;

$$\text{plim} \frac{1}{N} B_{3t} = \text{plim} \frac{1}{N} \text{ (t-th column of } \underline{X}'M_3 \ (\iota \otimes I_T))$$

$$= \text{plim} \frac{1}{N} \text{ (t-th column of } \underline{X}'(\frac{\iota_N \otimes \iota_T \iota_T'}{T}))$$

$$\text{(3.B.25)} \qquad = \text{constant, say } \bar{B}_{3t}.$$

As $\tilde{v}_t$ (s) are i.i.d. with mean zero and variance-covariance matrix Ω_ν, by applying Central Limit Theorem, we can say that the limiting distribution of (3.B.23) is normal with zero mean and the following variance-covariance matrix, if the limit exists :

$$\text{plim} \sum_t \begin{bmatrix} \frac{\sqrt{N}}{T} (\Omega_\mu + \Omega_\nu)^{-1} \\ \frac{1}{\sqrt{NT}} ((\Omega_\nu^{-1} \otimes \bar{B}_{2t}) + (\Omega_\mu + \Omega_\nu)^{-1} \otimes \bar{B}_{3t}) \end{bmatrix} \Omega_\nu \cdot$$

$$\begin{bmatrix} \frac{\sqrt{N}}{T} (\Omega_\mu + \Omega_\nu)^{-1} \vdots \frac{1}{\sqrt{NT}} (\Omega_\nu^{-1} \otimes \bar{B}_{2t}' + (\Omega_\mu + \Omega_\nu)^{-1} \otimes \bar{B}_{3t}') \end{bmatrix}$$

$$= \text{plim} \sum_t \begin{bmatrix} \frac{1}{T}(\frac{N}{T})(\Omega_\mu + \Omega_\nu)^{-1} \Omega_\nu (\Omega_\mu + \Omega_\nu)^{-1} & \frac{1}{T\sqrt{T}}(\Omega_\mu + \Omega_\nu)^{-1} \Omega_\nu (\Omega_\nu^{-1} \otimes \bar{B}_{2t}' + (\Omega_\mu + \Omega_\nu)^{-1} \otimes \bar{B}_{3t}') \\ \frac{1}{T\sqrt{T}}(\Omega_\nu^{-1} \otimes \bar{B}_{2t} + (\Omega_\mu + \Omega_\nu)^{-1} \otimes \bar{B}_{3t}) \Omega_\nu (\Omega_\mu + \Omega_\nu)^{-1} & \frac{1}{NT} (\Omega_\nu^{-1} \otimes \bar{B}_{2t} + (\Omega_\mu + \Omega_\nu)^{-1} \otimes \bar{B}_{3t}) \Omega_\nu \\ & (\Omega_\nu^{-1} \otimes \bar{B}_{2t}' + (\Omega_\mu + \Omega_\nu)^{-1} \otimes \bar{B}_{3t}') \end{bmatrix}$$

$$\text{(3.B.26)} \quad = \begin{bmatrix} (\Omega_\mu + \Omega_\nu)^{-1} \Omega_\nu (\Omega_\mu + \Omega_\nu)^{-1} & 0 \\ 0 & 0 \end{bmatrix}$$

as it can be easily verified that

$$\text{(3.B.27)} \quad \bar{B}_{it} \bar{B}_{jt}' = 0 \qquad i \neq j$$

$$\text{(3.B.28)} \quad \sum_t \text{plim} \frac{1}{NT} \bar{B}_{2t} \bar{B}_{2t}' = 0$$

$$\text{(3.B.29)} \quad \sum_t \text{plim} \frac{1}{NT} \bar{B}_{3t} \bar{B}_{3t}' = 0$$

c) <u>Limiting distribution of the third component of (3.B.14)</u> :

$$\sum_{\ell}\sum_{t}\begin{bmatrix}\frac{\sqrt{N}}{NT}\left(\frac{\hat{\Omega}_3}{T}\right)^{-1} \\ \frac{1}{\sqrt{NT}}\sum_i \hat{\Omega}_i^{-1} \otimes C_{i\ell t}\end{bmatrix}\tilde{\varepsilon}_{\ell t}$$

The limiting distribtution of the above vector is the same as that of

$$\sum_{\ell}\sum_{t}\operatorname{plim}\begin{bmatrix}\frac{\sqrt{N}}{NT}\left(\frac{\hat{\Omega}_3}{T}\right)^{-1} \\ \frac{1}{\sqrt{NT}}\sum_i \hat{\Omega}_i^{-1} \otimes C_{i\ell t}\end{bmatrix}\tilde{\varepsilon}_{\ell t}$$

$$\text{(3.B.30)} \quad = \sum_{\ell}\sum_{t}\begin{bmatrix}\frac{\sqrt{N}}{NT}(\Omega_\mu + \Omega_\nu)^{-1} \\ \frac{1}{\sqrt{NT}}\Omega_\varepsilon^{-1} \otimes \bar{C}_{1\ell t}\end{bmatrix}\tilde{\varepsilon}_{\ell t}$$

as

$$\operatorname{plim}\hat{\Omega}_i^{-1} = 0 \quad i=1,2,3 \; ; \; \operatorname{plim}\hat{\Omega}_4^{-1} = \Omega_4^{-1} = \Omega_\varepsilon^{-1} \; ;$$

$$\operatorname{plim} C_{1\ell t} = \operatorname{plim}\left((\ell t)\text{th column of } \underline{X}'M_1\right)$$

$$= \operatorname{plim}\begin{bmatrix}\frac{1}{T}\sum_t x_{2\ell t} - \frac{1}{NT}\sum_\ell\sum_t x_{2\ell t} \\ \vdots \\ \frac{1}{T}\sum_t x_{K\ell t} - \frac{1}{NT}\sum_\ell\sum_t x_{K\ell t}\end{bmatrix}$$

$$\text{(3.B.31)} \quad = \text{constant, say } \bar{C}_{1\ell t}$$

and similarly

$$\text{(3.B.32)} \quad \begin{cases}\operatorname{plim} C_{2\ell t} = \bar{C}_{2\ell t}\text{, a constant} \\ \operatorname{plim} C_{3\ell t} = \bar{C}_{3\ell t}\text{, a constant} \\ \operatorname{plim} C_{4\ell t} = \bar{C}_{4\ell t}\text{, a constant}\end{cases}$$

As $\tilde{\varepsilon}_{\ell t}$ (s) are i.i.d. with zero mean and variance-covariance matrix Ω_ε, the limiting distribution of (3.B.30) is normal with zero mean and variance-covariance matrix equal to

$$\text{plim} \sum_{\ell} \sum_{t} \begin{bmatrix} \frac{\sqrt{N}}{NT}(\Omega_{\mu}+\Omega_{\nu})^{-1} \\ \frac{1}{\sqrt{NT}}\,\Omega_{\varepsilon}^{-1} \otimes \bar{C}_{4\ell t} \end{bmatrix} \Omega_{\varepsilon} \begin{bmatrix} \frac{\sqrt{N}}{NT}(\Omega_{\mu}+\Omega_{\nu})^{-1} & \frac{1}{\sqrt{NT}}\,\Omega_{\varepsilon}^{-1} \otimes \bar{C}'_{4\ell t} \end{bmatrix}$$

$$\text{plim} \sum_{\ell} \sum_{t} \begin{bmatrix} \frac{1}{T}\frac{1}{NT}(\Omega_{\mu}+\Omega_{\nu})^{-1}\Omega_{\varepsilon}(\Omega_{\mu}+\Omega_{\nu})^{-1} & \frac{1}{NT\sqrt{T}}(\Omega_{\mu}+\Omega_{\nu})^{-1}\Omega_{\varepsilon}(\Omega_{\varepsilon}^{-1}\otimes \bar{C}'_{4\ell t}) \\ \frac{1}{NT\sqrt{T}}(\Omega_{\varepsilon}^{-1}\otimes \bar{C}_{4\ell t})\Omega_{\varepsilon}(\Omega_{\mu}+\Omega_{\nu})^{-1} & \frac{1}{NT}(\Omega_{\varepsilon}^{-1}\otimes \bar{C}_{4\ell t}\,\bar{C}'_{4\ell t}) \end{bmatrix}$$

$$(3.B.33) \quad = \begin{bmatrix} 0 & 0 \\ 0 & \Omega_{\varepsilon}^{-1} \otimes R \end{bmatrix}$$

as $\quad \text{plim} \sum_{\ell}\sum_{t} \frac{1}{NT}\,\bar{C}_{4\ell t}\,\bar{C}'_{4\ell t}$

$$= \text{plim} \sum_{\ell}\sum_{t} \frac{1}{NT}\,C_{4\ell t}\,C'_{4\ell t}$$

$$= \text{plim}\,\frac{1}{NT} \sum_{\ell}\sum_{t} C_{4\ell t}\,C'_{4\ell t}$$

$$= \text{plim}\,\frac{1}{NT}\,\underline{X}'M_4.M_4\underline{X}$$

$$(3.B.34) \quad = \text{plim}\,\frac{1}{NT}\,\underline{X}'M_4\underline{X} = R$$

Now, let us go back to the expression of $D^{\frac{1}{2}}\,D_2w$ of (3.B.14) (page 105). As its three components are independent and have normal limiting distributions (as we have seen above), the limiting distribution of $D^{\frac{1}{2}}\,D_2w$ is normal with mean and variance-covariance matrix equal to the sum of the component means and variance-covariance matrices respectively.

The components have all zero mean and thus the sum has also zero mean.

The variance-covariance matrix of the sum is the sum of matrices (3.B.19), (3.B.26) and (3.B.33), each of these matrices being partitioned into four blocks. The sum of the first blocks of the three matrices is given by :

$$(\Omega_{\mu}+\Omega_{\nu})^{-1}\,\Omega_{\mu}(\Omega_{\mu}+\Omega_{\nu})^{-1} + (\Omega_{\mu}+\Omega_{\nu})^{-1}\,\Omega_{\nu}(\Omega_{\mu}+\Omega_{\nu})^{-1} + 0$$

$$= (\Omega_{\mu}+\Omega_{\nu})^{-1}$$

The sum of the second blocks is given by :

$$0 + 0 + 0 = 0$$

The sum of the third blocks is also zero as the matrices are symmetric.

The sum of the fourth blocks is given by

$$0 + 0 + \Omega_\varepsilon^{-1} \otimes R = \Omega_\varepsilon^{-1} \otimes R$$

Thus the limiting distribution of $D^{\frac{1}{2}} D_2 w$ is normal with zero mean and the following variance-covariance matrix :

$$\text{(3.B.35)} \quad \begin{bmatrix} (\Omega_\mu + \Omega_\nu)^{-1} & 0 \\ 0 & (\Omega_\varepsilon^{-1} \otimes R) \end{bmatrix} = \bar{D}^{-1}$$ using the notation of (3.B.6)

Now, from (3.B.5)

$$D^{-\frac{1}{2}}(\hat{\tilde{\pi}}_{fGLS} - \tilde{\pi}) = [D^{\frac{1}{2}} D_1 D^{\frac{1}{2}}]^{-1} D^{\frac{1}{2}} D_2 w$$

and from (3.B.6)

$$\operatorname{plim} (D^{\frac{1}{2}} D_1 D^{\frac{1}{2}})^{-1} = \bar{D}$$

Therefore, $D^{-\frac{1}{2}}(\hat{\tilde{\pi}}_{fGLS} - \tilde{\pi})$ has a normal limiting distribution with zero mean and variance covariance matrix equal to

$$\bar{D}\,\bar{D}^{-1}\,\bar{D}' = \bar{D}$$

In other words,

$$\text{(3.B.36)} \begin{cases} \sqrt{N}(\hat{\pi}_{o,fGLS} - \pi_o) \sim N(0, (\Omega_\mu + \Omega_\nu)) \\ \sqrt{NT}(\hat{\pi}_{*,fGLS} - \pi_*) \sim N(0, \Omega_\varepsilon \otimes R^{-1}) \end{cases}$$

APPENDIX 3.C Limiting Distribution of the Reduced Form Maximum Likelihood Estimators

3.C.1 The Information Matrix

The bordered information matrix is obtained by taking the limit of the negative of the expectation of the matrix of second-order derivatives of the log-likelihood function i.e. limit of the so-called information matrix and bordering it with the gradients of the constraints of the maximisation problem. First we will derive the information matrix, then find its limit (dividing each block by the appropriate factor) and finally give the bordered information matrix.

In order to obtain the information matrix, first we need to write down the second-order differential of the loglikelihood function.

Let us recall that the first-order differential of log L is :

$$(3.C.1) \qquad d \log L = -\frac{1}{2} \sum_{i=1}^{4} m_i \operatorname{tr} \Omega_i^{-1} d\Omega_i - \sum_{i=1}^{4} \operatorname{tr} V'M_i dV\Omega_i^{-1} + \frac{1}{2} \sum_{i=1}^{4} \operatorname{tr} V'M_i V\Omega_i^{-1} d\Omega_i \, \Omega_i^{-1}$$

where $V = y - X \Pi$.

Therefore, using the results (3.92) given on page 66 , the second-order differential of log L can be written as follows :

$$(3.C.2) \quad d^2 \log L = \frac{1}{2} \sum_{i=1}^{4} m_i \operatorname{tr} \Omega_i^{-1} d\Omega_i \, \Omega_i^{-1} d\Omega_i + \sum_{i=1}^{4} \operatorname{tr} V'M_i \, dV \, \Omega_i^{-1} d\Omega_i \, \Omega_i^{-1}$$
$$- \sum_{i=1}^{4} \operatorname{tr} (dV)'M_i(dV) \, \Omega_i^{-1} - \sum_{i=1}^{4} \operatorname{tr} V'M_i \, V \, \Omega_i^{-1} d\Omega_i \, \Omega_i^{-1} d\Omega_i \, \Omega_i^{-1}$$
$$+ \sum_{i=1}^{4} \operatorname{tr} V'M_i \, dV \, \Omega_i^{-1} d\Omega_i \, \Omega_i^{-1}$$

Before taking its expectation, let us note the following results :

$$E(V'M_i V) = E[v_m' \, M_i \, v_{m'}] \quad , \; m,m'=1,\ldots,M$$
$$= E[\operatorname{tr} M_i \, v_{m'} \, v_m'], \; m,m'=1,..,M$$

$$= [\mathrm{tr}\; M_i\; \Omega_{mm'}],\quad m,m'=1,\ldots,M$$

$$= [\mathrm{tr}\; \omega_{imm'}\; M_i],\quad m,m'=1,\ldots,M$$

$$= [\omega_{imm'}\; m_i],\quad m,m'=1,\ldots,M$$

(3.C.3) $$= m_i\; \Omega_i$$

(3.C.4) $E(V'M_i dV) = -\,E(V'M_i\; X\, d\Pi) = 0$ using (3.A.19)

Using these results, it can be verified that

$$E(-d^2 \log L) = -\frac{1}{2}\sum_{i=1}^{4} m_i\;\mathrm{tr}\; d\Omega_i\;\Omega_i^{-1}\; d\Omega_i\;\Omega_i^{-1} + \sum_{i=1}^{4}\mathrm{tr}\;(d\Pi)'X'M_iX(d\Pi)\Omega_i^{-1}$$

$$-\sum_{i=1}^{4}\mathrm{tr}\; m_i\;\Omega_i\;\Omega_i^{-1}\; d\Omega_i\;\Omega_i^{-1}\; d\Omega_i\;\Omega_i^{-1}$$

(3.C.5) $$= \frac{1}{2}\sum_{i=1}^{4} m_i\;\mathrm{tr}\; d\Omega_i\;\Omega_i^{-1}\; d\Omega_i\;\Omega_i^{-1} + \sum_{i=1}^{4}\mathrm{tr}\;(d\Pi)'X'M_iX(d\Pi)\Omega_i^{-1}$$

Now, the above expression has to be written in vec form to obtain the information matrix. For that, we use the familiar vec-trace relationship :

$$\mathrm{tr}\; DB'A'C = (\mathrm{vec}\; A)'\;(B \otimes C)\;\mathrm{vec}\; D$$

Doing so, we can write :

(3.C.6) $$\mathrm{tr}\; d\Omega_i\Omega_i^{-1}d\Omega_i\Omega_i^{-1} = (d\;\mathrm{vec}\;\Omega_i)'(\Omega_i^{-1}\otimes\Omega_i^{-1})(d\;\mathrm{vec}\;\Omega_i)$$

and

(3.C.7) $$\mathrm{tr}(d\Pi)'X'M_iXd\Pi\Omega_i^{-1} = (d\;\mathrm{vec}(X\Pi))'(\Omega_i^{-1}\otimes M_i)(d\;\mathrm{vec}(X\Pi))$$

$$= d(\tilde{X}\tilde{\pi})'(\Omega_i^{-1}\otimes M_i)d(\tilde{X}\tilde{\pi}) \text{ as } \mathrm{vec}\; X\Pi = \tilde{X}\tilde{\pi}$$

$$= (d\tilde{\pi})'\tilde{X}'(\Omega_i^{-1}\otimes M_i)\;\tilde{X}\; d\tilde{\pi}$$

Note that we have expressed the second term of (3.C.5) in terms of $\tilde{\pi}$ as we wish to derive the limiting distribution of

$$\begin{bmatrix} \sqrt{N}\; I_M & 0 \\ 0 & \sqrt{NT}\; I_{M(K-1)} \end{bmatrix} \qquad (\hat{\tilde{\pi}}_{ML} - \tilde{\pi})$$

Thus, we have

$$(3.C.8)\quad E(-d^2\log L) = \frac{1}{2}\sum_{i=1}^{4} m_i(d\,\mathrm{vec}\,\Omega_i)'(\Omega_i^{-1}\otimes\Omega_i^{-1})(d\mathrm{vec}\,\Omega_i) + \sum_{i=1}^{4}(d\tilde{\pi})'\tilde{X}'(\Omega_i^{-1}\otimes M_i)\,\tilde{X}\,d\tilde{\pi}$$

Substituting formulae (3.93) in vec version in the above equation, we get

$$\begin{aligned}(3.C.9)\quad E(-d^2\log L) = {} & \frac{1}{2}\sum_{i=1}^{4} m_i(d\,\mathrm{vec}\,\Omega_\varepsilon)'(\Omega_i^{-1}\otimes\Omega_i^{-1})(d\,\mathrm{vec}\,\Omega_\varepsilon)\\ & + \frac{T^2}{2}\sum_{i=1,3} m_i(d\mathrm{vec}\,\Omega_\mu)'(\Omega_i^{-1}\otimes\Omega_i^{-1})(d\mathrm{vec}\,\Omega_\mu)\\ & + \frac{N^2}{2}\sum_{i=2,3} m_i(d\mathrm{vec}\,\Omega_\nu)'(\Omega_i^{-1}\otimes\Omega_i^{-1})(d\mathrm{vec}\,\Omega_\nu)\\ & + 2\,\frac{T}{2}\sum_{1,3} m_i\,(d\mathrm{vec}\Omega_\varepsilon)'(\Omega_i^{-1}\otimes\Omega_i^{-1})\,d\mathrm{vec}\,\Omega_\mu\\ & + 2\,\frac{N}{2}\sum_{2,3} m_i\,(d\mathrm{vec}\Omega_\varepsilon)'(\Omega_i^{-1}\otimes\Omega_i^{-1})\,d\mathrm{vec}\,\Omega_\nu\\ & + 2\,\frac{NT}{2} m_3\,(d\mathrm{vec}\Omega_\mu)'(\Omega_3^{-1}\otimes\Omega_3^{-1})\,d\mathrm{vec}\,\Omega_\nu\\ & + \sum_{i=1}^{4}(d\tilde{\pi})'\,\tilde{X}'(\Omega_i^{-1}\otimes M_i)\,\tilde{X}(d\tilde{\pi})\end{aligned}$$

Thus, the negative of expectation of the matrix of second-order derivatives of log L, i.e. the information matrix, is given by :

$$(3.C.10)\ \psi = \left[\begin{array}{c|c|c|c} \sum_{i=1}^{4}\tilde{X}'(\Omega_i^{-1}\otimes M_i)\tilde{X} & & 0 & \\ & \frac{1}{2}\sum_{i=1}^{4} m_i(\Omega_i^{-1}\otimes\Omega_i^{-1}) & & \text{(symmetric)} \\ 0 & & & \\ & \frac{T}{2}\sum_{1,3} m_i(\Omega_i^{-1}\otimes\Omega_i^{-1}) & \frac{T^2}{2}\sum_{1,3} m_i(\Omega_i^{-1}\otimes\Omega_i^{-1}) & \\ & \frac{N}{2}\sum_{2,3} m_i(\Omega_i^{-1}\otimes\Omega_i^{-1}) & \frac{NT}{2} m_3(\Omega_3^{-1}\otimes\Omega_3^{-1}) & \frac{N^2}{2}\sum_{2,3} m_i(\Omega_i^{-1}\otimes\Omega_i^{-1}) \end{array}\right]$$

3.C.2 Limit of the Information Matrix

The following results will be used in the calculations[1]:

$$(3.C.11)\quad \begin{cases} \frac{1}{NT}\underline{X}'M_4\underline{X} = \frac{1}{NT}\underline{X}'Q\underline{X} \to R & \text{(cf (3.A.12))} \\ \frac{1}{NT}\underline{X}'\iota \to r & \text{(cf (3.A.4))} \end{cases}$$

$$(3.C.12)\quad \Omega_i^{-1} \to 0 \quad \text{for } i=1,2,3$$

$$(3.C.13)\quad T\,\Omega_1^{-1} \to \left(\frac{\Omega_1}{T}\right)^{-1} = \left(\frac{\Omega_\varepsilon+T\Omega_\mu}{T}\right)^{-1} \to \Omega_\mu^{-1}$$

$$(3.C.14)\quad N\,\Omega_2^{-1} \to \left(\frac{\Omega_2}{N}\right)^{-1} = \left(\frac{\Omega_\varepsilon+N\Omega_\nu}{N}\right)^{-1} \to \Omega_\nu^{-1}$$

$$(3.C.15)\quad \begin{cases} T\,\Omega_3^{-1} \to \left(\frac{\Omega_3}{T}\right)^{-1} = \left(\frac{\Omega_\varepsilon+T\Omega_\mu+N\Omega_\nu}{T}\right)^{-1} \to (\Omega_\mu+\Omega_\nu)^{-1} \\ N\,\Omega_3^{-1} \to \left(\frac{\Omega_3}{N}\right)^{-1} = \left(\frac{\Omega_\varepsilon+T\Omega_\mu+N\Omega_\nu}{N}\right)^{-1} \to (\Omega_\mu+\Omega_\nu)^{-1} \end{cases}$$

assuming $\frac{N}{T} \to 1$.

Now, we have to take the limit of

$$\eta\;\psi\;\eta$$

where

$$(3.C.16)\quad \eta = \text{diag}\left[\frac{1}{\sqrt{N}} I_M, \frac{1}{\sqrt{NT}} I_{M(K-1)}, \frac{1}{\sqrt{NT}} I_{M^2}, \frac{1}{\sqrt{N}} I_{M^2}, \frac{1}{\sqrt{T}} I_{M^2}\right]$$

in order to finally determine the limiting distribution of the ML estimators in the form given on page 70 [2]. Let us find the limit of each block of $\eta\;\psi\;\eta$ separately. First,

---

1) The limit is taken as both N and T tend to infinity.

2) It can easily be verified that taking $\frac{1}{\sqrt{T}}$ I in place of the first $\frac{1}{\sqrt{N}}$ I in η will not alter any of the results that follow.

$$\begin{bmatrix} \frac{1}{\sqrt{N}} I & 0 \\ 0 & \frac{1}{\sqrt{NT}} I \end{bmatrix} \sum_{i=1}^{4} \tilde{X}'(\Omega_i^{-1} \otimes M_i)\tilde{X} \begin{bmatrix} \frac{1}{\sqrt{N}} I & 0 \\ 0 & \frac{1}{\sqrt{NT}} I \end{bmatrix}$$

$$= \begin{bmatrix} \frac{1}{\sqrt{N}} I & 0 \\ 0 & \frac{1}{\sqrt{NT}} I \end{bmatrix} \begin{bmatrix} \Omega_3^{-1}\, NT & \Omega_3^{-1} \otimes \iota'\underline{X} \\ \Omega_3^{-1} \otimes \underline{X}'\iota & \sum_i \Omega_i^{-1} \otimes \underline{X}'M_i\underline{X} \end{bmatrix} \begin{bmatrix} \frac{1}{\sqrt{N}} I & 0 \\ 0 & \frac{1}{\sqrt{NT}} I \end{bmatrix}$$

using (3.28), page 51

$$= \begin{bmatrix} \frac{1}{N}\Omega_3^{-1}\, NT & \frac{1}{N\sqrt{T}} \Omega_3^{-1} \otimes \iota'\underline{X} \\ \frac{1}{N\sqrt{T}} \Omega_3^{-1} \otimes \underline{X}'\iota & \frac{1}{NT} \sum_i \Omega_i^{-1} \otimes \underline{X}'M_i\underline{X} \end{bmatrix}$$

$$\rightarrow \begin{bmatrix} (\Omega_\mu + \Omega_\nu)^{-1} & 0 \\ 0 & \Omega_\varepsilon^{-1} \otimes R \end{bmatrix}$$ using (3.C.11), (3.C.12), (3.C.15)

Next,

$$\frac{1}{NT}\frac{1}{2}\sum_{i=1} m_i\,(\Omega_i^{-1} \otimes \Omega_i^{-1}) \rightarrow \frac{1}{2}\Omega_\varepsilon^{-1} \otimes \Omega_\varepsilon^{-1}$$

noting that

$$\frac{m_i}{NT} \rightarrow 0 \qquad \text{for } i=1,2,3 \text{ and } \frac{m_4}{NT} \rightarrow 1$$

and using (3.C.12).

Following a similar reasoning and making use of results (3.C.11) to (3.C.15), it can be easily verified that :

$$\frac{1}{N\sqrt{T}}\frac{T}{2}\sum_{1,3} m_i\,\Omega_i^{-1} \otimes \Omega_i^{-1} \rightarrow 0$$

$$\frac{1}{T\sqrt{N}}\frac{N}{2}\sum_{2,3} m_i\,\Omega_i^{-1} \otimes \Omega_i^{-1} \rightarrow 0$$

$$\frac{1}{N}\frac{T}{2}\sum_{1,3} m_i\,\Omega_i^{-1} \otimes \Omega_i^{-1} \rightarrow \frac{1}{2}\Omega_\mu^{-1} \otimes \Omega_\mu^{-1}$$

$$\frac{1}{\sqrt{NT}}\frac{NT}{2} m_3\,\Omega_3^{-1} \otimes \Omega_3^{-1} \rightarrow 0$$

and

$$\frac{1}{T}\frac{N^2}{2}\sum_{2,3} m_i\,\Omega_i^{-1} \otimes \Omega_i^{-1} \rightarrow \frac{1}{2}\Omega_\nu^{-1} \otimes \Omega_\nu^{-1}$$

Thus,

(3.C.17) $\lim \eta \psi \eta = \tilde{\psi} =$

$$\begin{bmatrix} (\Omega_\mu + \Omega_\nu)^{-1} & & & & \\ & \Omega_\varepsilon^{-1} \otimes R & & \bigcirc & \\ & & \frac{1}{2}\Omega_\varepsilon^{-1} \otimes \Omega_\varepsilon^{-1} & & \\ & \bigcirc & & \frac{1}{2}\Omega_\mu^{-1} \otimes \Omega_\mu^{-1} & \\ & & & & \frac{1}{2}\Omega_\nu^{-1} \otimes \Omega_\nu^{-1} \end{bmatrix}$$

3.C.3 The Limiting Distribution

Recalling the definition of the bordered information matrix given at the beginning of this Appendix, it is obvious that it is given by

(3.C.18) $$H = \begin{bmatrix} \tilde{\psi} & \tilde{G}' \\ \tilde{G} & 0 \end{bmatrix}$$

where

(3.C.19) $$\tilde{G} = \begin{bmatrix} 0 & 0 & C & 0 & 0 \\ 0 & 0 & 0 & C & 0 \\ 0 & 0 & 0 & 0 & C \end{bmatrix}$$

Replacing $\tilde{\psi}$ and $\tilde{G}$ in H by their respective expressions (3.C.17) and (3.C.19), the bordered information matrix (in limit) is written as:

$$(3.C.20)\quad H = \begin{bmatrix} (\Omega_\mu+\Omega_\nu)^{-1} & 0 & 0 & 0 & 0 & 0 & 0 & 0 \\ 0 & (\Omega_\varepsilon^{-1}\otimes R) & 0 & 0 & 0 & 0 & 0 & 0 \\ 0 & 0 & \frac{1}{2}(\Omega_\varepsilon^{-1}\otimes\Omega_\varepsilon^{-1}) & 0 & 0 & C' & 0 & 0 \\ 0 & 0 & 0 & \frac{1}{2}(\Omega_\mu^{-1}\otimes\Omega_\mu^{-1}) & 0 & 0 & C' & 0 \\ 0 & 0 & 0 & 0 & \frac{1}{2}(\Omega_\nu^{-1}\otimes\Omega_\nu^{-1}) & 0 & 0 & C' \\ 0 & 0 & C & 0 & 0 & 0 & 0 & 0 \\ 0 & 0 & 0 & C & 0 & 0 & 0 & 0 \\ 0 & 0 & 0 & 0 & C & 0 & 0 & 0 \end{bmatrix}$$

Due to the particularly simple form of H, the first two blocks on the main diagonal of its inverse are given by the inverse of the corresponding blocks. For the next three blocks on the main diagonal of the inverse of the information matrix, each block is the first block of the inverse of

$$(3.C.21)\quad \begin{bmatrix} \frac{1}{2}(\Omega_j^{-1}\otimes\Omega_j^{-1}) & C' \\ C & 0 \end{bmatrix}$$

using the fact that if we permute any two columns of a matrix and then find its inverse, we have to permute the two rows with the same indices in the inverse of the "permuted" matrix in order to get the inverse of the original matrix, and vice versa.

According to the usual inversion rule (see, for instance, Balestra [4] , p.10), the first block of the inverse of (3.C.21) can be expressed as

$$(3.C.22)\quad \tilde{F}\left[\tilde{F}'\,\frac{1}{2}\,(\Omega_j^{-1}\otimes\Omega_j^{-1})\,\tilde{F}\right]^{-1}\tilde{F}'$$

for $\tilde{F}$ such that $\tilde{F}\tilde{F}' = I - C'C = I - \frac{1}{2}(I-P) = \frac{1}{2}(I+P)$ where P is the commutation matrix. Note that $\frac{1}{2}(I+P)$ is idempotent.

From the properties of the commutation matrix, it is easily established that

(3.C.23) $\tilde{F}\tilde{F}'\,(2\,\Omega_j \otimes \Omega_j)\,\tilde{F}\tilde{F}' = \tilde{F}\tilde{F}'\,(2\,\Omega_j \otimes \Omega_j)$

From this we can write :

$$\tilde{F}\tilde{F}'\,(2\,\Omega_j \otimes \Omega_j)\,\tilde{F}\tilde{F}'(\tfrac{1}{2}\,\Omega_j^{-1} \otimes \Omega_j^{-1}) = \tilde{F}\tilde{F}'$$

or

$$\tilde{F}\tilde{F}'\,(2\,\Omega_j \otimes \Omega_j)\,\tilde{F}\left[\tilde{F}'(\tfrac{1}{2}\,\Omega_j^{-1} \otimes \Omega_j^{-1})\,\tilde{F}\right] = \tilde{F}\;,\quad (\tilde{F}'\tilde{F}=I)$$

or

$$\tilde{F}\tilde{F}'\,(2\,\Omega_j \otimes \Omega_j)\,\tilde{F} = \tilde{F}\left[\tilde{F}'(\tfrac{1}{2}\,\Omega_j^{-1} \otimes \Omega_j^{-1})\,\tilde{F}\right]^{-1}$$

or

(3.C.24) $\frac{1}{2}(I+P)(2\,\Omega_j \otimes \Omega_j)\,\frac{1}{2}(I+P) = \tilde{F}\left[\tilde{F}'(\tfrac{1}{2}\,\Omega_j^{-1} \otimes \Omega_j^{-1})\,\tilde{F}\right]^{-1}\tilde{F}'$

Thus we can conclude that :

(3.C.25) $\begin{cases} \sqrt{N}^{1)}\,(\hat{\pi}_{o,ML} - \pi_o) \sim N\left(0\,,\,(\Omega_\mu + \Omega_\nu)\right) \\ \sqrt{NT}\,(\hat{\pi}_{*,ML} - \pi_*) \sim N\left(0\,,\,\Omega_\varepsilon \otimes R^{-1}\right) \end{cases}$

(3.C.26) $\begin{cases} \sqrt{NT}\,\mathrm{vec}(\hat{\Omega}_{\varepsilon,ML} - \Omega_\varepsilon) \sim N\left(0\,,\,\frac{1}{2}(I+P)(2\,\Omega_\varepsilon \otimes \Omega_\varepsilon)\,\frac{1}{2}(I+P)\right) \\ \sqrt{N}\,\mathrm{vec}(\hat{\Omega}_{\mu,ML} - \Omega_\mu) \sim N\left(0\,,\,\frac{1}{2}(I+P)(2\,\Omega_\mu \otimes \Omega_\mu)\,\frac{1}{2}(I+P)\right) \\ \sqrt{T}\,\mathrm{vec}(\hat{\Omega}_{\nu,ML} - \Omega_\nu) \sim N\left(0\,,\,\frac{1}{2}(I+P)(2\,\Omega_\nu \otimes \Omega_\nu)\,\frac{1}{2}(I+P)\right) \end{cases}$

CHAPTER 4

ESTIMATION OF THE STRUCTURAL FORM - PART 1

4.1 Generalised Two Stage Least Squares - A Single Equation Method

4.1.1 Estimation in the Case of Known Variance Components

Let us consider the m-th structural equation, over-identified, and rewrite it as follows :

(4.1) $$y_m = Y_m \gamma_m + X_m \beta_m + u_m$$

where y_m is the NT x 1 vector of the values taken by the jointly dependent variable explained by the equation and Y_m is the NT x $\tilde{M}_m$ matrix of the values taken by the $\tilde{M}_m$ jointly dependent explanatory variables present in the equation ; X_m is the NT x K_m matrix of the values taken by the K_m explanatory predetermined variables of the equation ; γ_m and β_m are the corresponding coefficient vectors of dimensions $\tilde{M}_m$ x 1 and K_m x 1 respectively. We can also write (4.1) as :

(4.2) $$y_m = Z_m \alpha_m + u_m$$

where

(4.3) $$Z_m = [Y_m \; X_m] \; ; \; \alpha_m' = [\gamma_m' \; \beta_m']$$

Since, in general, in a structural equation, the explanatory variables are correlated with the errors, the equation cannot be as such estimated by GLS. One way of overcoming this problem is to use the instrumental variables (IV) method, which consists in premultiplying the equation by a matrix of instruments that are uncorrelated with the errors and then estimating it by GLS. In Theil's two stage least squares estimation[1)] for homoscedastic and non-auto-correlated structural errors, the X matrix is taken to be the instrumental variables matrix i.e. equation (4.2) is premultiplied by X' to get :

---

1) See Theil [51], page 451 e.g.

(4.4) $X'y_m = X'Z_m \alpha_m + X'u_m$

with

$$E(X'u_m u_m' X') = \sigma^2_{mm} (X'X)$$

and GLS is applied to (4.4) which gives :

$$\alpha^e_m = [Z_m' X(X'X)^{-1}X'Z_m]^{-1} Z_m'X(X'X)^{-1}X'y_m$$

the Aitken estimator of α_m .

For our model, we develop a more general method which selects the "best" instruments from a class of instruments of the form FX. The term "best" is used to mean that F is chosen in such a way as to minimise the trace and determinant of, and the quadratic form in, the asymptotic variance-covariance matrix of the resulting GLS estimator. Let us now expand the above idea in formal terms.

So, first we premultiply (4.2) by X'F', F being a nonsingular matrix (to be determined) and apply GLS on the transformed equation to obtain an estimator of α_m denoted as α^E_m :

$$\alpha^E_m = [Z_m'FX(X'F'\Sigma_{mm}FX)^{-1}X'F'Z_m]^{-1}Z_m'FX(X'F'\Sigma_{mm}FX)^{-1}X'F'y_m$$

The asymptotic variance-covariance matrix of α^E_m can be written as

(4.5) $[\Pi^{*\prime}_m X'F X(X'F'\Sigma_{mm}FX)^{-1}X'F'X \Pi^*_m]^{-1}$

where $\Pi^*_m = [\Pi_m H_m]$, Π_m is the $K x \tilde{M}_m$ matrix containing the $\tilde{M}_m$ coefficient vectors of X in the $\tilde{M}_m$ reduced form equations explaining Y_m i.e.

(4.6) $Y_m = X \Pi_m + V_m$,

H_m is a KxK_m selection matrix such that

(4.7) $X H_m = X_m$;

where the limit value of $\frac{1}{NT} V_m' FX$ is assumed to be zero, V_m representing the $NTx\tilde{M}_m$ matrix of disturbances of the $\tilde{M}_m$ reduced form equations explaining Y_m ; and where the limit values of $\frac{1}{NT}$ X'FX and $\frac{1}{NT}$ $X'F'\Sigma_{mm}FX$ have to be inserted.

It can be shown[1] (see Balestra [5] , p. 101) that $F = \Sigma_{mm}^{-1}$ minimises the trace of (4.5); minimises the determinant of (4.5) and gives the minimal positive definite matrix of the form, in the sense that any other transformation would lead to an asymptotic variance-covariance matrix of the coefficient estimator that, when subtracted from this one, would give a non-positive definite matrix.

Thus the generalised two stage least squares estimator of α_m is given by :

(4.8)

$$\hat{\alpha}_{m,G2SLS} = [Z_m'\Sigma_{mm}^{-1}X(X'\Sigma_{mm}^{-1}X)^{-1}X'\Sigma_{mm}^{-1}Z_m]^{-1} Z_m'\Sigma_{mm}^{-1}X(X'\Sigma_{mm}^{-1}X)^{-1}X'\Sigma_{mm}^{-1}y_m$$

The consistency of $\hat{\alpha}_{m,G2SLS}$ is shown in Appendix 4.C (page 149).

4.1.2 Estimation in the Case of Unknown Variance Components

In the case of unknown variance components, which is the most frequent case in practice, they have to be estimated before applying the generalised two stage least squares. Thus we may speak of feasible generalised two stage least squares while referring to the methods developed below.

Here again, the variance components are estimated by Analysis of Variance (AOV) of the errors of the equation. The AOV estimates (cf. [1]) of $\sigma_{\varepsilon mm}$, $\sigma_{\mu mm}$ and $\sigma_{\nu mm}$ are as follows :

$$(4.9) \quad \begin{cases} \tilde{\sigma}_{\varepsilon mm} = \frac{1}{(N-1)(T-1)} u_m' Q u_m \\ \tilde{\sigma}_{\mu mm} = \frac{1}{T} \left[\frac{1}{N-1} u_m' M_1 u_m - \tilde{\sigma}_{\varepsilon mm}\right] \\ \tilde{\sigma}_{\nu mm} = \frac{1}{N} \left[\frac{1}{T-1} u_m' M_2 u_m - \tilde{\sigma}_{\varepsilon mm}\right] \end{cases}$$

---

1) At least when X contains observations on exogenous variables only.

The resulting estimators of the eigenvalues σ_{1mm}, σ_{2mm}, σ_{3mm} and σ_{4mm} are :

$$(4.10)\quad \begin{cases} \tilde{\sigma}_{1mm} = \frac{1}{N-1} u_m' M_1 u_m \\ \tilde{\sigma}_{2mm} = \frac{1}{T-1} u_m' M_2 u_m \\ \tilde{\sigma}_{3mm} = \tilde{\sigma}_{1mm} + \tilde{\sigma}_{2mm} - \tilde{\sigma}_{\varepsilon mm} \\ \tilde{\sigma}_{4mm} = \tilde{\sigma}_{\varepsilon mm} \end{cases}$$

But the vector of errors u_m is not observable and has to be estimated in turn. Here we will present two alternate ways of estimating u_m using two different (2SLS) covariance estimators of α_m and show when they both lead to the same result.

Method 1 :

Let us rewrite the m-th structural equation to be estimated :

$$(4.11)\quad y_m = Y_m \gamma_m + X_m \beta_m + u_m$$

Suppose we have a consistent estimator of Π_m denoted by $\hat{\Pi}_m$, for instance the estimator given by the asymptotically efficient feasible GLS method developed in the previous chapter. But OLS estimator will also do, as the only condition imposed on $\hat{\Pi}_m$ is consistency i.e.

$$(4.12)\quad \text{plim}\ \hat{\Pi}_m = \Pi_m$$

Then we can write :

$$(4.13)\quad y_m = \hat{Y}_m \gamma_m + X_m \beta_m + u_m + (Y_m - \hat{Y}_m) \gamma_m$$

or

$$(4.14)\quad y_m = \hat{Z}_m \alpha_m + u_m + (Y_m - \hat{Y}_m) \gamma_m$$

where

$$(4.15)\quad \hat{Y}_m = X \hat{\Pi}_m$$

and

(4.16) $$Z_m = [\hat{Y}_m \; X_m]$$

If X_m contains the vector ι_{NT} as its first column i.e. if the structural equation (4.1) has a constant term say a_m which we assume to be the case here, we can rewrite (4.11) as

(4.17) $$y_m = Y_m \gamma_m + \iota_{NT} a_m + X^*_m b_m + u_m$$

writing

(4.18) $$X_m = [\iota_{NT} \; X^*_m] \qquad \beta_m = \begin{bmatrix} a_m \\ b_m \end{bmatrix}$$

or

(4.19) $$y_m = \iota_{NT} a_m + Z^*_m \alpha^*_m + u_m$$

where

(4.20) $$Z^*_m = [Y_m \; X^*_m] \qquad \alpha^*_m = \begin{bmatrix} \gamma_m \\ b_m \end{bmatrix}$$

Now, in the same way, equation (4.14) can be rewritten as

(4.21) $$y_m = \iota_{NT} a_m + \hat{Z}^*_m \alpha^*_m + u_m + (Y_m - \hat{Y}_m) \gamma_m$$

where

(4.22) $$\hat{Z}^*_m = [\hat{Y}_m \; X^*_m]$$

Applying the covariance transformation Q to (4.21) yields:

(4.23) $$Q\, y_m = Q\, \hat{Z}^*_m \alpha^*_m + Q\, u_m + Q(Y_m - \hat{Y}_m) \gamma_m$$

where $Q = I_{NT} - \frac{A}{T} - \frac{B}{N} + \frac{J_{NT}}{NT}$ and $Q\, \iota_{NT} = 0$.

Performing OLS on (4.23) we get a covariance estimator of α^*_m denoted as $\hat{\alpha}^*_{m,cov}$:

(4.24) $$\hat{\alpha}^*_{m,cov} = (\hat{Z}^{*\prime}_m Q\, \hat{Z}^*_m)^{-1} \hat{Z}^{*\prime}_m Q\, y_m$$

The constant term can be estimated by :

(4.25) $\hat{a}_{m,cov} = \frac{1}{NT} \iota'_{NT} (y_m - Z^*_m \hat{\alpha}^*_{m,cov})$

These estimators are consistent and their consistency is proved in detail in Appendix 4.A. (page 134).

Thus we have a first set of consistent estimators of the structural coefficients vector α_m which is

(4.26) $$\hat{\alpha}_{m,cov} = \begin{bmatrix} \hat{\gamma}_{m,cov} \\ \hat{a}_{m,cov} \\ \hat{b}_{m,cov} \end{bmatrix}$$

with $\hat{\gamma}_{m,cov}$, $\hat{b}_{m,cov}$ being the subvectors that form $\hat{\alpha}^*_{m,cov}$.

But it is not interesting to stop at this point as no information is acquired on variance components and also because the generalised two stage least squares estimator of (4.8) possesses additional properties of optimality as mentioned on page 122.

In case the constant term is absent from the structural equation (i.e. $a_m = 0$), $\hat{Z}^*_m = \hat{Z}_m$ and $\alpha^*_m = \alpha_m$; and making these substitutions in (4.24) yields the corresponding covariance estimator of α_m , $\hat{\alpha}_{m,cov}$.

Now, u_m can be predicted by

(4.27) $\hat{u}_m = y_m - Z^*_m \hat{\alpha}^*_{m,cov} - \iota_{NT} \hat{a}_{m,cov}$

Replacing u_m by $\hat{u}_m$ in (4.9) we get the following estimates of variance components :

(4.28) $$\begin{cases} \hat{\sigma}_{\varepsilon mm} = \frac{1}{(N-1)(T-1)} \hat{u}'_m Q \hat{u}_m \\ \hat{\sigma}_{\mu mm} = \frac{1}{T} \left[\frac{1}{N-1} \hat{u}'_m M_1 \hat{u}_m - \hat{\sigma}_{\varepsilon mm}\right] \\ \hat{\sigma}_{\nu mm} = \frac{1}{N} \left[\frac{1}{T-1} \hat{u}'_m M_2 \hat{u}_m - \hat{\sigma}_{\varepsilon mm}\right] \end{cases}$$

The estimators of the characteristic roots of Σ_{mm} namely, $\hat{\sigma}_{1mm}$, $\hat{\sigma}_{2mm}$, $\hat{\sigma}_{3mm}$ and $\hat{\sigma}_{4mm}$ are given by replacing u_m by $\hat{u}_m$ in (4.10), i.e. :

$$(4.29)\quad \begin{cases} \hat{\sigma}_{1mm} = \frac{1}{N-1}\hat{u}'_m M_1 \hat{u}_m \\ \hat{\sigma}_{2mm} = \frac{1}{T-1}\hat{u}'_m M_2 \hat{u}_m \\ \hat{\sigma}_{3mm} = \hat{\sigma}_{1mm} + \hat{\sigma}_{2mm} - \hat{\sigma}_{\varepsilon mm} \\ \hat{\sigma}_{4mm} = \hat{\sigma}_{\varepsilon mm} \end{cases}$$

Once again, there is no guarantee that $\hat{\sigma}_{3mm}$ will be positive and in case it is not so, centered data can be used to eliminate it from the expression of the (feasible) GLS estimator.

Method 2 :

Let us consider the structural equation in the form

$$y_m = \iota_{NT}\, a_m + Z^*_m \alpha^*_m + u_m$$

Premultiplying it by $\underline{X}'Q$ we get :

$$(4.30)\quad \underline{X}'Q\, y_m = \underline{X}'Q\, Z^*_m \alpha^*_m + \underline{X}'Q\, u_m \qquad \text{as } Q\,\iota_{NT} = 0$$

Now, applying GLS on (4.30) we get what we call the covariance 2SLS estimator of α^*_m :

$$(4.31)\quad \hat{\alpha}^*_{m,C2SLS} = (Z^{*\prime}_m Q\underline{X}(\underline{X}'Q\underline{X})^{-1}\underline{X}'Q\, Z^*_m)^{-1}\, Z^{*\prime}_m Q\underline{X}(\underline{X}'Q\underline{X})^{-1}\underline{X}'Qy_m$$

The constant term a_m is estimated as :

$$(4.32)\quad \hat{a}_{m,C2SLS} = \frac{1}{NT}\,\iota'(y_m - Z^*_m\, \hat{\alpha}^*_{m,C2SLS})$$

The above estimators are also consistent as shown in Appendix 4.A. (page 137). Further, straight calculation shows that when we use $\hat{\Pi}_{m,cov}$ for estimating Π_m in Method 1,

$$\hat{\alpha}^*_{m,cov} = \hat{\alpha}^*_{m,C2SLS}$$

Writing

$$(4.33)\quad \hat{\alpha}_{m,C2SLS} = \begin{bmatrix} \hat{a}_{m,C2SLS} \\ \hat{\alpha}^*_{m,C2SLS} \end{bmatrix}$$

we can compute the residuals of the equation as follows :

$$(4.34)\quad \hat{\hat{u}}_m = y_m - Z_m\, \hat{\alpha}_{m,C2SLS}$$

Substituting $\hat{\hat{u}}_m$ for u_m in equations (4.9), we get a second set of estimators of variance components and making the same substitution in (4.10) gives a second set of estimators of the characteristic roots or eigenvalues. For simplicity of notations, we will denote both sets of variance components estimators the same way, specifying, whenever necessary, which one we are referring to.

A detailed proof of the consistency, under some general assumptions, of both the sets of estimators of eigenvalues and variance components is given in Appendix 4.B. (page 139). The following observation may be made at this point :

When N <u>and</u> T go to infinity :

$$(4.35)\quad \begin{cases} \text{plim}\ \hat{\sigma}_{1mm} = \text{plim}\ \sigma_{1mm} = \infty\ ;\ \text{plim}\ \hat{\sigma}_{2mm} = \text{plim}\ \sigma_{2mm} = \infty\ ; \\ \text{plim}\ \hat{\sigma}_{3mm} = \text{plim}\ \sigma_{3mm} = \infty\ ;\ \text{plim}\ \hat{\sigma}_{4mm} = \text{plim}\ \sigma_{4mm} = \sigma_{4mm} \\ \text{plim}\ \dfrac{1}{\hat{\sigma}_{1mm}} = \text{plim}\ \dfrac{1}{\sigma_{1mm}} = 0;\ \text{plim}\ \dfrac{1}{\hat{\sigma}_{2mm}} = \text{plim}\ \dfrac{1}{\sigma_{2mm}} = 0\ ; \\ \text{plim}\ \dfrac{1}{\hat{\sigma}_{3mm}} = \text{plim}\ \dfrac{1}{\sigma_{3mm}} = 0;\ \text{plim}\ \dfrac{1}{\hat{\sigma}_{4mm}} = \text{plim}\ \dfrac{1}{\sigma_{4mm}} = \dfrac{1}{\sigma_{4mm}} \end{cases}$$

whether the eigenvalues are estimated by Method 1 or Method 2.

Thus the estimator of the variance-covariance matrix Σ_{mm} is given by

$$\hat{\Sigma}_{mm} = \hat{\sigma}_{\mu mm}\ A + \hat{\sigma}_{\nu mm}\ B + \hat{\sigma}_{\varepsilon mm}\ I_{NT}$$

or equivalently

$$(4.36)\quad \hat{\Sigma}_{mm} = \hat{\sigma}_{1mm}\ M_1 + \hat{\sigma}_{2mm}\ M_2 + \hat{\sigma}_{3mm}\ M_3 + \hat{\sigma}_{4mm}\ M_4\ ,$$

the latter expression being more useful for calculating $\hat{\Sigma}^{-1}_{mm}$.

This leads to the following feasible G2SLS estimator of α_m :

$$(4.37)\quad \hat{\alpha}_{m,fG2SLS} = [Z'_m\hat{\Sigma}^{-1}_{mm}X(X'\hat{\Sigma}^{-1}_{mm}X)^{-1}X'\hat{\Sigma}^{-1}_{mm}Z_m]^{-1}$$
$$Z'_m\hat{\Sigma}^{-1}_{mm}X(X'\hat{\Sigma}^{-1}_{mm}X)^{-1}\ X'\hat{\Sigma}^{-1}_{mm}y_m$$

The above estimator is consistent, its consistency being shown in Appendix 4.C. (page 149).

Using the consistency results regarding the variance components estimators proved in Appendix 4.B., it can be easily verified that both the pure G2SLS and the feasible G2SLS estimators are asymptotically equivalent. Hence we will derive the asymptotic distribution of only the feasible G2SLS estimator.

As done in Appendix 4.C. (see (4.C.2)), in presence of a constant term in the structural equation, we reformulate the equation as :

$$y_m = \tilde{Z}_m \tilde{\alpha}_m + u_m$$

with $\tilde{Z}_m = [\iota \;\; Z^*_m]$ and $\tilde{\alpha}'_m = [a_m \;\; \alpha^{*\prime}_m]$

Then the expression of the fG2SLS estimator, $\hat{\tilde{\alpha}}_{m,fG2SLS}$, also changes slightly, though both the earlier expression (4.37) and the modified one given in (4.C.4) (page 149) are just two different arrangements of the elements of the same vector.

Now, let us derive the limiting distribution of

$$(4.38) \qquad \begin{bmatrix} \sqrt{N} & 0 \\ 0 & \sqrt{NT}\, I_{\tilde{M}_m+K_m-1} \end{bmatrix} (\hat{\tilde{\alpha}}_{m,fG2SLS} - \tilde{\alpha}_m)$$

i.e. the constant term estimator is multiplied by only $\sqrt{N}$ 1) whereas the remaining coefficients estimators are multiplied as usual by $\sqrt{NT}$.

The derivation of the limiting distribution is given in detail in Appendix 4.D. It is obtained that $\bar{D}_2^{-\frac{1}{2}}(\hat{\tilde{\alpha}}_{m,fG2SLS} - \tilde{\alpha}_m)$ has a normal limiting distribution with zero mean and variance covariance matrix equal to

$$(4.39) \qquad (\tilde{P}'_m \tilde{R}_m \tilde{P}_m)^{-1}$$

where

---

1) or only $\sqrt{T}$ (see footnote on page 100)

$$(4.40) \qquad \bar{D}_2 = \begin{bmatrix} \frac{1}{N} & 0 \\ 0 & \frac{1}{NT} I_{\tilde{M}_m+K_m-1} \end{bmatrix}$$

$$(4.41) \qquad \tilde{P}_m = \left[\begin{array}{c|c} 1 & 0 \\ \hline 0 & \Pi'_{*m} \\ 0 & H^{*\prime}_m \end{array}\right]$$

$$(4.42) \qquad \tilde{R}_m = \begin{bmatrix} \frac{1}{\sigma_{\mu mm}+\sigma_{\nu mm}} & 0 \\ 0 & \frac{1}{\sigma_{\varepsilon mm}} R \end{bmatrix}$$

In other words, omitting the subscript 'fG2SLS', we have

$$(4.43) \qquad \begin{cases} \sqrt{N} \; (\hat{a}_m - a_m) \sim N\,(0, \sigma_{1mm} + \sigma_{\nu mm}) \\ \sqrt{NT} \; (\hat{\alpha}^*_m - \alpha^*_m) \sim N\left(0, \sigma_{\varepsilon mm} \begin{bmatrix} \Pi'_{*m} \\ H^{*\prime}_m \end{bmatrix} R \, [\Pi_{*m} \; H^*_m]\right) \end{cases}$$

4.2 Generalised Three Stage Least Squares - A System Method

The G2SLS method developed in the previous section estimates only one equation at a time. In this section, a method of estimating simultaneously all the structural (first-order) parameters will be presented. This method is a generalisation of the three stage least squares estimation of Theil and Zellner (see [57]). The extension from G2SLS to a generalised 3SLS is done in a similar way to that from classical 2SLS to 3SLS, adopting it to the specifications of the model.

In the G2SLS method of Section 4.1, the structural equation to be estimated, say the m-th one, is first premultiplied by $X'\Sigma^{-1}_{mm}$ to give :

$$(4.44) \qquad X'\Sigma^{-1}_{mm} \, y_m = X'\Sigma^{-1}_{mm} \, Z_m \, \alpha_m + X'\Sigma^{-1}_{mm} \, u_m$$

Then α_m is estimated by performing GLS on (4.44) . Let us apply similar transformations to all the equations of the system :

$$(4.45)\quad \begin{cases} X'\Sigma_{11}^{-1} y_1 = X'\Sigma_{11}^{-1} Z_1 \alpha_1 + X'\Sigma_{11}^{-1} u_1 \\ X'\Sigma_{22}^{-1} y_2 = X'\Sigma_{22}^{-1} Z_2 \alpha_2 + X'\Sigma_{22}^{-1} u_2 \\ \quad\vdots \\ X'\Sigma_{MM}^{-1} y_M = X'\Sigma_{MM}^{-1} Z_M \alpha_M + X'\Sigma_{MM}^{-1} u_M \end{cases}$$

We can write the above set of equations more compactly as :

$$(4.46)\quad \mathcal{X}'\Sigma_*^{-1} y = \mathcal{X}'\Sigma_*^{-1} Z_*\alpha + \mathcal{X}'\Sigma_*^{-1} u$$

where

$$(4.47)\quad \mathcal{X} = \begin{bmatrix} X & & 0 \\ & \ddots & \\ 0 & & X \end{bmatrix} ;\; Z_* = \begin{bmatrix} Z_1 & & 0 \\ & \ddots & \\ 0 & & Z_M \end{bmatrix} ;\; \Sigma_* = \begin{bmatrix} \Sigma_{11} & & 0 \\ & \ddots & \\ 0 & & \Sigma_{MM} \end{bmatrix}$$

and

$$(4.48)\quad E(\mathcal{X}'\Sigma_*^{-1} u\, u'\, \Sigma_*^{-1}\mathcal{X}) = \mathcal{X}'\, \Sigma_*^{-1}\, E(u\, u')\, \Sigma_*^{-1}\mathcal{X} = \mathcal{X}'\, \Sigma_*^{-1}\, \Sigma\, \Sigma_*^{-1}\mathcal{X}$$

Applying GLS on (4.46) we obtain the generalised 3SLS estimator :

$$(4.49)\quad \hat{\alpha}_{G3SLS} = \left[Z_*'\Sigma_*^{-1}\mathcal{X}(\mathcal{X}'\Sigma_*^{-1}\Sigma\Sigma_*^{-1}\mathcal{X})^{-1}\mathcal{X}'\Sigma_*^{-1}Z_*\right]^{-1} Z_*'\Sigma_*^{-1}\mathcal{X}(\mathcal{X}'\Sigma_*^{-1}\Sigma\Sigma_*^{-1}\mathcal{X})^{-1}\mathcal{X}'\Sigma_*^{-1}y$$

Again the problem of estimating $\Sigma_{mm'}$, $m,m'=1,\ldots,M$ has to be solved before applying the above formula. In the fG2SLS method of Section 4.1.2 , first, the residuals of the equation under consideration, say the m-th one, u_m, are predicted using covariance estimators of α_m and then, the variance components of the variance-covariance matrix Σ_{mm} are estimated by AOV methods. For our present purpose, the same residuals can be used to consistently estimate the covariance components of the matrix of covariance $\Sigma_{mm'}$ between errors of two different equations, m and m' .

Let us recall that the matrix of covariance between errors of any two different equations, say m and m', is

$$(4.50)\quad \Sigma_{mm'} = \sigma_{\mu mm'} A + \sigma_{\nu mm'} B + \sigma_{\varepsilon mm'} I_{NT}$$

or

(4.51) $\Sigma_{mm'} = \sigma_{1mm'} M_1 + \sigma_{2mm'} M_2 + \sigma_{3mm'} M_3 + \sigma_{4mm'} M_4$

The errors of the two equations, u_m and $u_{m'}$, can be predicted either by Method 1 (cf. (4.27)) or Method 2 (cf. (4.34)). Then the estimators of covariance components of $\Sigma_{mm'}$ can be constructed as follows (choosing say $\hat{u}_m$ (Method 1) for predicting u_m) :

(4.52)
$$\begin{cases} \hat{\sigma}_{1mm'} = \frac{1}{N-1} \hat{u}_m' M_1 \hat{u}_{m'} \\ \hat{\sigma}_{2mm'} = \frac{1}{T-1} \hat{u}_m' M_2 \hat{u}_{m'} \\ \hat{\sigma}_{4mm'} = \frac{1}{(N-1)(T-1)} \hat{u}_m' Q \hat{u}_{m'} \\ \hat{\sigma}_{3mm'} = \hat{\sigma}_{1mm'} + \hat{\sigma}_{2mm'} - \hat{\sigma}_{4mm'} \end{cases}$$

and

(4.53)
$$\begin{cases} \hat{\sigma}_{\varepsilon mm'} = \hat{\sigma}_{4mm'} \\ \hat{\sigma}_{\mu mm'} = \frac{1}{T} (\hat{\sigma}_{1mm'} - \hat{\sigma}_{\varepsilon mm'}) \\ \hat{\sigma}_{\nu mm'} = \frac{1}{N} (\hat{\sigma}_{2mm'} - \hat{\sigma}_{\varepsilon mm'}) \end{cases}$$

Thus $\Sigma_{mm'}$ is estimated by :

(4.54) $\hat{\Sigma}_{mm'} = \hat{\sigma}_{1mm'} M_1 + \hat{\sigma}_{2mm'} M_2 + \hat{\sigma}_{3mm'} M_3 + \hat{\sigma}_{4mm'} M_4$

Carrying out the same procedure for all different pairs of equations, we can form :

(4.55) $\hat{\Sigma} = [\hat{\Sigma}_{mm'}]$ $m,m'=1,\ldots,M$;

(4.56) $\hat{\Sigma}_* = \text{diag}(\hat{\Sigma}_{11}, \hat{\Sigma}_{22}, \ldots, \hat{\Sigma}_{MM})$

Substituting $\hat{\Sigma}$ for Σ and $\hat{\Sigma}_*$ for Σ_* in (4.49) yields a feasible G3SLS estimator that can be written as :

(4.57) $\hat{\alpha}_{fG3SLS} = \left[Z_*'\hat{\Sigma}_*^{-1}X(X'\hat{\Sigma}_*^{-1}\hat{\Sigma}\hat{\Sigma}_*^{-1}X)^{-1}X'\hat{\Sigma}_*^{-1}Z_*\right]^{-1} Z_*'\hat{\Sigma}_*^{-1}X(X'\hat{\Sigma}_*^{-1}\hat{\Sigma}\hat{\Sigma}_*^{-1}X)^{-1}X'\hat{\Sigma}_*^{-1}y$

Using $\hat{\hat{u}}_m$ instead of $\hat{u}_m$ for estimating Σ and Σ_* will lead to another feasible G3SLS estimator of α, say $\hat{\hat{\alpha}}_{fG3SLS}$.

There exists a third possible way of estimating $\Sigma_{mm'}$, $m,m'=1,\ldots,M$. Instead of using the residuals of the covariance models, we can compute the residuals of the feasible G2SLS estimation :

(4.58) $\hat{\hat{\hat{u}}}_m = y_m - Z_m \hat{\alpha}_{m,fG2SLS}$, $m=1,\ldots,M$

and replace $\hat{u}_m$ by $\hat{\hat{\hat{u}}}_m$ and $\hat{u}_{m'}$ by $\hat{\hat{\hat{u}}}_{m'}$ in (4.52) to get a new set of estimates of the covariance components and hence that of $\Sigma_{mm'}$, say $\hat{\hat{\hat{\Sigma}}}_{mm'}$. Using the fG2SLS residuals of all the different pairs of equations we can estimate $\Sigma_{mm'}$, $m,m'=1,\ldots M$ and obtain $\hat{\hat{\hat{\Sigma}}}$ and $\hat{\hat{\hat{\Sigma}}}_*$. Replacing Σ, Σ_* by $\hat{\hat{\hat{\Sigma}}}$, $\hat{\hat{\hat{\Sigma}}}_*$ respectively in (4.57) yields another feasible G3SLS estimator of α that can be denoted by $\hat{\hat{\hat{\alpha}}}_{fG3SLS}$.

Combining the different results of Appendices 4.A., 4.B. and 4.C. that lead to consistency of $\hat{\alpha}_{m,fG2SLS}$, it can be easily shown that $\hat{\alpha}_{G3SLS}$, $\hat{\alpha}_{fG3SLS}$, $\hat{\hat{\alpha}}_{fG3SLS}$ and $\hat{\hat{\hat{\alpha}}}_{fG3SLS}$ are all consistent. We will not give any separate proof of consistency of the different fG3SLS estimators as it will largely be a repetition of the earlier proofs and will only add unnecessary volume without containing any new result.

The limiting distribution of the feasible G3SLS estimator is derived in detail in Appendix 4.E. It is shown that $\tilde{D}^{-\frac{1}{2}}(\hat{\hat{\alpha}}_{fG3SLS} - \tilde{\alpha})$ has a normal limiting distribution with zero mean and variance-covariance matrix equal to

$$(4.59) \quad \begin{bmatrix} \Sigma_\mu + \Sigma_\nu & 0 \\ 0 & [\bar{\Pi}'(\Sigma_\varepsilon^{-1} \otimes R)\ \bar{\Pi}]^{-1} \end{bmatrix}$$

where

$$(4.60) \qquad \tilde{D} = \begin{bmatrix} \frac{1}{N} I_M & 0 \\ 0 & \frac{1}{NT} I_{\Sigma_m (\tilde{M}_m + K_m - 1)} \end{bmatrix}$$

$$(4.61) \qquad \bar{\Pi} = \text{diag}\,([\Pi_{*m} \; H^*_m]) \qquad m=1,\ldots,M$$

In other words, omitting the subscript 'fG3SLS',

$$(4.62) \quad \begin{cases} \sqrt{N}\;(\hat{a} - a) \sim N\,(0, \Sigma_\mu + \Sigma_\nu) \\ \sqrt{NT}\;(\hat{\alpha}^* - \alpha^*) \sim N\,(0, [\,\bar{\Pi}'(\Sigma_\varepsilon^{-1} \otimes R)\bar{\Pi}]^{-1}) \end{cases}$$

APPENDIX 4.A : Proof of the Consistency of the 2SLS Covariance Estimators $\hat{\alpha}_{m,cov}$ and $\hat{\alpha}_{m,C2SLS}$

4.A.1 Consistency of $\hat{\alpha}_{m,cov}$

From (4.24), (4.25) we have

$$\hat{\alpha}^*_{m,cov} = (\hat{Z}^{*\prime}_m \; Q \; \hat{Z}^*_m)^{-1} \; \hat{Z}^{*\prime}_m \; Q \; y_m$$

$$\hat{a}_{m,cov} = \frac{1}{NT} \iota'_{NT} \; (y_m - Z^*_m \; \hat{\alpha}^*_{m,cov})$$

By substituting $y_m = \iota_{NT} \; a_m + \hat{Z}^*_m \; \alpha^*_m + u_m + (Y_m - \hat{Y}_m) \; \gamma_m$ (see (4.21)), we can write :

$$(4.A.1) \quad \hat{\alpha}^*_{m,cov} = \alpha^*_m + (\hat{Z}^{*\prime}_m Q \hat{Z}^*_m)^{-1} \hat{Z}^{*\prime}_m Q u_m + (\hat{Z}^{*\prime}_m Q \hat{Z}^*_m)^{-1} \hat{Z}^{*\prime}_m Q (Y_m - \hat{Y}_m) \gamma_m$$

Let us calculate the probability limits of these three terms one by one :

a) (4.A.2) $\quad \text{plim} \; \alpha^*_m = \alpha^*_m$

b) $\quad \text{plim} \; (\hat{Z}^{*\prime}_m \; Q \; \hat{Z}^*_m)^{-1} \; \hat{Z}^{*\prime}_m \; Q \; u_m$

$$= \text{plim} \left[\frac{1}{NT} \begin{pmatrix} \hat{Y}'_m \\ X^{*\prime}_m \end{pmatrix} Q \; (\hat{Y}_m \;\; X^*_m) \right]^{-1} \frac{1}{NT} \begin{pmatrix} \hat{Y}'_m \\ X^{*\prime}_m \end{pmatrix} Q \; u_m$$

$$(4.A.3) \quad = \text{plim} \left\{ \frac{1}{NT} \begin{bmatrix} \hat{Y}'_m \; Q \; \hat{Y}_m & \hat{Y}'_m \; Q \; X^*_m \\ X^{*\prime}_m Q \; \hat{Y}_m & X^{*\prime}_m Q \; X^*_m \end{bmatrix} \right\}^{-1} \frac{1}{NT} \begin{bmatrix} \hat{Y}'_m \; Q \; u_m \\ X^{*\prime}_m Q \; u_m \end{bmatrix}$$

Before continuing, let us recall that

$$\hat{Y}_m = X \; \hat{\Pi}_m$$

where Π_m is a consistent estimator of Π_m i.e.

(4.A.4) $\quad \text{plim} \; \hat{\Pi}_m = \Pi_m$

Now, let us expand the blocks appearing inside the double brackets in (4.A.3).

$$\text{plim} \frac{1}{NT} \hat{Y}'_m \; Q \; \hat{Y}_m = \text{plim} \frac{1}{NT} \hat{\Pi}'_m \; X'Q \; X \; \hat{\Pi}_m$$

$$= \Pi'_m \; R_* \; \Pi_m$$

where

$$(4.A.5)\quad R_* = \begin{bmatrix} 0 & 0 \\ 0 & R \end{bmatrix} = \text{plim}\ \frac{1}{NT}\ X'Q\ X \qquad \text{using (3.A.12) and } \iota'Q = 0$$

Hence,

$$(4.A.6)\quad \text{plim}\ \frac{1}{NT}\ \hat{Y}_m'\ Q\ \hat{Y}_m = \Pi_{*m}'\ R\ \Pi_{*m}$$

writing

$$(4.A.7)\quad \Pi_m' = [\Pi_{om}\quad \Pi_{*m}']$$

Next,

$$(4.A.8)\quad \text{plim}\ \frac{1}{NT}\ \hat{Y}_m'Q\ X_m^* = \text{plim}\ \frac{1}{NT}\ \Pi_{*m}'\ \underline{X}'Q\ \underline{X}\ H_m^*$$

using $X = [\iota\ \underline{X}]$, (4.A.7), and writing

$$(4.A.9)\quad X_m^* = \underline{X}\ H_m^* \quad ,$$

H_m^* being an appropriate selection matrix.

Using (3.A.12) (page 76), we can write

$$(4.A.10)\quad \text{plim}\ \frac{1}{NT}\ \hat{Y}_m'\ Q\ X_m^* = \Pi_{*m}'\ R\ H_m^*$$

Next,

$$\text{plim}\ \frac{1}{NT}\ X_m^{*\prime}Q\ X_m^* = \text{plim}\ \frac{1}{NT}\ H_m^{*\prime}\underline{X}'Q\ \underline{X}\ H_m^* \quad \text{using (4.A.9)}$$

$$(4.A.11)\qquad = H_m^{*\prime}\ R\ H_m^* \quad \text{using (3.A.12)}$$

$$\text{Thus}\quad \text{plim}\ (\frac{1}{NT}\ \hat{Z}_m^{*\prime}Q\ \hat{Z}_m^*)^{-1} = \begin{bmatrix} \Pi_{*m}'\ R\ \Pi_{*m} & \Pi_{*m}'\ R\ H_m^* \\ H_m^{*\prime}\ R\ \Pi_{*m} & H_m^{*\prime}\ R\ H_m^* \end{bmatrix}^{-1}$$

$$(4.A.12)\qquad = \left(\begin{bmatrix} \Pi_{*m}' \\ H_m^{*\prime} \end{bmatrix} R\ [\Pi_{*m}\ \ H_m^*] \right)^{-1}$$

By verifying that the rank of $[\Pi_{*m}\ H_m^*] = K_m + \tilde{M}_m - 2$, using the identification requirement, namely that $K \geqslant K_m + \tilde{M}_m$, it can be seen that

$$(4.A.13)\quad \begin{bmatrix} \Pi_{*m}' \\ H_m^{*\prime} \end{bmatrix} R\ [\Pi_{*m}\ \ H_m^*] \text{ is non-singular}$$

Now, the other two blocks of expression (4.A.3) can be shown to tend to zero in plim :

$$(4.A.14)\quad \text{plim}\ \frac{1}{NT}\ \hat{Y}_m' Q u_m = \text{plim}\ \frac{1}{NT}\ \hat{\Pi}_m' X' Q u_m = \Pi_{*m}'\ \text{plim}\ \frac{1}{NT}\ \underline{X}' Q u_m = 0$$

(using (3.A.18))

$$(4.A.15)\quad \text{plim}\ \frac{1}{NT}\ X_m^{*\prime} Q u_m = \text{plim}\ \frac{1}{NT}\ H_m^{*\prime}\ \underline{X}' Q u_m = 0 \text{ (using (3.A.18))}$$

Hence

$$(4.A.16)\quad \text{plim}\ \frac{1}{NT}\ \hat{Z}_m^{*\prime}\ Q\ u_m = 0$$

and thus

$$(4.A.17)\quad \text{plim}\ (\hat{Z}_m^{*\prime} Q\ \hat{Z}_m^{*})^{-1} \hat{Z}_m^{*\prime} Q\ u_m = 0 \text{ (using (4.A.12),(4.A.16))}$$

c) $\text{plim}\ (\hat{Z}_m^{*\prime}\ Q\ \hat{Z}_m^{*})^{-1}\ \hat{Z}_m^{*\prime}\ Q\ (Y_m - \hat{Y}_m)\ \gamma_m$

$$= \text{plim}\ (\frac{1}{NT}\ \hat{Z}_m^{*\prime}\ Q\ \hat{Z}_m^{*})^{-1}\ \frac{1}{NT}\ \hat{Z}_m^{*\prime}\ Q\ (X\ \Pi_m + V_m - X\ \hat{\Pi}_m)\ \gamma_m$$

Now, $\text{plim}\ (\frac{1}{NT}\ \hat{Z}_m^{*\prime}\ Q\ \hat{Z}_m^{*})^{-1}$ is given in (4.A.12)

and

$$\text{plim}\ \frac{1}{NT}\ \hat{Z}_m^{*\prime}\ Q(X\ (\Pi_m - \hat{\Pi}_m) + V_m)\ \gamma_m$$

$$= \text{plim}\ \frac{1}{NT} \begin{bmatrix} \hat{Y}_m' Q\ X\ (\Pi_m - \hat{\Pi}_m)\ \gamma_m + \hat{Y}_m'\ Q\ V_m\ \gamma_m \\ X_m^{*\prime} Q\ X\ (\Pi_m - \hat{\Pi}_m)\ \gamma_m + X_m^{*\prime}\ Q\ V_m\ \gamma_m \end{bmatrix}$$

$$= \text{plim} \begin{bmatrix} \frac{1}{NT}\ \hat{\Pi}_{*m}' \underline{X}' Q\ \underline{X}\ (\Pi_{*m} - \hat{\Pi}_{*m})\gamma_m + \frac{1}{NT}\ \hat{\Pi}_{*m}'\ \underline{X}' Q\ V_m\ \gamma_m \\ \frac{1}{NT}\ H_m^{*\prime} \underline{X}' Q\ \underline{X}\ (\Pi_{*m} - \hat{\Pi}_{*m})\gamma_m + \frac{1}{NT}\ H_m^{*\prime}\ \underline{X}' Q\ V_m\ \gamma_m \end{bmatrix}$$

$= 0$ using (4.A.4), (3.A.12), (3.A.19).

Thus

$$(4.A.18)\quad \text{plim}\ (\hat{Z}_m^{*\prime}\ Q\ \hat{Z}_m^{*})^{-1}\ \hat{Z}_m^{*\prime}\ Q\ (Y_m - \hat{Y}_m)\ \gamma_m = 0$$

Combining (4.A.1), (4.A.2), (4.A.17) and (4.A.18) we get

$$(4.A.19)\quad \text{plim}\ \hat{\alpha}_{m,cov}^{*} = \alpha_m^{*}$$

Finally,

$$\text{plim } \hat{a}_{m,cov} = \text{plim } \frac{1}{NT} \iota'_{NT} (y_m - Z^*_m \hat{\alpha}^*_{m,cov})$$
$$= \text{plim } \frac{1}{NT} \iota'_{NT} (y_m - Z^*_m \alpha^*_m) \text{ using (4.A.19)}$$
$$= \text{plim } \frac{1}{NT} \iota'_{NT} (\iota_{NT} a_m + u_m) \text{ using (4.19)}$$
$$(4.A.20) \qquad = a_m$$

4.A.2 Consistency of $\hat{\alpha}_{m,C2SLS}$

From (4.30), (4.31) we have

$$\hat{\alpha}^*_{m,C2SLS} = \left[Z^{*\prime}_m Q\, \underline{X}(\underline{X}'Q\, \underline{X})^{-1}\underline{X}'Q\, Z^*_m\right]^{-1} Z^{*\prime}_m Q\, \underline{X}(\underline{X}'Q\, \underline{X})^{-1}\underline{X}'Q\, y_m$$

and

$$\hat{a}_{m,C2SLS} = \frac{1}{NT} \iota' (y_m - Z^*_m \hat{\alpha}^*_{m,cov})$$

Substituting (4.17) for y_m in the above expressions, we get

$$(4.A.21) \quad \hat{\alpha}^*_{m,C2SLS} = \alpha^*_m + (Z^{*\prime}_m Q\underline{X}(\underline{X}'Q\underline{X})^{-1}\underline{X}'QZ^*_m)^{-1} Z^{*\prime}_m Q\underline{X}(\underline{X}'Q\underline{X})^{-1}\underline{X}'Q\, u_m$$

Now,

$$\text{plim } \frac{1}{NT} Z^{*\prime}_m Q\, \underline{X} = \text{plim } \frac{1}{NT} \begin{bmatrix} Y'_m Q\, \underline{X} \\ X^{*\prime}_m Q\, \underline{X} \end{bmatrix}$$
$$= \text{plim } \frac{1}{NT} \begin{bmatrix} \Pi'_m X'Q\, \underline{X} + V'_m Q\, \underline{X} \\ H^{*\prime}_m \underline{X}\, Q\, \underline{X} \end{bmatrix}$$
$$(4.A.22) \qquad = \begin{bmatrix} \Pi'_{*m} R \\ H^{*\prime}_m R \end{bmatrix} \quad ;$$

$$(4.A.23) \quad \text{plim } (\frac{1}{NT} \underline{X}'Q\, \underline{X})^{-1} = R^{-1} \qquad \text{from (3.A.12)}$$

and

$$(4.A.24) \quad \text{plim } \frac{1}{NT} \underline{X}'Q\, u_m = 0 \qquad \text{from (3.A.18)}$$

Substituting these results in (4.A.21) and simplifying, we get

(4.A.25) $\text{plim}\ \hat{\alpha}^*_{m,C2SLS} = \alpha^*_m$

The consistency of $\hat{a}_{m,C2SLS}$ is verified as follows :

$$\text{plim}\ \hat{a}_{m,C2SLS} = \text{plim}\ \frac{1}{NT}\ \iota'(y_m - Z^*_m\ \hat{\alpha}^*_{m,C2SLS})$$

$$= \text{plim}\ \frac{1}{NT}\ \iota'(y_m - Z^*_m\ \alpha^*_{m,C2SLS}) \text{ using (4.A.25)}$$

$$= \text{plim}\ \frac{1}{NT}\ \iota'(\iota\ a_m + u_m) \text{ from (4.19)}$$

(4.A.26) $$= a_m$$

APPENDIX 4.B : Proof of the Consistency of AOV Estimators of Eigenvalues and Variance Components of Σ_{mm}

4.B.1 Method 1

Consistency of $\hat{\sigma}_{\varepsilon mm}$:

We have, from (4.28) :

$$(4.B.1) \quad \hat{\sigma}_{\varepsilon mm} = \frac{1}{(N-1)(T-1)} \hat{u}_m' Q \hat{u}_m$$

where

$$(4.B.2) \quad \hat{u}_m = Y_m - Z_m \hat{\alpha}_{m,cov}$$

with

$$(4.B.3) \quad \begin{aligned} Z_m &= [\, Y_m \; \iota_{NT} \; X_m^* \,] \\ \hat{\alpha}_{m,cov}' &= [\, \hat{\gamma}_{m,cov}' \; \hat{a}_{m,cov} \; \hat{b}_{m,cov}' \,] \end{aligned}$$

In the following proofs, let us omit the subscript "cov" in the symbols of the estimators for the sake of simplicity of notations.

Substituting (4.B.2) in (4.B.1), we get :

$$\text{plim } \hat{\sigma}_{\varepsilon mm} = \text{plim } \frac{1}{(N-1)(T-1)} \left[(y_m - Z_m \hat{\alpha}_m)' \, Q (y_m - Z_m \hat{\alpha}_m) \right]$$

$$= \text{plim } \frac{1}{(N-1)(T-1)} \left[(Z_m(\alpha_m - \hat{\alpha}_m) + u_m)' Q (Z_m(\alpha_m - \hat{\alpha}_m) + u_m) \right]$$

replacing y_m by $Z_m \alpha_m + u_m$ (from (4.2))

$$(4.B.4) \quad = \text{plim } \frac{1}{(N-1)(T-1)} \left[(\alpha_m - \hat{\alpha}_m)' Z_m' Q \, Z_m (\alpha_m - \hat{\alpha}_m) + 2(\alpha_m - \hat{\alpha}_m)' Z_m' \, Q \, u_m + u_m' \, Q \, u_m \right]$$

Let us consider these three terms one by one.

a) $\quad \text{plim } \frac{1}{(N-1)(T-1)} (\alpha_m - \hat{\alpha}_m)' \, Z_m' \, Q \, Z_m \, (\alpha_m - \hat{\alpha}_m)$

Now,

$$Z_m' \, Q \, Z_m = \begin{bmatrix} Y_m' \\ X_m' \end{bmatrix} Q \begin{bmatrix} Y_m & X_m \end{bmatrix}$$

$$= \begin{bmatrix} \Pi_m' X' + V_m' \\ H_m' X' \end{bmatrix} Q \begin{bmatrix} X \Pi_m + V_m & X H_m \end{bmatrix}$$

using (4.6), (4.7)

$$= \begin{bmatrix} \Pi_m'X'QX\Pi_m + V_m'QX\Pi_m + \Pi_m'X'QV_m + V_m'QV_m & \Pi_m'X'QXH_m + V_m'QXH_m \\ H_m'X'QX\Pi_m + H_m'X'QV_m & H_m'X'QXH_m \end{bmatrix}$$

To calculate $\text{plim} \frac{1}{(N-1)(T-1)} Z_m' Q Z_m$ we need the following limits :

(4.B.5) $\text{plim} \frac{1}{(N-1)(T-1)} \Pi_m' X'Q X \Pi_m = \Pi_m' R_* \Pi_m$ using (4.A.5)

(4.B.6) $\text{plim} \frac{1}{(N-1)(T-1)} V_m' Q X \Pi_m = 0$

using $X=[\iota \ \underline{X}]$, $Q\iota=0$ and (3.A.19)

It follows from (3.A.32) that :

(4.B.7) $\text{plim} \frac{1}{(N-1)(T-1)} V_m' Q V_m$ = a finite matrix, say $\Omega_{(m)}$

Next,

(4.B.8) $\text{plim} \frac{1}{(N-1)(T-1)} H_m' X'Q X \Pi_m = H_m' R_* \Pi_m$ using (4.A.5)

(4.B.9) $\text{plim} \frac{1}{(N-1)(T-1)} H_m' X'Q V_m = 0$

using $X=[\iota \ \underline{X}]$, $Q\iota=0$ and (3.A.19)

(4.B.10) $\text{plim} \frac{1}{(N-1)(T-1)} H_m' X'Q X H_m = H_m' R_* H_m$ using (4.A.5)

With the above results, we obtain

(4.B.11) $$\text{plim} \frac{1}{(N-1)(T-1)} Z_m'Q Z_m = \left[\begin{array}{c|c} \Pi_m' R_* \Pi_m + \Omega_{(m)} & \Pi_m' R_* H_m \\ H_m' R_* \Pi_m & H_m' R_* H_m \end{array}\right]$$

And as

$$\text{plim} (\alpha_m - \hat{\alpha}_m) = 0$$

we can conclude that

(4.B.12) $$\text{plim} \frac{1}{(N-1)(T-1)} (\alpha_m - \hat{\alpha}_m)'Z_m'Q Z_m(\alpha_m - \hat{\alpha}_m) = 0$$

b) $$\text{plim} \frac{1}{(N-1)(T-1)} (\alpha_m - \hat{\alpha}_m)' Z_m' Q\, u_m$$

Now,

$$Z_m' Q\, u_m = \begin{bmatrix} Y_m' \\ X_m' \end{bmatrix} Q\, u_m$$

$$= \begin{bmatrix} (\Pi_m' X' + V_m')\, Q\, u_m \\ H_m' X' Q\, u_m \end{bmatrix} \quad \text{using (4.6),(4.7)}$$

$$= \begin{bmatrix} \Pi_m' X' Q\, u_m + \bar{T}_m' U' Q\, u_m \\ H_m' X' Q\, u_m \end{bmatrix}$$

using $V_m = U\,\bar{T}_m$ (cf. (3.37), page 53)

We have

$$\text{plim} \frac{1}{(N-1)(T-1)} \Pi_m' X' Q u_m = 0 \text{ using } \iota' Q = 0 \text{ and (3.A.18)}$$

and

$$\text{plim} \frac{1}{(N-1)(T-1)} H_m' X' Q u_m = 0 \text{ using } \iota' Q = 0 \text{ and (3.A.18)}$$

It can be deduced from result (4.B.29) of c) below that

$$\text{plim} \frac{1}{(N-1)(T-1)} U' Q\, u_m = \text{a finite vector, say } \sigma_{(m)}$$

Hence we can write

$$\text{plim} \frac{1}{(N-1)(T-1)} \bar{T}_m' U' Q\, u_m = \bar{T}_m' \sigma_{(m)}$$

Thus

$$\text{plim} \frac{1}{(N-1)(T-1)} Z_m' Q\, u_m = \begin{bmatrix} \bar{T}_m' \sigma_{(m)} \\ 0 \end{bmatrix}$$

and hence

$$\text{(4.B.13)} \quad \text{plim} \frac{1}{(N-1)(T-1)} (\alpha_m - \hat{\alpha}_m)' Z_m' Q\, u_m = 0 \text{ as plim } \alpha_m - \hat{\alpha}_m = 0$$

c) $$\text{plim} \frac{1}{(N-1)(T-1)} u_m' Q\, u_m$$

By noting that

$$Q\, u_m = Q\, \varepsilon_m$$

and that

$$Q = Q^2$$

we have

$$(4.B.14)\quad \operatorname{plim} \frac{1}{(N-1)(T-1)} u_m' Q u_m = \operatorname{plim} \frac{1}{(N-1)(T-1)} \varepsilon_m' Q \varepsilon_m$$

$$(4.B.15)\quad = \operatorname{plim} \frac{1}{(N-1)(T-1)} \left[\varepsilon_m' \varepsilon_m - \varepsilon_m' \frac{A}{T} \varepsilon_m - \varepsilon_m' \frac{B}{N} \varepsilon_m + \varepsilon_m' \frac{J_{NT}}{NT} \varepsilon_m\right]$$

using the definition of Q given in (3.26), page 51 .

Let us calculate the probability limits of the above four terms, one by one. The following assumptions will be used for this purpose :

$$(4.B.16)\quad E(\varepsilon_{mit}) = 0 \qquad \text{(cf. (3.16), page 49)}$$

$$(4.B.17)\quad E(\varepsilon_{mit}\varepsilon_{mjs}) = \delta_{ij}\,\delta_{ts}\,\sigma_{\varepsilon mm} \qquad \text{(cf. (3.17), page 50)}$$

(4.B.18) ε_{mit} and ε_{mjs} are <u>independent</u> for $i \neq j$ or $t \neq s$ or both.

Then, $$\operatorname{plim} \frac{1}{(N-1)(T-1)} \varepsilon_m' \varepsilon_m = \operatorname{plim} \frac{1}{(N-1)(T-1)} \sum_i \sum_t \varepsilon_{mit}^2$$

$$(4.B.19)\quad = \sigma_{\varepsilon mm}$$

using (L-1-9) of Lemma L-1 (page 81), under assumptions (4.B.16), (4.B.17), (4.B.18) and other general assumptions concerning higher-order moments.

Next,

$$(4.B.20)\quad \operatorname{plim} \frac{1}{(N-1)(T-1)} \varepsilon_m' \frac{A}{T} \varepsilon_m = \operatorname{plim} \frac{1}{(N-1)(T-1)} \frac{1}{T} \sum_i \left(\sum_t \varepsilon_{mit}\right)^2$$

$$= \operatorname{plim} \frac{T}{(N-1)(T-1)} \sum_i \varepsilon_{mi.}^2$$

where

$$(4.B.21)\quad \varepsilon_{mi.} = \frac{1}{T} \sum_t \varepsilon_{mit}$$

Using the following results :

$$(4.B.22)\quad \begin{cases} E(\varepsilon_{mi.}) = 0 \\ E(\varepsilon_{mi.}\ \varepsilon_{mj.}) = \delta_{ij} \dfrac{\sigma_{\varepsilon mm}}{T} \\ \text{independence between } \varepsilon_{mi.} \text{ and } \varepsilon_{mj.} \text{ for } i \neq j \ ; \end{cases}$$

and other assumptions concerning higher-order moments, we can say that

$$(4.B.23)\quad \underset{N\to\infty}{\text{plim}} \frac{1}{N-1} \sum_i \varepsilon^2_{mi.} = \frac{\sigma_{\varepsilon mm}}{T}$$

according to (L-1-9) of Lemma L-1 (page 81).

Thus

$$\underset{\substack{N\to\infty \\ T\to\infty}}{\text{plim}} \frac{1}{(N-1)(T-1)} \varepsilon'_m \frac{A}{T} \varepsilon_m = \underset{\substack{N\to\infty \\ T\to\infty}}{\text{plim}} \frac{1}{(N-1)(T-1)} T \sum_i \varepsilon^2_{mi.}$$

$$= \underset{T\to\infty}{\text{plim}} \frac{T}{T-1} \frac{1}{T} \sigma_{\varepsilon mm} \qquad \text{using (4.B.23)}$$

$$= \text{plim} \frac{1}{T-1} \sigma_{\varepsilon mm}$$

$$(4.B.24)\qquad = 0$$

Similarly, it can be shown that

$$(4.B.25)\quad \text{plim} \frac{1}{(N-1)(T-1)} \varepsilon'_m \frac{B}{N} \varepsilon_m = 0$$

Finally,

$$\text{plim} \frac{1}{(N-1)(T-1)} \varepsilon'_m \frac{J_{NT}}{NT} \varepsilon_m = \text{plim} \frac{1}{(N-1)(T-1)} \frac{1}{NT} \Big(\sum_i \sum_t \varepsilon_{mit}\Big)\Big(\sum_j \sum_s \varepsilon_{mjs}\Big)$$

$$= \text{plim} \frac{NT}{(N-1)(T-1)} \varepsilon^2_{m..}$$

where

$$(4.B.26)\quad \varepsilon_{m..} = \frac{1}{NT} \sum_i \sum_t \varepsilon_{mit}$$

Applying (L-1-16) of Lemma L-1 (page 82), we have

$$(4.B.27)\quad \text{plim}\ \varepsilon^2_{m..} = 0$$

Thus

$$(4.B.28)\quad \operatorname{plim} \frac{1}{(N-1)(T-1)} \varepsilon_m' \frac{J_{NT}}{NT} \varepsilon_m = 0$$

Hence, from (4.B.14), (4.B.15), (4.B.19), (4.B.24), (4.B.25) and (4.B.28), we have

$$(4.B.29)\quad \operatorname{plim} \frac{1}{(N-1)(T-1)} u_m' \, Q \, u_m = \sigma_{\varepsilon mm}$$

Therefore, combining (4.B.4), (4.B.12), (4.B.13) and (4.B.29) we get

$$(4.B.30)\quad \operatorname{plim} \hat{\sigma}_{\varepsilon mm} = \sigma_{\varepsilon mm}$$

<u>Consistency of $\hat{\sigma}_{1mm}$</u> :

From (4.29), we have

$$\hat{\sigma}_{1mm} = \frac{1}{N-1} \hat{u}_m' \, M_1 \, \hat{u}_m$$

$$(4.B.31)\quad = \frac{1}{N-1} \left[Z_m(\alpha_m - \hat{\alpha}_m) + u_m \right]' M_1 \left[Z_m(\alpha_m - \hat{\alpha}_m) + u_m \right]$$

Noting that

$\operatorname{plim}_{N\to\infty} \frac{1}{N-1} \underline{X}' \, M_1 \, \underline{X}$ is a finite non-singular matrix (cf. (3.A.9))

$\operatorname{plim}_{N\to\infty} \frac{1}{N-1} \underline{X}' \, M_1 \, u_m = 0$ (cf. (3.A.15))

and using the consistency of $\hat{\alpha}_m$, it can be easily verified that :

$$(4.B.32)\quad \operatorname{plim}_{N\to\infty} \hat{\sigma}_{1mm} = \operatorname{plim}_{N\to\infty} \frac{1}{N-1} u_m' \, M_1 \, u_m$$

Replacing u_m by $(I_N \otimes \iota_T) \mu^m + (\iota_N \otimes I_T) \nu^m + \varepsilon_m$ in (4.B.32) and noting that

$$M_1 \, (\iota_N \otimes I_T) = 0$$

we can write

(4.B.33) $$\operatorname*{plim}_{N\to\infty} \hat{\sigma}_{1mm} = \operatorname*{plim}_{N\to\infty} \frac{1}{N-1} ((I_N \otimes \iota_T)\mu^m + \varepsilon_m)' M_1 ((I_N \otimes \iota_T)\mu^m + \varepsilon_m)$$

$$= \operatorname*{plim}_{N\to\infty} \frac{1}{N-1} \Big[(\mu^m)'(I_N \otimes \iota_T') M_1 (I_N \otimes \iota_T)\mu^m + 2(\mu^m)'(I_N \otimes \iota_T') M_1 \varepsilon_m + \varepsilon_m' M_1 \varepsilon_m \Big]$$

Let us look at these three terms one by one.

a) $$\operatorname*{plim}_{N\to\infty} \frac{1}{N-1} (\mu^m)'(I_N \otimes \iota_T') M_1 (I_N \otimes \iota_T)\mu^m$$

$$= \operatorname*{plim}_{N\to\infty} \frac{1}{N-1} T \sum_i (\mu_{mi} - \mu_{m.})^2$$ using the definition of M_1 and carrying out the multiplication

where $\mu_{m.} = \frac{1}{N}\sum_i \mu_{mi}$

Applying (L-1-12) of Lemma L-1 (page 81), we get :

(4.B.34) $$\operatorname*{plim}_{N\to\infty} \frac{1}{N-1} \sum_i (\mu_{mi} - \mu_{m.})^2 = \sigma_{\mu mm}$$

Thus

(4.B.35) $$\operatorname*{plim}_{N\to\infty} \frac{1}{N-1} (\mu^m)'(I_N \otimes \iota_T') M_1 (I_N \otimes \iota_T)\mu^m = T\sigma_{\mu mm}$$

b) $$\operatorname*{plim}_{N\to\infty} \frac{1}{N-1} (\mu^m)'(I_N \otimes \iota_T') M_1 \varepsilon_m$$

$$= \operatorname*{plim}_{N\to\infty} \frac{1}{N-1} T \sum_i (\mu_{mi} - \mu_{m.})(\varepsilon_{mi.} - \varepsilon_{m..})$$

(see page 86 for the derivation of a similar result obtained in the treatment of the second term of (3.A.34))

Applying (L-1-14) of Lemma L-1 (page 82), we obtain

$$= \operatorname*{plim}_{N\to\infty} \frac{1}{N-1} \sum_i (\mu_{mi} - \mu_{m.})(\varepsilon_{mi.} - \varepsilon_{m..}) = 0$$

Thus

(4.B.36) $$\operatorname*{plim}_{N\to\infty} \frac{1}{N-1} (\mu^m)'(I_N \otimes \iota_T') M_1 \varepsilon_m = 0$$

c) $$\operatorname*{plim}_{N\to\infty} \frac{1}{N-1} \varepsilon_m' M_1 \varepsilon_m$$

$$= \operatorname*{plim}_{N\to} \frac{1}{N-1} T \sum_i (\varepsilon_{mi.} - \varepsilon_{m..})^2$$

(using the definition of M_1 and expanding the quadratic form)

Applying (L-1-12) of Lemma L-1 (page 81), we obtain

$$\operatorname*{plim}_{N\to\infty} \frac{1}{N-1} \sum_i (\varepsilon_{mi.} - \varepsilon_{m..})^2 = \frac{\sigma_{\varepsilon mm}}{T}$$

Thus

(4.B.37) $$\operatorname*{plim}_{N\to\infty} \frac{1}{N-1} \varepsilon_m' M_1 \varepsilon_m = \sigma_{\varepsilon mm}$$

Finally, (4.B.35), (4.B.36) and (4.B.37) lead to

(4.B.38) $$\operatorname*{plim}_{N\to\infty} \hat{\sigma}_{1mm} = T \sigma_{\mu mm} + \sigma_{\varepsilon mm} = \sigma_{1mm}$$

Before letting T tend to infinity, it is necessary to redefine the concept of consistency in this case, as the true value, $\sigma_{1mm} = T \sigma_{\mu mm} + \sigma_{\varepsilon mm}$, is itself a linear function of T and thus becomes infinitely large as $T\to\infty$. Here, it seems more appropriate to adopt the following definition of consistency :

<u>Definition</u> : $\hat{\lambda}(T)$ is a consistent estimator of $\lambda(T)$ if

(4.B.39) $$\operatorname*{plim}_{T\to\infty} \left(\frac{\hat{\lambda}}{\lambda} \right) = 1$$

In our case, $\hat{\sigma}_{1mm}$ is a consistent estimator of σ_{1mm} as we can write, starting from (4.B.38), that :

(4.B.40) $$\operatorname*{plim}_{\substack{N\to\infty \\ T\to\infty}} \frac{\hat{\sigma}_{1mm}}{\sigma_{1mm}} = \operatorname*{plim}_{T\to\infty} \frac{T \sigma_{\mu mm} + \sigma_{\varepsilon mm}}{T \sigma_{\mu mm} + \sigma_{\varepsilon mm}} = 1$$

Note also that

(4.B.41) $$\operatorname*{plim}_{\substack{N\to\infty \\ T\to\infty}} \left(\frac{1}{\hat{\sigma}_{1mm}} \right) = 0$$

Consistency of $\hat{\sigma}_{2mm}$ and $\hat{\sigma}_{3mm}$ can be proved similarly using the definition (4.B.39). Further, we also have, as before :

(4.B.42) $$\underset{\substack{N\to\infty \\ T\to\infty}}{\text{plim}} \left(\frac{1}{\hat{\sigma}_{2mm}} \right) = 0$$

and

(4.B.43) $$\underset{\substack{N\to\infty \\ T\to\infty}}{\text{plim}} \left(\frac{1}{\hat{\sigma}_{3mm}} \right) = 0$$

Consistency of $\hat{\sigma}_{\mu mm}$:

We have :

$$\hat{\sigma}_{\mu mm} = \frac{1}{T} (\hat{\sigma}_{1mm} - \hat{\sigma}_{\varepsilon mm})$$

and hence

$$\underset{\substack{N\to\infty \\ T\to\infty}}{\text{plim}} \hat{\sigma}_{\mu mm} = \underset{\substack{N\to\infty \\ T\to\infty}}{\text{plim}} \frac{1}{T} (\hat{\sigma}_{1mm} - \hat{\sigma}_{\varepsilon mm}) = \underset{\substack{N\to\infty \\ T\to\infty}}{\text{plim}} \frac{1}{T} \hat{\sigma}_{1mm} - \underset{\substack{N\to\infty \\ T\to\infty}}{\text{plim}} \frac{1}{T} \hat{\sigma}_{\varepsilon mm}$$

(4.B.44) $$= \underset{T\to\infty}{\text{plim}} \frac{1}{T} (T \sigma_{\mu mm} + \sigma_{\varepsilon mm}) = \sigma_{\mu mm}$$

Consistency of $\hat{\sigma}_{\nu mm}$ is proved similarly.

Finally, let us make note of the following results :

$$\underset{\substack{N\to\infty \\ T\to\infty}}{\text{plim}} \left(\frac{\hat{\sigma}_{1mm}}{T}\right) = \underset{T\to\infty}{\text{plim}} \left(\frac{T \sigma_{\mu mm} + \sigma_{\varepsilon mm}}{T} \right) \quad \text{from (4.B.38)}$$

(4.B.45) $$= \sigma_{\mu mm}$$

Similarly, it can be verified that

(4.B.46) $$\underset{\substack{N\to\infty \\ T\to\infty}}{\text{plim}} \frac{\hat{\sigma}_{2mm}}{N} = \sigma_{\nu mm}$$

and

(4.B.47) $$\underset{\substack{N\to\infty \\ T\to\infty}}{\text{plim}} \left(\frac{\hat{\sigma}_{3mm}}{T} \right) = \underset{\substack{N\to\infty \\ T\to\infty}}{\text{plim}} \left(\frac{\hat{\sigma}_{3mm}}{N} \right) = \sigma_{\mu mm} + \sigma_{\nu mm}$$

4.B.2 Method 2

By noting that all the above proofs of Section 4.B.1 can be rewritten replacing $\hat{\alpha}_{m,cov}$ by $\hat{\alpha}_{m,C2SLS}$ without changing the results, the consistency of the estimators of eigenvalues and variance compoments by method 2 follows automatically.

APPENDIX 4.C : Proof of the Consistency of the Feasible (and pure) G2SLS Estimator

Let us recall that the equation to be estimated is

(4.C.1) $y_m = Z_m \alpha_m + u_m$

and that the feasible G2SLS procedure consists in first premultiplying (4.C.1) by $X'\hat{\Sigma}_{mm}^{-1}$ and then applying GLS. But before doing so, let us reformulate the structural equation by separating the constant term from the other coefficients, as we are going to deal with limits and some of them do not exist if we maintain the present formulation.

Thus we will rewrite equation (4.C.1) as :

$$y_m = (Y_m \ \iota \ X_m^*) \begin{pmatrix} \gamma_m \\ a_m \\ b_m \end{pmatrix} + u_m \qquad \text{(cf. (4.17))}$$

$$= \iota \, a_m + (Y_m \ X_m^*) \begin{pmatrix} \gamma_m \\ b_m \end{pmatrix} + u_m$$

$$= \iota \, a_m + Z_m^* \alpha_m^* + u_m$$

$$= (\iota \ Z_m^*) \begin{pmatrix} a_m \\ \alpha_m^* \end{pmatrix} + u_m$$

or

(4.C.2) $y_m = \tilde{Z}_m \tilde{\alpha}_m + u_m$

with

(4.C.3) $\tilde{Z}_m = [\iota \ Z_m^*]$ and $\tilde{\alpha}_m = (a_m \ \alpha_m^{*\prime})$

Now, premultiplying (4.C.2) by $X'\hat{\Sigma}_{mm}^{-1}$ and applying GLS we obtain

(4.C.4) $$\hat{\tilde{\alpha}}_{m,fG2SLS} = \left[\tilde{Z}_m' \hat{\Sigma}_{mm}^{-1} X (X' \hat{\Sigma}_{mm}^{-1} X)^{-1} X' \hat{\Sigma}_{mm}^{-1} \tilde{Z}_m\right]^{-1} \tilde{Z}_m' \hat{\Sigma}_{mm}^{-1} X (X' \hat{\Sigma}_{mm}^{-1} X)^{-1} X \hat{\Sigma}_{mm}^{-1} y_m$$

Substituting (4.C.2) in (4.C.4) and simplifying we get

$$(4.C.5)\quad \hat{\hat{\alpha}}_{m,fG2SLS} = \tilde{\alpha}_m + \left[\tilde{Z}_m' \hat{\Sigma}_{mm}^{-1} X (X'\hat{\Sigma}_{mm}^{-1}X)^{-1} X'\hat{\Sigma}_{mm}^{-1}\tilde{Z}_m\right]^{-1} \tilde{Z}_m' \hat{\Sigma}_{mm}^{-1} X (X'\hat{\Sigma}_{mm}^{-1}X)^{-1} X'\hat{\Sigma}_{mm}^{-1} u_m$$

or

$$(4.C.6)\quad = \tilde{\alpha}_m + \left[(\bar{D}_2 \tilde{Z}_m' \hat{\Sigma}_{mm}^{-1} X)(\bar{D}_1 X'\hat{\Sigma}_{mm}^{-1} X)^{-1} \bar{D}_1 X'\hat{\Sigma}_{mm}^{-1}\tilde{Z}_m\right]^{-1}$$

$$(\bar{D}_2 \tilde{Z}_m' \hat{\Sigma}_{mm}^{-1} X)(\bar{D}_1 X'\hat{\Sigma}_{mm}^{-1} X)^{-1} \bar{D}_1 X'\hat{\Sigma}_{mm}^{-1} u_m$$

where

$$(4.C.7)\quad \bar{D}_1 = \begin{bmatrix} \frac{1}{N} & 0 \\ 0 & \frac{1}{NT} I_{K-1} \end{bmatrix} \quad \text{and } \bar{D}_2 = \begin{bmatrix} \frac{1}{N} & 0 \\ 0 & \frac{1}{NT} I_{\tilde{M}_m+K_m-1} \end{bmatrix}$$

Here, since we have assumed that X contains ι_{NT} as its first column, we have taken care to premultiply $X'\hat{\Sigma}_{mm}^{-1}X$ by $\bar{D}_1$, instead of simply multiplying it by $\frac{1}{NT}$, as plim $\frac{1}{NT}X'\hat{\Sigma}_{mm}^{-1}X$ is singular in this case ; similarly we consider $\bar{D}_2\tilde{Z}_m'\hat{\Sigma}_{mm}^{-1}X$ or $\bar{D}_1X'\hat{\Sigma}_{mm}^{-1}\tilde{Z}_m$ instead of $\frac{1}{NT}Z_m'\hat{\Sigma}_{mm}^{-1}X$.

Now,

$$(4.C.8)\quad \text{plim } \bar{D}_2\tilde{Z}_m'\hat{\Sigma}_{mm}^{-1}X = \text{plim} \begin{bmatrix} \frac{1}{N} & 0 \\ 0 & \frac{1}{NT}I \end{bmatrix} \begin{bmatrix} \iota'\hat{\Sigma}_{mm}^{-1}\iota & \iota'\hat{\Sigma}_{mm}^{-1}\underline{X} \\ Z_m^{*\prime}\hat{\Sigma}_{mm}^{-1}\iota & Z_m^{*\prime}\hat{\Sigma}_{mm}^{-1}\underline{X} \end{bmatrix}$$

$$= \text{plim} \left[\begin{array}{c|c} \frac{1}{N}\frac{NT}{\hat{\sigma}_{3mm}} & \frac{1}{N}\frac{1}{\hat{\sigma}_{3mm}}\iota'\underline{X} \\ \frac{1}{NT}\frac{1}{\hat{\sigma}_{3mm}} Z_m^{*\prime}\iota & \sum_j \frac{1}{NT}\frac{1}{\hat{\sigma}_{jmm}} Z_m^{*\prime}M_j\underline{X} \end{array}\right]$$

using (3.22) and (3.28)

Let us take the limits of the four blocks one by one.

(i)

$$(4.C.9)\quad \frac{1}{NT}\frac{NT}{\hat{\sigma}_{3mm}} = \frac{1}{\left(\frac{\hat{\sigma}_{3mm}}{T}\right)} \dashrightarrow \frac{1}{\sigma_{\mu mm}+\sigma_{\nu mm}} \qquad \text{using (4.B.47)}$$

(ii)

$$(4.C.10)\quad \frac{1}{N}\frac{1}{\hat{\sigma}_{3mm}}\,\iota'\underline{X} = \frac{1}{\left(\frac{\hat{\sigma}_{3mm}}{T}\right)}\,\frac{1}{NT}\,\iota'\underline{X} \;\text{-->}\; \frac{1}{\sigma_{\mu mm}+\sigma_{\nu mm}}\,r'$$

using (4.B.47) and (3.A.4)

(iii)

$$(4.C.11)\quad \frac{1}{NT}\frac{1}{\hat{\sigma}_{3mm}}\,Z_m^{*\prime}\iota = \frac{1}{T}\,\frac{1}{\left(\frac{\hat{\sigma}_{3mm}}{T}\right)}\,\frac{1}{NT}\,Z_m^{*\prime}\iota \;\text{-->}\; 0$$

as $\frac{\hat{\sigma}_{3mm}}{T}$ --> $\sigma_{\mu mm}+\sigma_{\nu mm}$ (cf. (4.B.47)

$$\text{and}\quad \frac{1}{NT}\,Z_m^{*\prime}\iota = \frac{1}{NT}\,[Y_m\; X_m^*]'\iota$$

$$= \frac{1}{NT}\begin{bmatrix} \Pi_m'X'\iota + V_m'\iota \\ X_m^{*\prime}\iota \end{bmatrix}$$

$$= \frac{1}{NT}\begin{bmatrix} \Pi_m'\begin{bmatrix}\iota'\iota \\ \underline{X}'\iota\end{bmatrix} + V_m'\iota \\ X_m^{*\prime}\iota \end{bmatrix}$$

$$= \frac{1}{NT}\begin{bmatrix} \Pi_m'\begin{bmatrix}NT \\ \underline{X}'\iota\end{bmatrix} + V_m'\iota \\ H_m^{*\prime}\underline{X}'\iota \end{bmatrix}$$

using (4.A.9)

$$(4.C.12)\quad \text{->}\; \frac{1}{NT}\begin{bmatrix} \Pi_m' & \begin{bmatrix}1 \\ r\end{bmatrix} \\ H_m^{*\prime} & r \end{bmatrix}$$

using (3.A.4)

(iv)

$$\frac{1}{NT}\sum_j \frac{1}{\hat{\sigma}_{jmm}}\,Z_m^{*\prime}\,M_j\,\underline{X} = \frac{1}{NT}\sum_j \frac{1}{\hat{\sigma}_{jmm}}\begin{bmatrix} Y_m' & M_j\,\underline{X} \\ X_m^{*\prime} & M_j\,\underline{X} \end{bmatrix}$$

$$= \frac{1}{NT}\sum_j \frac{1}{\hat{\sigma}_{jmm}}\begin{bmatrix} \Pi_m'X'M_j\underline{X} + V_m'M_j\underline{X} \\ X_m^{*\prime}\,M_j\,\underline{X} \end{bmatrix}$$

We have :

$$\frac{1}{NT}\,\underline{X}'\,M_j\,\underline{X} \dashrightarrow \text{constant} \quad \forall j \qquad \text{(cf. Section 3.A.1)}$$

$$\frac{1}{NT}\,V_m'\,M_j\,\underline{X} \dashrightarrow 0 \quad \forall j \qquad \text{(cf. Section 3.A.1)}$$

$$\frac{1}{\hat{\sigma}_{jmm}} \to 0 \text{ for } j=1,2,3 \text{ (cf. (4.B.41),(4.B.42),(4.B.43))}$$

and

$$\frac{1}{\hat{\sigma}_{4mm}} \to \frac{1}{\sigma_{4mm}} = \frac{1}{\sigma_{\varepsilon mm}} \qquad \text{(cf. (4.B.30))}$$

Thus

$$\frac{1}{NT}\sum_j \frac{1}{\hat{\sigma}_{jmm}}\,\Pi_m'\,\,X'M_j\underline{X} \to \frac{1}{\sigma_{4mm}}\,\Pi_m'\,\text{plim}\,X'M_4\underline{X}$$

$$= \frac{1}{\sigma_{4mm}}\,\text{plim}\,\left[\Pi_{om}\;\Pi_{*m}'\right]\begin{bmatrix}\iota'M_4\underline{X}\\ \underline{X}'M_4\underline{X}\end{bmatrix}$$

(4.C.13) $$= \frac{1}{\sigma_{4mm}}\,\Pi_{*m}'\,R \quad \text{as} \quad \iota'M_4 = 0, \quad \frac{1}{NT}\,\underline{X}'M_4\underline{X} \to R$$

(cf. (3.A.12)

and $$\frac{1}{NT}\sum_j \frac{1}{\hat{\sigma}_{jmm}}\,X_m^{*\prime}\,M_j\,\underline{X} \to \frac{1}{\sigma_{4mm}}\,\text{plim}\,\frac{1}{NT}\,X_m^{*\prime}\,M_j\,\underline{X}$$

$$= \frac{1}{\sigma_{4mm}}\,\text{plim}\,\frac{1}{NT}\,H_m^{*\prime}\,\underline{X}'\,M_4\,\underline{X}$$

using (4.A.9)

(4.C.14) $$= \frac{1}{\sigma_{4mm}}\,H_m^{*\prime}\,R \qquad \text{using (3.A.12)}$$

Therefore,

(4.C.15) $$\bar{D}_2\,\tilde{Z}_m'\,\hat{\Sigma}_{mm}^{-1}\,X = \begin{bmatrix} \dfrac{1}{\sigma_{\mu mm}+\sigma_{\nu mm}} & \dfrac{1}{\sigma_{\mu mm}+\sigma_{\nu mm}}\,r' \\ 0 & \dfrac{1}{\sigma_{\varepsilon mm}}\begin{bmatrix}\Pi_{*m}'\\ H_m^{*\prime}\end{bmatrix} R \end{bmatrix}$$

using (4.C.9), (4.C.10), (4.C.11), (4.C.13), (4.C.14).

Let us now determine plim $\bar{D}_1 X' \hat{\Sigma}_{mm}^{-1} X$.

$$\bar{D}_1 X'\hat{\Sigma}_{mm}^{-1} X = \begin{bmatrix} \frac{1}{N} & 0 \\ 0 & \frac{1}{NT}I \end{bmatrix} \begin{bmatrix} \iota'\hat{\Sigma}_{mm}^{-1}\iota & \iota'\hat{\Sigma}_{mm}^{-1}\underline{X} \\ \underline{X}'\hat{\Sigma}_{mm}^{-1}\iota & \underline{X}'\hat{\Sigma}_{mm}^{-1}\underline{X} \end{bmatrix}$$

$$= \begin{bmatrix} \frac{1}{N}\frac{NT}{\hat{\sigma}_{3mm}} & \frac{1}{N}\frac{1}{\hat{\sigma}_{3mm}}\iota'\underline{X} \\ \frac{1}{NT}\frac{1}{\hat{\sigma}_{3mm}}\underline{X}'\iota & \frac{1}{NT}\sum_j \frac{1}{\hat{\sigma}_{jmm}}\underline{X}'M_j\underline{X} \end{bmatrix}$$

using (3.22) and (3.28)

(4.C.16)
$$\rightarrow \begin{bmatrix} \frac{1}{\sigma_{\mu mm}+\sigma_{\nu mm}} & \frac{1}{\sigma_{\mu mm}+\sigma_{\nu mm}} r' \\ 0 & \frac{1}{\sigma_{\varepsilon mm}} R \end{bmatrix}$$

using (4.B.47), (4.B.30) and (3.A.12)

and hence

(4.C.17)
$$\text{plim } (\bar{D}_1 X' \hat{\Sigma}_{mm}^{-1} X)^{-1} = \begin{bmatrix} \frac{1}{\sigma_{\mu mm}+\sigma_{\nu mm}} & \frac{1}{\sigma_{\mu mm}+\sigma_{\nu mm}} r' \\ 0 & \frac{1}{\sigma_{\varepsilon mm}} R \end{bmatrix}^{-1}$$

Next, we need plim $\bar{D}_1 X' \hat{\Sigma}_{mm}^{-1} \tilde{Z}_m$.

$$\bar{D}_1 X' \hat{\Sigma}_{mm}^{-1} \tilde{Z}_m = \begin{bmatrix} \frac{1}{N} & 0 \\ 0 & \frac{1}{NT}I \end{bmatrix} \begin{bmatrix} \iota'\hat{\Sigma}_{mm}^{-1}\iota & \iota'\hat{\Sigma}_{mm}^{-1}Z_m^* \\ \underline{X}'\hat{\Sigma}_{mm}^{-1}\iota & \underline{X}'\hat{\Sigma}_{mm}^{-1}Z_m^* \end{bmatrix}$$

$$= \begin{bmatrix} \frac{1}{N}\frac{NT}{\hat{\sigma}_{3mm}} & \frac{1}{N}\frac{1}{\hat{\sigma}_{3mm}}\iota'Z_m^* \\ \frac{1}{NT}\frac{1}{\hat{\sigma}_{3mm}}\iota'\underline{X} & \frac{1}{NT}\sum_j \frac{1}{\hat{\sigma}_{jmm}}\underline{X}'M_jZ_m^* \end{bmatrix}$$

(4.C.18)
$$\rightarrow \begin{bmatrix} \frac{1}{\sigma_{\mu mm}+\sigma_{\nu mm}} & \frac{1}{\sigma_{\mu mm}+\sigma_{\nu mm}} \begin{bmatrix} \Pi_m' \\ H_m^{*\prime} \end{bmatrix} \begin{bmatrix} 1 \\ r \\ r \end{bmatrix} \\ 0 & \frac{1}{\sigma_{\varepsilon mm}} \begin{bmatrix} \Pi_{*m}' \\ H_m^{*\prime} \end{bmatrix} R \end{bmatrix}$$

using (4.B.47), (4.C.12), (4.C.13), (4.C.14).

Finally, plim $\bar{D}_1 \; X' \; \hat{\Sigma}_{mm}^{-1} \; u_m$ has to be calculated.

$$\bar{D}_1 \; X'\hat{\Sigma}_{mm}^{-1}u_m = \begin{bmatrix} \frac{1}{N} & 0 \\ 0 & \frac{1}{NT}I \end{bmatrix} \begin{bmatrix} \iota'\hat{\Sigma}_{mm}^{-1} \; u_m \\ \underline{X}'\hat{\Sigma}_{mm}^{-1} \; u_m \end{bmatrix}$$

$$= \begin{bmatrix} \frac{1}{N} \frac{1}{\hat{\sigma}_{3mm}} \iota' u_m \\ \frac{1}{NT} \sum_j \frac{1}{\hat{\sigma}_{jmm}} \underline{X}' \; M_j \; u_m \end{bmatrix}$$

$$= \begin{bmatrix} \frac{1}{\left(\frac{\hat{\sigma}_{3mm}}{T}\right)} \frac{1}{NT} \iota' \; u_m \\ \sum_j \frac{1}{\hat{\sigma}_{jmm}} \frac{1}{NT} \underline{X}' \; M_j \; u_m \end{bmatrix}$$

(4.C.19) $= 0$ using (4.B.47), plim $\frac{1}{NT} \iota' u_m = 0$ and (3.A.15) to (3.A.18)

Therefore, combining (4.C.15), (4.C.17), (4.C.18) and (4.C.19) we get :

(4.C.20) plim $\hat{\tilde{\alpha}}_{m,fG2SLS} = \tilde{\alpha}_m + 0 = \tilde{\alpha}_m$

Thus we have shown that $\hat{\alpha}_{m,fG2SLS}$ is consistent.

Note that the same proof can be repeated with the real values of the variance components instead of their estimators thereby proving the consistency of the pure G2SLS estimator.

APPENDIX 4.D Limiting Distribution of the Feasible G2SLS Estimator

Let us recall that the expression of the feasible G2SLS estimator $\hat{\tilde{\alpha}}_{m,fG2SLS}$ is given in (4.C.5):

$$(4.D.1)\quad \hat{\tilde{\alpha}}_{m,fG2SLS} = \tilde{\alpha}_m + \left[\tilde{Z}_m' \hat{\Sigma}_{mm}^{-1}X(X'\hat{\Sigma}_{mm}^{-1}X)^{-1}X'\hat{\Sigma}_{mm}^{-1}\tilde{Z}_m\right]^{-1} \tilde{Z}_m' \hat{\Sigma}_{mm}^{-1}X(X'\hat{\Sigma}_{mm}^{-1}X)^{-1}X'\hat{\Sigma}_{mm}^{-1}u_m$$

By using the notation $\bar{D}_2$ of (4.C.7), i.e.

$$\bar{D}_2 = \begin{bmatrix} \frac{1}{N} & 0 \\ 0 & \frac{1}{NT} I_{M_m + K_m - 1} \end{bmatrix}$$

we can write (4.38) as $\bar{D}_2^{-\frac{1}{2}}(\hat{\tilde{\alpha}}_m - \tilde{\alpha}_m)$[1] and using (4.D.1) we can say that its limiting distribution is the same as that of

$$\bar{D}_2^{-\frac{1}{2}} \operatorname{plim}\left[\tilde{Z}_m' \hat{\Sigma}_{mm}^{-1} X (X' \hat{\Sigma}_{mm}^{-1} X)^{-1} X' \hat{\Sigma}_{mm}^{-1} \tilde{Z}_m\right]^{-1}$$
$$\tilde{Z}_m' \hat{\Sigma}_{mm}^{-1} X (X' \hat{\Sigma}_{mm}^{-1} X)^{-1} X' \hat{\Sigma}_{mm}^{-1} u_m$$

or that of

$$(4.D.2)\quad \operatorname{plim}\left[\bar{D}_2^{\frac{1}{2}}\tilde{Z}_m'\hat{\Sigma}_{mm}^{-1} X \bar{D}_1^{\frac{1}{2}} (\bar{D}_1^{\frac{1}{2}}X'\hat{\Sigma}_{mm}^{-1} X \bar{D}_1^{\frac{1}{2}})^{-1} \bar{D}_1^{\frac{1}{2}}X'\hat{\Sigma}_{mm}^{-1}\tilde{Z}_m\bar{D}_2^{\frac{1}{2}}\right]^{-1}$$
$$\operatorname{plim} \bar{D}_2^{\frac{1}{2}}\tilde{Z}_m'\hat{\Sigma}_{mm}^{-1}X\bar{D}_1^{\frac{1}{2}} (\bar{D}_1^{\frac{1}{2}}X'\hat{\Sigma}_{mm}^{-1}X\bar{D}_1^{\frac{1}{2}})^{-1} \bar{D}_1^{\frac{1}{2}}X'\hat{\Sigma}_{mm}^{-1}u_m$$

---

1) We will omit writing the subscript fG2SLS each time till the end of this appendix, in order to simplify notations.

Now,

$$\text{plim}\ \bar{D}_2^{\frac{1}{2}}\ \tilde{Z}'_m\ \hat{\Sigma}^{-1}_{mm}\ X\ \bar{D}_1^{\frac{1}{2}}$$

$$= \text{plim} \begin{bmatrix} \frac{1}{\sqrt{N}} & 0 \\ 0 & \frac{1}{\sqrt{NT}} I \end{bmatrix} \begin{bmatrix} \iota'\hat{\Sigma}^{-1}_{mm}\iota & \iota'\hat{\Sigma}^{-1}_{mm}\underline{X} \\ Z^{*\prime}_m\hat{\Sigma}^{-1}_{mm}\iota & Z^{*\prime}_m\hat{\Sigma}^{-1}_{mm}\underline{X} \end{bmatrix} \begin{bmatrix} \frac{1}{\sqrt{N}} & 0 \\ 0 & \frac{1}{\sqrt{NT}} I \end{bmatrix}$$

$$= \text{plim} \begin{bmatrix} \frac{1}{N}\ \frac{NT}{\hat{\sigma}_{3mm}} & \frac{1}{\sqrt{T}}\ \frac{1}{NT}\ \frac{1}{\left(\frac{\hat{\sigma}_{3mm}}{T}\right)}\ \iota'\underline{X} \\ \frac{1}{\sqrt{T}}\ \frac{1}{NT}\ \frac{1}{\left(\frac{\hat{\sigma}_{3mm}}{T}\right)}\ Z^{*\prime}_m\iota & \frac{1}{NT}\sum_j \frac{1}{\hat{\sigma}_{jmm}}\ Z^{*\prime}_m M_j\underline{X} \end{bmatrix} \quad \text{using (3.22) and (3.28)}$$

$$= \begin{bmatrix} \frac{1}{\sigma_{\mu mm}+\sigma_{\nu mm}} & 0 \\ 0 & \frac{1}{\sigma_{\varepsilon mm}} \begin{bmatrix} \Pi'_{*m} \\ H^{*\prime}_m \end{bmatrix} R \end{bmatrix} \quad \text{using (4.B.47), (4.C.13),(4.C.14)}$$

$$= \left[\begin{array}{c|c} 1 & 0 \\ \hline 0 & \Pi'_{*m} \\ 0 & H^{*\prime}_m \end{array}\right] \begin{bmatrix} \frac{1}{\sigma_{\mu mm}+\sigma_{\nu mm}} & 0 \\ 0 & \frac{1}{\sigma_{\varepsilon mm}} R \end{bmatrix}$$

$$\text{(4.D.3)} \qquad = \tilde{P}'_m\ \tilde{R}_m$$

with obvious notations for $\tilde{P}_m$ and $\tilde{R}_m$.

$$\text{plim}\ \bar{D}_1^{\frac{1}{2}} X'\hat{\Sigma}^{-1}_{mm} X\ \bar{D}_1^{\frac{1}{2}} = \text{plim} \begin{bmatrix} \frac{1}{N}\ \iota'\hat{\Sigma}^{-1}_{mm}\iota & \frac{1}{N\sqrt{T}}\ \iota'\hat{\Sigma}^{-1}_{mm}\ \underline{X} \\ \frac{1}{N\sqrt{T}}\ \underline{X}'\hat{\Sigma}^{-1}_{mm}\iota & \frac{1}{NT}\ \underline{X}'\hat{\Sigma}^{-1}_{mm}\ \underline{X} \end{bmatrix}$$

$$= \text{plim} \begin{bmatrix} \frac{1}{\left(\frac{\hat{\sigma}_{3mm}}{T}\right)} & \frac{1}{\sqrt{T}}\ \frac{1}{NT}\ \frac{1}{\left(\frac{\hat{\sigma}_{3mm}}{T}\right)}\ \iota'\underline{X} \\ \frac{1}{\sqrt{T}}\ \frac{1}{NT}\ \frac{1}{\left(\frac{\hat{\sigma}_{3mm}}{T}\right)}\ \underline{X}'\iota & \frac{1}{NT}\sum_j \frac{1}{\hat{\sigma}_{jmm}}\ \underline{X}'M_j\underline{X} \end{bmatrix}$$

using (3.22) and (3.28)

$$= \begin{bmatrix} \frac{1}{\sigma_{\mu mm}+\sigma_{\nu mm}} & 0 \\ 0 & \frac{1}{\sigma_{\varepsilon mm}} R \end{bmatrix} \quad \text{using (4.B.47), (4.B.41), (4.B.42), (4.B.43), (4.B.30) and (3.A.12)}$$

$$\text{(4.D.4)} \qquad = \tilde{R}_m$$

Thus the limiting distribution of $\bar{D}_2^{-\frac{1}{2}}\,(\hat{\tilde{\alpha}}_m - \tilde{\alpha}_m)$ is the same as that of

(4.D.5) $$[\tilde{P}_m' \, \tilde{R}_m \, \tilde{P}_m]^{-1} \quad \tilde{P}_m' \quad \bar{D}_1^{\frac{1}{2}} \, X'\hat{\Sigma}_{mm}^{-1} \, u_m$$

Now,

$$\bar{D}_1^{\frac{1}{2}} \, X' \, \hat{\Sigma}_{mm}^{-1} \, u_m = \begin{bmatrix} \frac{1}{\sqrt{N}} & 0 \\ 0 & \frac{1}{\sqrt{NT}} I \end{bmatrix} \begin{bmatrix} \iota' \, \hat{\Sigma}_{mm}^{-1} \, u_m \\ \underline{X}' \, \hat{\Sigma}_{mm}^{-1} \, u_m \end{bmatrix}$$

$$= \begin{bmatrix} \frac{1}{\sqrt{N}} & 0 \\ 0 & \frac{1}{\sqrt{NT}} I \end{bmatrix} \begin{bmatrix} \frac{1}{\hat{\sigma}_{3mm}} \quad \iota' \, u_m \\ \sum_{\ell} \frac{1}{\hat{\sigma}_{\ell mm}} \, \underline{X}' M_{\ell} \, u_m \end{bmatrix}$$

$$= \begin{bmatrix} \frac{1}{\sqrt{N}} \frac{1}{\hat{\sigma}_{3mm}} \, \iota' u_m \\ \frac{1}{\sqrt{NT}} \sum_{\ell} \frac{1}{\hat{\sigma}_{\ell mm}} \, \underline{X}' M_{\ell} \, u_m \end{bmatrix}$$

$$= \begin{bmatrix} \sqrt{N} \, \frac{1}{\left(\frac{\hat{\sigma}_{3mm}}{T}\right)} \, \frac{1}{NT} \, \iota' u_m \\ \frac{1}{\sqrt{NT}} \sum_{\ell} \frac{1}{\hat{\sigma}_{\ell mm}} \, \underline{X}' M_{\ell} \, u_m \end{bmatrix}$$

$$= \begin{bmatrix} \sqrt{N} \, \frac{1}{\left(\frac{\hat{\sigma}_{3mm}}{T}\right)} \left(\frac{1}{N} \sum_i \mu_{mi} + \frac{1}{T} \sum_t \nu_{mt} + \frac{1}{NT} \sum_i \sum_t \varepsilon_{mit}\right) \\ \frac{1}{\sqrt{NT}} \sum_{\ell} \frac{1}{\hat{\sigma}_{\ell mm}} \left(\underline{X}' M_{\ell} \, (I \otimes \iota) \mu^m + \underline{X}' M_{\ell} \, (\iota \otimes I) \nu^m + \underline{X}' M_{\ell} \, \varepsilon_m\right) \end{bmatrix}$$

(4.D.6) $$= \sum_i \begin{bmatrix} \frac{1}{\sqrt{N}} \, \frac{1}{\left(\frac{\hat{\sigma}_{3mm}}{T}\right)} \\ \frac{1}{\sqrt{NT}} \sum_{\ell} \frac{1}{\hat{\sigma}_{\ell mm}} A_{\ell i} \end{bmatrix} \mu_{mi} + \sum_t \begin{bmatrix} \frac{\sqrt{N}}{T} \, \frac{1}{\left(\frac{\hat{\sigma}_{3mm}}{T}\right)} \\ \frac{1}{\sqrt{NT}} \sum_{\ell} \frac{1}{\hat{\sigma}_{\ell mm}} B_{\ell t} \end{bmatrix} \nu_{mt}$$

$$+ \sum_i \sum_t \begin{bmatrix} \frac{\sqrt{N}}{NT} \, \frac{1}{\left(\frac{\hat{\sigma}_{3mm}}{T}\right)} \\ \frac{1}{\sqrt{NT}} \sum_{\ell} \frac{1}{\hat{\sigma}_{\ell mm}} \, C_{\ell it} \end{bmatrix} \varepsilon_{mit}$$

where $A_{\ell i}$, $B_{\ell t}$ and $C_{\ell it}$ are defined in (3.B.11), page 104 and verify (3.B.15).

Thus $\bar{D}_1^{-\frac{1}{2}} X'\hat{\Sigma}_{mm}^{-1} u_m$ has been split into three independent components. Next, let us derive the limiting distribution of each component.

i) First component :

$$\sum_i \begin{bmatrix} \frac{1}{\sqrt{N}} & \left(\frac{\hat{\sigma}_{3mm}}{T}\right)^{-1} \\ \frac{1}{\sqrt{NT}} \sum_\ell & \frac{1}{\hat{\sigma}_{\ell mm}} A_{\ell i} \end{bmatrix} \mu_{mi}$$

The limiting distribution of the above variable is the same as that of

$$\sum_i \operatorname{plim} \begin{bmatrix} \frac{1}{\sqrt{N}} & \left(\frac{\hat{\sigma}_{3mm}}{T}\right)^{-1} \\ \frac{1}{\sqrt{NT}} & \sum_{\ell=1,3} \left(\frac{\hat{\sigma}_{\ell mm}}{T}\right)^{-1} \frac{A_{\ell i}}{T} \end{bmatrix} \mu_{mi}$$

as $A_\ell = 0$, $\ell=2,4$ (cf. (3.B.15))

$$(4.D.7) \quad = \sum_i \begin{bmatrix} \frac{1}{\sqrt{N}} (\sigma_{\mu mm} + \sigma_{\nu mm})^{-1} \\ \frac{1}{\sqrt{NT}} \left(\frac{1}{\sigma_{\mu mm}} \bar{A}_{1i} + \frac{1}{\sigma_{\mu mm}+\sigma_{\nu mm}} \bar{A}_{3i}\right) \end{bmatrix} \mu_{mi}$$

where $\bar{A}_{\ell i} = \operatorname{plim} \frac{1}{T} A_{\ell i}$, $i=1,3$ (see (3.B.17),(3.B.18), page 106) and using (4.B.45), (4.B.47).

As the μ_{mi}(s) are i.i.d. with zero mean and variance $\sigma_{\mu mm}$, applying the Central Limit Theorem, we can conclude that (4.D.7) has a normal limiting distribution with zero mean and variance

$$\sigma_{\mu mm} \operatorname{plim} \sum_i \begin{bmatrix} \frac{1}{\sqrt{N}} \frac{1}{\sigma_{\mu mm}+\sigma_{\nu mm}} \\ \frac{1}{\sqrt{NT}}\left(\frac{1}{\sigma_{\mu mm}} \bar{A}_{1i} + \frac{1}{\sigma_{\mu mm}+\sigma_{\nu mm}} \bar{A}_{3i}\right) \end{bmatrix} \begin{bmatrix} \frac{1}{\sqrt{N}} \frac{1}{\sigma_{\mu mm}+\sigma_{\nu mm}} \\ \frac{1}{\sqrt{NT}}\left(\frac{1}{\sigma_{\mu mm}} \bar{A}_{1i} + \frac{1}{\sigma_{\mu mm}+\sigma_{\nu mm}} \bar{A}_{3i}\right) \end{bmatrix}$$

$$= \sigma_{\mu mm} \cdot \operatorname{plim} \sum_i \begin{bmatrix} \frac{1}{N} \left(\frac{1}{\sigma_{\mu mm}+\sigma_{\nu mm}}\right)^2 & \frac{1}{N\sqrt{T}} \frac{1}{\sigma_{\mu mm}+\sigma_{\nu mm}} \left(\frac{1}{\sigma_{\mu mm}} \bar{A}'_{1i} + \frac{1}{\sigma_{\mu mm}+\sigma_{\nu mm}} \bar{A}'_{3i}\right) \\ \frac{1}{N\sqrt{T}} \frac{1}{\sigma_{\mu mm}+\sigma_{\nu mm}}\left(\frac{1}{\sigma_{\mu mm}} \bar{A}_{1i} + \frac{1}{\sigma_{\mu mm}+\sigma_{\nu mm}} \bar{A}_{3i}\right) & \frac{1}{NT}\left(\left(\frac{1}{\sigma_{\mu mm}}\right)^2 \bar{A}_{1i}\bar{A}'_{1i} + \left(\frac{1}{\sigma_{\mu mm}+\sigma_{\nu mm}}\right)^2 \bar{A}_{3i}\bar{A}'_{3i}\right) \end{bmatrix}$$

using (3.B.20)

$$(4.D.8) \quad = \sigma_{\mu mm} \begin{bmatrix} \frac{1}{(\sigma_{\mu mm}+\sigma_{\nu mm})^2} & 0 \\ 0 & 0 \end{bmatrix}$$

using (3.B.21),(3.B.22) (pages 106-107)

ii) Second component :

$$\sum_t \begin{bmatrix} \frac{\sqrt{N}}{T} \left(\frac{\hat{\sigma}_{3mm}}{T}\right)^{-1} \\ \frac{1}{\sqrt{NT}} \sum_{\ell=2,3} \left(\frac{\hat{\sigma}_{\ell mm}}{N}\right)^{-1} \frac{B_{\ell t}}{N} \end{bmatrix} \nu_{mt} \qquad \text{as } B_\ell = 0 \text{ , } \ell=1,4 \text{ (cf. (3.B.15))}$$

As in (i), the limiting distribution of the above variable is the same as that of

$$\text{(4.D.9)} \qquad \sum_t \begin{bmatrix} \frac{\sqrt{N}}{T} (\sigma_{\mu mm} + \sigma_{\nu mm})^{-1} \\ \frac{1}{\sqrt{NT}} \left(\frac{1}{\sigma_{\nu mm}} \bar{B}_{2t} + \frac{1}{\sigma_{\mu mm}+\sigma_{\nu mm}} \bar{B}_{3t} \right) \end{bmatrix} \nu_{mt}$$

using (3.B.24), (3.B.25) and (4.B.46), (4.B.47)

As the ν_{mt}(s) are i.i.d $(0, \sigma_{\nu mm})$, applying the Central Limit Theorem, we can say that the limiting distribution of (4.D.9) is normal with zero mean and variance

$$\sigma_{\nu mm} \operatorname{plim} \sum_t \begin{bmatrix} \frac{\sqrt{N}}{T} (\sigma_{\mu mm} + \sigma_{\nu mm})^{-1} \\ \frac{1}{\sqrt{NT}} (\frac{1}{\sigma_{\nu mm}} \bar{B}_{2t} + \frac{1}{\sigma_{\mu mm}+\sigma_{\nu mm}} \bar{B}_{3t}) \end{bmatrix} \begin{bmatrix} \frac{\sqrt{N}}{T} (\sigma_{\mu mm} + \sigma_{\nu mm})^{-1} \\ \frac{1}{\sqrt{NT}} (\frac{1}{\sigma_{\nu mm}} \bar{B}_{2t} + \frac{1}{\sigma_{\mu mm}+\sigma_{\nu mm}} \bar{B}_{3t}) \end{bmatrix}$$

$$\text{(4.D.10)} \qquad = \sigma_{\nu mm} \begin{bmatrix} (\sigma_{\mu mm} + \sigma_{\nu mm})^{-2} & 0 \\ 0 & 0 \end{bmatrix} \qquad \text{using (3.B.27), (3.B.28) and (3.B.29)}$$

(iii) Third component :

$$\sum_i \sum_t \begin{bmatrix} \frac{\sqrt{N}}{T} \left(\frac{\hat{\sigma}_{3mm}}{T}\right)^{-1} \\ \frac{1}{\sqrt{NT}} \sum_\ell \frac{1}{\hat{\sigma}_{\ell mm}} C_{\ell it} \end{bmatrix} \varepsilon_{mit}$$

This is asymptotically equivalent to

$$\text{(4.D.11)} \qquad \sum_i \sum_t \begin{bmatrix} \frac{\sqrt{N}}{NT} (\sigma_{\mu mm} + \sigma_{\nu mm})^{-1} \\ \frac{1}{\sqrt{NT}} \frac{1}{\sigma_{\varepsilon mm}} \bar{C}_{\varepsilon it} \end{bmatrix} \varepsilon_{mit}$$

using $\operatorname{plim} C_{\ell it} = \bar{C}_{\ell it}$ (cf. (3.B.31), (4.B.47) and (4.35))

Since the ε_{mit}(s) are i.i.d. $(0,\sigma_{\varepsilon mm})$, applying the Central Limit Theorem, we can say that the limiting distribution of (4.D.11) is normal with zero mean and the following variance:

$$\sigma_{\varepsilon mm} \operatorname{plim} \sum_{i,t} \begin{bmatrix} \frac{\sqrt{N}}{NT}(\sigma_{\mu mm}+\sigma_{\nu mm})^{-1} \\ \frac{1}{\sqrt{NT}}\frac{1}{\sigma_{\varepsilon mm}}\bar{C}_{\varepsilon it} \end{bmatrix} \begin{bmatrix} \frac{\sqrt{N}}{NT}(\sigma_{\mu mm}+\sigma_{\nu mm})^{-1} & \frac{1}{\sqrt{NT}}\sigma_{\varepsilon mm}^{-1}\bar{C}'_{\varepsilon it} \end{bmatrix}$$

$$=\sigma_{\varepsilon mm}\sum_{i,t} \begin{bmatrix} \frac{1}{T}\frac{1}{NT}(\sigma_{\mu mm}+\sigma_{\nu mm})^{-2} & \frac{1}{NT\sqrt{T}}(\sigma_{\mu mm}+\sigma_{\nu mm})^{-1}\sigma_{\varepsilon mm}^{-1}\bar{C}'_{\varepsilon it} \\ \frac{1}{NT\sqrt{T}}\frac{1}{\sigma_{\varepsilon mm}}\bar{C}_{\varepsilon it}(\sigma_{\mu mm}+\sigma_{\nu mm})^{-1} & \frac{1}{NT}\left(\frac{1}{\sigma_{\varepsilon mm}}\right)^2\bar{C}_{\varepsilon it}\bar{C}'_{\varepsilon it} \end{bmatrix}$$

$$(4.D.12) \quad = \sigma_{\varepsilon mm}\begin{bmatrix} 0 & 0 \\ 0 & \frac{1}{(\sigma_{\varepsilon mm})^2}R \end{bmatrix} \qquad \text{using (3.B.34)}$$

Thus the limiting distribution of $\bar{D}_1^{-\frac{1}{2}} X'\hat{\Sigma}_{mm}^{-1} u_m$ is normal with zero mean and variance-covariance matrix equal to the sum of (4.D.8), (4.D.10) and (4.D.12) i.e.

$$(4.D.13) \quad \begin{bmatrix} \frac{1}{\sigma_{\mu mm}+\sigma_{\nu mm}} & 0 \\ 0 & \frac{1}{\sigma_{\varepsilon mm}}R \end{bmatrix} = \tilde{R}_m$$

Therefore, the limiting distribution of $\bar{D}_2^{-\frac{1}{2}}(\hat{\tilde{\alpha}}_{m,fG2SLS}-\tilde{\alpha}_m)$ is normal with zero mean and the following variance-covariance matrix :

$$(\tilde{P}'_m \tilde{R}_m \tilde{P}_m)^{-1} \tilde{P}'_m \tilde{R}_m \tilde{P}_m (\tilde{P}'_m \tilde{R}_m \tilde{P}_m)^{-1}$$

$$(4.D.14) \quad = (\tilde{P}'_m \tilde{R}_m \tilde{P}_m)^{-1}$$

APPENDIX 4.E Limiting Distribution of the Feasible G3SLS Estimator

Let us substitute (4.46) in the expression of the feasible G3SLS estimator (4.57) and simplify it to get :

$$(4.E.1)\quad \hat{\alpha}_{fG3SLS} - \alpha = \left[Z_*'\hat{\Sigma}_*^{-1} \mathcal{X} (\mathcal{X}'\hat{\Sigma}_*^{-1}\hat{\Sigma}\hat{\Sigma}_*^{-1} \mathcal{X})^{-1} \mathcal{X}'\hat{\Sigma}_*^{-1} Z_* \right]^{-1} Z_*'\hat{\Sigma}_*^{-1}\mathcal{X}(\mathcal{X}'\hat{\Sigma}_*^{-1}\hat{\Sigma}\hat{\Sigma}_*^{-1}\mathcal{X})^{-1} \mathcal{X}'\hat{\Sigma}_*^{-1}u$$

As done in the case of the reduced form, we will reformulate the system (4.45) separating all the constant terms from the other coefficients :

$$(4.E.2)\quad X'\Sigma_{mm}^{-1}y_m = X'\Sigma_{mm}^{-1} [Y_m \;\; \iota \;\; X_m^*] \begin{bmatrix} \gamma_m^* \\ a_m \\ b_m \end{bmatrix} + X'\Sigma_{mm}^{-1} u_m \qquad m=1,\dots,M$$

or

$$(4.E.3)\quad X'\Sigma_{mm}^{-1}y_m = X'\Sigma_{mm}^{-1} \iota\, a_m + X'\Sigma_{mm}^{-1} Z_m^* \alpha_m^* + X'\Sigma_{mm}^{-1} u_m \qquad m=1,\dots,M$$

or

$$\begin{bmatrix} \iota' \\ \underline{X}' \end{bmatrix} \Sigma_{mm}^{-1}y_m = \begin{bmatrix} \iota' \\ \underline{X}' \end{bmatrix} \Sigma_{mm}^{-1} \iota\, a_m + \begin{bmatrix} \iota' \\ \underline{X}' \end{bmatrix} \Sigma_{mm}^{-1} Z_m^* \alpha_m^* + \begin{bmatrix} \iota' \\ \underline{X}' \end{bmatrix} \Sigma_{mm}^{-1} u_m \qquad m=1,\dots,M$$

or

$$(4.E.4)\quad \begin{bmatrix} I \otimes \iota' \\ I \otimes \underline{X}' \end{bmatrix} \Sigma_*^{-1} y = \begin{bmatrix} I \otimes \iota' \\ I \otimes \underline{X}' \end{bmatrix} \Sigma_*^{-1}(I \otimes \iota)a + \begin{bmatrix} I \otimes \iota' \\ I \otimes \underline{X}' \end{bmatrix} \Sigma_*^{-1} Z^* \alpha^* + \begin{bmatrix} I \otimes \iota' \\ I \otimes \underline{X}' \end{bmatrix} \Sigma_*^{-1} u$$

$$(4.E.5)\quad \text{where} \quad \begin{cases} a' = [a_1 \; \dots \; a_M] \\ \alpha^{*\prime} = [\alpha_1^{*\prime} \dots \alpha_M^{*\prime}] \\ Z^* = \begin{bmatrix} Z_1^* & & 0 \\ & \ddots & \\ 0 & & Z_M^* \end{bmatrix} \end{cases}$$

or

$$(4.E.6) \qquad \tilde{X}' \Sigma_*^{-1} y = \tilde{X}' \Sigma_*^{-1} \tilde{Z} \tilde{\alpha} + \tilde{X}' \Sigma_*^{-1} u$$

denoting

$$(4.E.7) \quad \begin{cases} \tilde{X} = [I \otimes \iota \quad I \otimes \underline{X}] \\ \tilde{Z} = [I \otimes \iota \quad Z^*] \\ \tilde{\alpha} = \begin{bmatrix} a \\ \alpha^* \end{bmatrix} \end{cases}$$

Thus, the feasible G3SLS is given by :

$$(4.E.8) \qquad \hat{\tilde{\alpha}}_{fG3SLS} = \left[\tilde{Z}'\hat{\Sigma}_*^{-1}\tilde{X}(\tilde{X}'\hat{\Sigma}_*^{-1}\hat{\Sigma}\hat{\Sigma}_*^{-1}\tilde{X})^{-1}\tilde{X}'\hat{\Sigma}_*^{-1}\tilde{Z}\right]^{-1} \tilde{Z}' \hat{\Sigma}_*^{-1}\tilde{X}(\tilde{X}'\hat{\Sigma}_*^{-1}\hat{\Sigma}\hat{\Sigma}_*^{-1}\tilde{X})^{-1}\tilde{X}' \hat{\Sigma}_*^{-1}y$$

Substituting (4.E.6) in (4.E.8) we get

$$(4.E.9) \qquad \hat{\tilde{\alpha}}_{fG3SLS} - \tilde{\alpha} = \left[\tilde{Z}'\hat{\Sigma}_*^{-1}\tilde{X}(\tilde{X}' \hat{\Sigma}_*^{-1} \hat{\Sigma} \hat{\Sigma}_*^{-1}\tilde{X})^{-1}\tilde{X}' \hat{\Sigma}_*^{-1}\tilde{Z}\right]^{-1} \tilde{Z}'\hat{\Sigma}_*^{-1}\tilde{X}(\tilde{X}'\hat{\Sigma}_*^{-1}\hat{\Sigma}\hat{\Sigma}_*^{-1}\tilde{X})^{-1}\tilde{X}'\hat{\Sigma}_*^{-1}u$$

As before, we will derive the limiting distribution of

$$\begin{bmatrix} \sqrt{N}\, I_M & 0 \\ 0 & \sqrt{NT}\, I_{M(\tilde{M}_m+K_m-1)} \end{bmatrix} (\hat{\tilde{\alpha}}_{fG3SLS} - \tilde{\alpha})$$

to avoid problems of singularity of limits of certain matrices. Denoting

$$(4.E.10) \qquad \tilde{D} = \begin{bmatrix} \frac{1}{N} I_M & 0 \\ 0 & \frac{1}{NT} I_{\sum_m(\tilde{M}_m+K_m-1)} \end{bmatrix}$$

and

$$(4.E.11) \qquad D^* = \begin{bmatrix} \frac{1}{N} I_M & 0 \\ & \frac{1}{NT} I_{M(K-1)} \end{bmatrix}$$

we can write[1)]

$$\tilde{D}^{-\frac{1}{2}}(\hat{\tilde{\alpha}}-\tilde{\alpha}) = \left[\tilde{D}^{\frac{1}{2}}\tilde{Z}'\hat{\Sigma}_*^{-1}\tilde{X}D^{*\frac{1}{2}}(D^{*\frac{1}{2}}\tilde{X}'\hat{\Sigma}_*^{-1}\hat{\Sigma}\hat{\Sigma}_*^{-1}\tilde{X}D^{*\frac{1}{2}})^{-1}D^{*\frac{1}{2}}\tilde{X}'\hat{\Sigma}_*^{-1}\tilde{Z}\tilde{D}^{\frac{1}{2}}\right]^{-1}$$

$$\tilde{D}^{\frac{1}{2}}\tilde{Z}'\hat{\Sigma}_*^{-1}\tilde{X}D^{*\frac{1}{2}}(D^{*\frac{1}{2}}\tilde{X}'\hat{\Sigma}_*^{-1}\hat{\Sigma}\hat{\Sigma}_*^{-1}\tilde{X}D^{*\frac{1}{2}})^{-1}D^{*\frac{1}{2}}\tilde{X}'\hat{\Sigma}_*^{-1}u \qquad (4.E.12)$$

Now, let us calculate, one by one, the plim of the matrices premultiplying $D^{*\frac{1}{2}}\ \tilde{X}'\ \hat{\Sigma}_*^{-1}\ u$ in (4.E.12).

(1) plim $\tilde{D}^{\frac{1}{2}}\ \tilde{Z}'\ \hat{\Sigma}_*^{-1}\ \tilde{X}\ D^{*\frac{1}{2}}$

$$= \text{plim} \begin{bmatrix} \frac{1}{\sqrt{N}}(I\otimes\iota') \\ \frac{1}{\sqrt{NT}}Z^{*'} \end{bmatrix} \begin{bmatrix} \hat{\Sigma}_{11}^{-1} & & 0 \\ & \ddots & \\ 0 & & \hat{\Sigma}_{MM}^{-1} \end{bmatrix} \begin{bmatrix} \frac{1}{\sqrt{N}}(I\otimes\iota) & \frac{1}{\sqrt{NT}}(I\otimes\underline{X}) \end{bmatrix}$$

using (4.E.7)

$$= \text{plim} \begin{bmatrix} \text{diag}\ (\frac{1}{N}\ \iota'\hat{\Sigma}_{mm}^{-1}\ \iota) & \text{diag}\ (\frac{1}{N\sqrt{T}}\ \iota'\hat{\Sigma}_{mm}^{-1}\ \underline{X}) \\ \text{diag}\ (\frac{1}{N\sqrt{T}}\ \underline{X}'\hat{\Sigma}_{mm}^{-1}\ \iota) & \text{diag}\ (\frac{1}{NT}\ \underline{X}'\hat{\Sigma}_{mm}^{-1}\ \underline{X}) \end{bmatrix}$$

$$= \text{plim} \begin{bmatrix} \text{diag}\ (\frac{1}{N}\ \frac{1}{\hat{\sigma}_{3mm}}\ NT) & \text{diag}\ (\frac{1}{N\sqrt{T}}\ \frac{1}{\hat{\sigma}_{3mm}}\ \iota'\underline{X}) \\ \text{diag}\ (\frac{1}{N\sqrt{T}}\ \frac{1}{\hat{\sigma}_{3mm}}\ Z_m^{*'}\iota) & \text{diag}\ (\frac{1}{NT}\sum_i \frac{1}{\hat{\sigma}_{imm}}\ Z_m^{*'}M_i\underline{X}) \end{bmatrix}$$

using (3.24) and (3.28)

$$= \begin{bmatrix} \text{diag}\ (\frac{1}{\sigma_{\mu mm}+\sigma_{\nu mm}}) & 0 \\ 0 & \text{diag}\left(\frac{1}{\sigma_{4mm}}\begin{bmatrix}\Pi_{*m}' \\ H_m^{*'}\end{bmatrix} R\right) \end{bmatrix}$$

using (4.B.47),(4.B.41),(4.B.42), (4.B.43) and (4.C.13), (4.C.14).

$$= \begin{bmatrix} I & 0 \\ 0 & \text{diag}\ \left(\begin{bmatrix}\Pi_{*m}' \\ H_m^{*'}\end{bmatrix} R\right) \end{bmatrix} \begin{bmatrix} \text{diag}(\frac{1}{\sigma_{\mu mm}+\sigma_{\nu mm}}) & 0 \\ 0 & \text{diag}(\frac{1}{\sigma_{\varepsilon mm}}\ I) \end{bmatrix}$$

$$(4.E.13) = \tilde{P}'\ \Lambda_1$$

with obvious notations for $\tilde{P}$ and Λ_1.

---

1) We will omit writing the subscript fG3SLS till the end of this derivation to avoid making the notations more cumbersome.

(2) $\quad \text{plim } D^{*\frac{1}{2}} \tilde{X}'\hat{\Sigma}_*^{-1} \hat{\Sigma} \hat{\Sigma}_*^{-1} \tilde{X} D^{*\frac{1}{2}}$

$$(4.E.14) = \text{plim} \begin{bmatrix} \left[\frac{1}{N} \iota'\hat{\Sigma}_{mm}^{-1}\hat{\Sigma}_{m\ell}\hat{\Sigma}_{\ell\ell}^{-1}\iota\right]_{m,\ell=1,\dots,M} & \left[\frac{1}{N\sqrt{T}} \iota'\hat{\Sigma}_{mm}^{-1}\hat{\Sigma}_{m\ell}\hat{\Sigma}_{\ell\ell}^{-1}\underline{X}\right]_{m,\ell=1,\dots,M} \\ \left[\frac{1}{N\sqrt{T}} \underline{X}'\hat{\Sigma}_{mm}^{-1}\hat{\Sigma}_{m\ell}\hat{\Sigma}_{\ell\ell}^{-1}\iota\right]_{m,\ell=1,\dots,M} & \left[\frac{1}{NT} \underline{X}'\hat{\Sigma}_{mm}^{-1}\hat{\Sigma}_{m\ell}\hat{\Sigma}_{\ell\ell}^{-1}\underline{X}\right]_{m,\ell=1,\dots,M} \end{bmatrix}$$

Now,

$$\text{plim } \frac{1}{N} \iota' \hat{\Sigma}_{mm}^{-1} \hat{\Sigma}_{m\ell} \hat{\Sigma}_{\ell\ell}^{-1}\iota = \text{plim } \frac{1}{N} \sum_i \frac{\hat{\sigma}_{im\ell}}{\hat{\sigma}_{imm}\hat{\sigma}_{i\ell\ell}} \iota' M_i \iota \quad \text{using (3.24)}$$

$$= \text{plim } \frac{1}{N} \frac{\hat{\sigma}_{3m\ell}}{\hat{\sigma}_{3mm}\hat{\sigma}_{3\ell\ell}} NT \quad \text{using (3.28)}$$

$$= \text{plim } \frac{\left(\frac{\hat{\sigma}_{3m\ell}}{T}\right)}{\left(\frac{\hat{\sigma}_{3mm}}{T}\right)\left(\frac{\hat{\sigma}_{3\ell\ell}}{T}\right)}$$

$$= \frac{(\sigma_{\mu m\ell} + \sigma_{\nu m\ell})}{(\sigma_{\mu mm}+\sigma_{\nu mm})(\sigma_{\mu\ell\ell}+\sigma_{\nu\ell\ell})} \quad \text{using (4.B.47)}$$

$$(4.E.15) \qquad = \bar{\sigma}_{m\ell}, \text{ say };$$

$$\text{plim } \frac{1}{N\sqrt{T}} \iota'\hat{\Sigma}_{mm}^{-1}\hat{\Sigma}_{m\ell}\hat{\Sigma}_{\ell\ell}^{-1}\underline{X} = \text{plim } \frac{1}{N\sqrt{T}} \sum_i \frac{\hat{\sigma}_{im\ell}}{\hat{\sigma}_{imm}\hat{\sigma}_{i\ell\ell}} \iota' M_i \underline{X} \quad \text{using (3.24)}$$

$$= \text{plim } \frac{1}{N\sqrt{T}} \frac{\hat{\sigma}_{3m\ell}}{\hat{\sigma}_{3mm}\hat{\sigma}_{3\ell\ell}} \iota'\underline{X} \quad \text{using (3.28)}$$

$$= \text{plim } \frac{1}{\sqrt{T}} \frac{1}{NT} \iota'\underline{X} \frac{\left(\frac{\hat{\sigma}_{3m\ell}}{T}\right)}{\left(\frac{\hat{\sigma}_{3mm}}{T}\right)\left(\frac{\hat{\sigma}_{3\ell\ell}}{T}\right)}$$

$$(4.E.16) \qquad = 0 \quad \text{(cf. (3.A.4)) (cf. (4.E.15) above)}$$

and

$$\text{plim } \frac{1}{NT} \underline{X}'\hat{\Sigma}_{mm}^{-1}\hat{\Sigma}_{m\ell}\hat{\Sigma}_{\ell\ell}^{-1}\underline{X} = \text{plim } \sum_i \frac{\hat{\sigma}_{im\ell}}{\hat{\sigma}_{imm}\hat{\sigma}_{i\ell\ell}} \underline{X}' M_i \underline{X} \quad \text{using (3.24)}$$

$$(4.E.17) \qquad = \frac{\sigma_{4m\ell}}{\sigma_{4mm}\sigma_{4\ell\ell}} R \quad \text{using (3.A.12)}$$

Thus, the plim in (4.E.14) equals

$$\begin{bmatrix} [\bar{\sigma}_{m\ell}]_{m,\ell=1,\dots,M} & 0 \\ 0 & \left[\dfrac{\sigma_{4m\ell}}{\sigma_{4mm}\sigma_{4\ell\ell}}\right] R_{m,\ell=1,\dots,M} \end{bmatrix}$$

$$= \begin{bmatrix} \text{diag}\left(\frac{1}{\sigma_{\mu mm}+\sigma_{\nu mm}}\right) & 0 \\ 0 & \text{diag}\left(\frac{1}{\sigma_{\varepsilon mm}} I\right) \end{bmatrix} \begin{bmatrix} [\sigma_{\mu m\ell}+\sigma_{\nu m\ell}]_{m,\ell=1,\dots,M} & 0 \\ 0 & [\sigma_{\varepsilon m\ell} \; R]_{m,\ell=1,\dots,M} \end{bmatrix}$$
$$\begin{bmatrix} \text{diag}\left(\frac{1}{\sigma_{\mu\ell\ell}+\sigma_{\nu\ell\ell}}\right) & 0 \\ 0 & \text{diag}\left(\frac{1}{\sigma_{\varepsilon\ell\ell}} I\right) \end{bmatrix}$$

(4.E.18) $= \Lambda_1 \; \Lambda_2 \; \Lambda_1$ using (4.E.13) and defining Λ_2 appropriately

Hence

(4.E.19) $\text{plim } D^{*\frac{1}{2}} \; \tilde{X}' \; \hat{\Sigma}_*^{-1} \; \hat{\Sigma} \; \hat{\Sigma}_*^{-1} \; \tilde{X} \; D^* \; = (\Lambda_1 \; \Lambda_2 \; \Lambda_1)$

Thus, the limiting distribution of $\tilde{D}^{-\frac{1}{2}}(\hat{\tilde{\alpha}} - \tilde{\alpha})$ is the same as that of $[\tilde{P}'\Lambda_1(\Lambda_1 \; \Lambda_2 \; \Lambda_1)^{-1}\Lambda_1\tilde{P}]^{-1} \; \tilde{P}'\Lambda_1(\Lambda_1 \; \Lambda_2 \; \Lambda_1)^{-1}D^{*\frac{1}{2}}\tilde{X}'\hat{\Sigma}_*^{-1} \; u$.

Now,

$$D^{*\frac{1}{2}}\tilde{X}' \; \hat{\Sigma}_*^{-1}u = \begin{bmatrix} \left[\frac{1}{\sqrt{N}} \; \iota'\hat{\Sigma}_{mm}^{-1}u_m\right]_{m=1,\dots,M} \\ \left[\frac{1}{\sqrt{NT}} \; \underline{X}'\hat{\Sigma}_{mm}^{-1} \; u_m\right]_{m=1,\dots,M} \end{bmatrix}$$

$$= \begin{bmatrix} \left[\frac{1}{\sqrt{N}} \frac{1}{\hat{\sigma}_{3mm}} \; \iota' u_m\right]_{m=1,\dots,M} \\ \left[\frac{1}{\sqrt{NT}} \sum_j \frac{1}{\hat{\sigma}_{jmm}} \; \underline{X}'M_j \; u_m\right]_{m=1,\dots,M} \end{bmatrix}$$

using (3.24), (3.28)

$$= \begin{bmatrix} \left[\sqrt{N} \frac{1}{\left(\frac{\hat{\sigma}_{3mm}}{T}\right)} \frac{1}{NT} \; \iota' u_m\right]_{m=1,\dots,M} \\ \left[\frac{1}{\sqrt{NT}} \sum_j \frac{1}{\hat{\sigma}_{jmm}} \; \underline{X}'M_j \; u_m\right]_{m=1,\dots,M} \end{bmatrix}$$

$$= \sum_i \begin{bmatrix} \left[\frac{1}{\sqrt{N}} \frac{1}{\left(\frac{\hat{\sigma}_{3mm}}{T}\right)} \mu_{mi} \right]_{m=1,\dots,M} \\ \left[\frac{1}{\sqrt{NT}} \sum_j \frac{1}{\hat{\sigma}_{jmm}} A_{ji} \mu_{mi} \right]_{m=1,\dots,M} \end{bmatrix} + \sum_t \begin{bmatrix} \left[\frac{\sqrt{N}}{T} \frac{1}{\left(\frac{\hat{\sigma}_{3mm}}{T}\right)} \nu_{mt} \right]_{m=1,\dots,M} \\ \left[\frac{1}{\sqrt{NT}} \sum_j \frac{1}{\hat{\sigma}_{jmm}} B_{jt} \nu_{mt} \right]_{m=1,\dots,M} \end{bmatrix}$$

$$+ \sum_i \sum_t \begin{bmatrix} \left[\frac{\sqrt{N}}{\sqrt{NT}} \frac{1}{\left(\frac{\hat{\sigma}_{3mm}}{T}\right)} \varepsilon_{mit} \right]_{m=1,\dots,M} \\ \left[\frac{1}{\sqrt{NT}} \sum_j \frac{1}{\hat{\sigma}_{jmm}} C_{jit} \varepsilon_{mit} \right]_{m=1,\dots,M} \end{bmatrix}$$

using (4.D.5)

$$(4.E.20) \quad = \sum_i \begin{bmatrix} \operatorname{diag}\left(\frac{1}{\sqrt{N}} \frac{1}{\left(\frac{\hat{\sigma}_{3mm}}{T}\right)} \right) \\ \operatorname{diag}\left(\frac{1}{\sqrt{NT}} \sum_j \frac{1}{\hat{\sigma}_{jmm}} A_{ji} \right) \end{bmatrix} \begin{bmatrix} \mu_{1i} \\ \vdots \\ \mu_{Mi} \end{bmatrix} + \sum_t \begin{bmatrix} \operatorname{diag}\left(\frac{\sqrt{N}}{T} \frac{1}{\left(\frac{\hat{\sigma}_{3mm}}{T}\right)} \right) \\ \operatorname{diag}\left(\frac{1}{NT} \sum_j \frac{1}{\hat{\sigma}_{jmm}} B_{jt} \right) \end{bmatrix} \begin{bmatrix} \nu_{1t} \\ \vdots \\ \nu_{Mt} \end{bmatrix}$$

$$+ \sum_i \sum_t \begin{bmatrix} \operatorname{diag}\left(\frac{\sqrt{N}}{NT} \frac{1}{\left(\frac{\hat{\sigma}_{3mm}}{T}\right)} \right) \\ \operatorname{diag}\left(\frac{1}{\sqrt{NT}} \sum_j \frac{1}{\hat{\sigma}_{jmm}} C_{jit} \right) \end{bmatrix} \begin{bmatrix} \varepsilon_{1it} \\ \vdots \\ \varepsilon_{Mit} \end{bmatrix}$$

The expression (4.E.20) above consists of three independent components which we will now examine one by one :

(i) First component :

$$(4.E.21) \quad \sum_i \begin{bmatrix} \operatorname{diag}\left(\frac{1}{\sqrt{N}} \frac{1}{\left(\frac{\hat{\sigma}_{3mm}}{T}\right)} \right) \\ \operatorname{diag}\left(\frac{1}{\sqrt{NT}} \sum_j \frac{1}{\hat{\sigma}_{jmm}} A_{ji} \right) \end{bmatrix} \mu_i$$

The limiting distribution of (4.E.21) is the same as that of

$$(4.E.22) \quad \sum_i \begin{bmatrix} \operatorname{diag}\left(\frac{1}{\sqrt{N}} \frac{1}{(\sigma_{\mu mm} + \sigma_{\nu mm})} \right) \\ \operatorname{diag} \frac{1}{\sqrt{NT}} \left(\frac{1}{\sigma_{\mu mm}} \bar{A}_{1i} + \frac{1}{\sigma_{\mu mm} + \sigma_{\nu mm}} \bar{A}_{3i} \right) \end{bmatrix} \mu_i$$

where $\bar{A}_{ji} = \operatorname{plim} \frac{1}{T} A_{ji}$, $j=1,3$ (see (3.B.17),(3.B.18))

Since the μ_i(s) are i.i.d $(0, \Sigma_\mu)$, applying the Central Limit Theorem, the limiting distribution of (4.E.22) is normal with zero mean and variance-covariance matrix :

$$= \operatorname{plim} \sum_i \begin{bmatrix} \operatorname{diag} \left(\frac{1}{\sqrt{N}} \frac{1}{\sigma_{\mu mm}+\sigma_{\nu mm}}\right) \\ \operatorname{diag} \frac{1}{\sqrt{NT}} \left(\frac{1}{\sigma_{\mu mm}} \bar{A}_{1i} + \frac{1}{\sigma_{\mu mm}+\sigma_{\nu mm}} \bar{A}_{3i}\right) \end{bmatrix} \Sigma_\mu$$

$$\left[\operatorname{diag} \left(\frac{1}{\sqrt{N}} \frac{1}{\sigma_{\mu mm}+\sigma_{\nu mm}}\right) \vdots \operatorname{diag} \frac{1}{\sqrt{NT}} \left(\frac{1}{\sigma_{\mu mm}} \bar{A}'_{1i} + \frac{1}{\sigma_{\mu mm}+\sigma_{\nu mm}} \bar{A}'_{3i}\right) \right]$$

$$(4.E.23) \quad = \begin{bmatrix} \left[\frac{\sigma_{\mu m\ell}}{(\sigma_{\mu mm}+\sigma_{\nu mm})(\sigma_{\mu \ell\ell}+\sigma_{\nu \ell\ell})} \right]_{m,\ell=1,\dots,M} & 0 \\ 0 & 0 \end{bmatrix}$$

using the various properties of A_{ji} and $\bar{A}_{ji}$, j=1,2,3,4 derived in Appendix 3.B.

(ii) Second component :

$$(4.E.24) \quad \sum_t \begin{bmatrix} \operatorname{diag} \left(\frac{\sqrt{N}}{T} \frac{1}{\left(\frac{\hat{\sigma}_{3mm}}{T}\right)} \right) \\ \operatorname{diag} \left(\frac{1}{\sqrt{NT}} \sum_j \frac{1}{\hat{\sigma}_{jmm}} B_{jt} \right) \end{bmatrix} \nu_t$$

The limiting distribution of (4.E.24) is the same as that of

$$(4.E.25) \quad \sum_t \begin{bmatrix} \operatorname{diag} \left(\frac{\sqrt{N}}{T} \frac{1}{\sigma_{\mu mm}+\sigma_{\nu mm}} \right) \\ \operatorname{diag} \left(\frac{1}{\sqrt{NT}} \left(\frac{1}{\sigma_{\mu mm}} \bar{B}_{2t} + \frac{1}{\sigma_{\mu mm}+\sigma_{\nu mm}} \bar{B}_{3t} \right) \right) \end{bmatrix} \nu_t$$

By proceeding as in (i), since the ν_t(s) are i.i.d. $(0,\Sigma_\nu)$, we can apply the Central Limit Theorem and conclude that the limiting distribution of (4.E.24) is normal with zero mean and the following variance-covariance matrix.

$$(4.E.26) \quad \begin{bmatrix} \left[\frac{\sigma_{\nu m\ell}}{(\sigma_{\mu mm}+\sigma_{\nu mm})(\sigma_{\mu \ell\ell}+\sigma_{\nu \ell\ell})} \right]_{m,\ell=1,\dots,M} & 0 \\ 0 & 0 \end{bmatrix}$$

(iii) Third component :

$$(4.E.27)\quad \sum_i \sum_t \begin{bmatrix} \text{diag} \; \dfrac{\sqrt{N}}{NT} \dfrac{1}{\dfrac{\hat{\sigma}_{3mm}}{T}} \\ \text{diag} \; \dfrac{1}{\sqrt{NT}} \sum_j \dfrac{1}{\hat{\sigma}_{jmm}} C_{jit} \end{bmatrix} \varepsilon_{it}$$

Similarly to (i) and (ii), it can be verified that the limiting distribution of (4.E.27) is normal with zero mean and variance-covariance matrix

$$(4.E.28)\quad \begin{bmatrix} 0 & 0 \\ 0 & \left[\dfrac{\sigma_{\varepsilon m\ell}}{\sigma_{\varepsilon mm}\,\sigma_{\varepsilon \ell\ell}} R \right]_{m,\ell=1,..,M} \end{bmatrix}$$

Combining the results of (i), (ii) and (iii) above, we can say that the limiting distribution of $\tilde{D}^{\frac{1}{2}} \tilde{X}' \hat{\Sigma}_*^{-1} u$ is normal with zero mean and variance-covariance matrix equal to the sum of (4.E.23),(4.E.26) and (4.E.28) i.e.

$$\begin{bmatrix} \left[\dfrac{\sigma_{\mu m\ell}+\sigma_{\nu m\ell}}{(\sigma_{\mu mm}+\sigma_{\nu mm})(\sigma_{\mu\ell\ell}+\sigma_{\nu\ell\ell})} \right]_{m,\ell=1,...,M} & 0 \\ 0 & \left[\dfrac{\sigma_{\varepsilon m\ell}}{\sigma_{\varepsilon mm}\sigma_{\varepsilon\ell\ell}} R \right]_{m,\ell=1,...,M} \end{bmatrix}$$

(4.E.29) $= \Lambda_1 \, \Lambda_2 \, \Lambda_1$ using the notation of (4.E.13), page 163 and that of (4.E.18), page 165.

Therefore, the limiting distribution of $\tilde{D}^{-\frac{1}{2}}(\hat{\tilde{\alpha}} - \tilde{\alpha})$ is normal with zero mean and the following covariance matrix

$$[\tilde{P}'\Lambda_1(\Lambda_1\Lambda_2\Lambda_1)^{-1}\Lambda_1\tilde{P}]^{-1}\; \tilde{P}'\Lambda_1(\Lambda_1\Lambda_2\Lambda_1)^{-1}\Lambda_1\Lambda_2\Lambda_1(\Lambda_1\Lambda_2\Lambda_1)^{-1}\Lambda_1\tilde{P}\; [\tilde{P}'\Lambda_1(\Lambda_1\Lambda_2\Lambda_1)^{-1}\Lambda_1\tilde{P}]^{-1}$$

$$= [\tilde{P}'\Lambda_1(\Lambda_1\Lambda_2\Lambda_1)^{-1}\Lambda_1\tilde{P}]^{-1}$$

$$= [\tilde{P}'\Lambda_1\Lambda_1^{-1}\Lambda_2^{-1}\Lambda_1^{-1}\Lambda_1\tilde{P}]^{-1} \quad \text{as } \Lambda_1 \text{ and } \Lambda_2 \text{ are non-singular}$$

$$(4.E.30)\quad = (\tilde{P}'\Lambda_2^{-1}\tilde{P})^{-1}$$

i.e.

$$\left\{\begin{bmatrix} I & 0 \\ 0 & \text{diag}\left(\begin{bmatrix} \Pi'_{*m} \\ H^{*'}_{m} \end{bmatrix} R\right) \end{bmatrix} \begin{bmatrix} [\sigma_{\mu m\ell}+\sigma_{\nu m\ell}]_{m,\ell=1,\dots,M} & 0 \\ 0 & [\sigma_{\varepsilon m\ell}\, R]_{m,\ell=1,\dots,M} \end{bmatrix}^{-1} \begin{bmatrix} I & 0 \\ 0 & \text{diag}\left((\Pi_{*m} H^{*}_{m})\ R\right) \end{bmatrix}\right\}^{-1}$$

replacing notation definitions (4.E.13) and (4.E.18).

$$= \begin{bmatrix} (\Sigma_{\mu}+\Sigma_{\nu})^{-1} & 0 \\ 0 & \bar{\Pi}'(\Sigma_{\varepsilon}^{-1} \otimes R)\, \bar{\Pi} \end{bmatrix}^{-1}$$

(4.E.31) denoting $\bar{\Pi} = \text{diag}\left([\Pi_{*m}\ H^{*}_{m}]\right)$

$$(4.E.32) \quad = \begin{bmatrix} (\Sigma_{\mu}+\Sigma_{\nu}) & 0 \\ 0 & \left[\bar{\Pi}'(\Sigma_{\varepsilon}^{-1} \otimes R)\, \bar{\Pi}\right]^{-1} \end{bmatrix}$$

CHAPTER 5

ESTIMATION OF THE STRUCTURAL FORM - PART 2

5.1 Full Information Maximum Likelihood (FIML) Estimation of the Structural Form

In this section, we will develop the maximum likelihood estimation of the constrained structural form. As in the case of the reduced form, it is extremely difficult to get analytical results since the normal equations can be solved only by numerical methods. To this end, we adapt a procedure suggested by Pollock [37] for the classical simultaneous equation model.

The starting point is the log-likelihood function expressed in terms of the structural parameters. Let us recall that the structural form is written, in matrix notation, as (see (3.11), page 49) :

(5.1) $$Y \Gamma + X B + U = 0$$

Let us rewrite (5.1) in a slightly different manner, by separating the constant term of each equation, from the remaining terms :

(5.2) $$Y \Gamma + \iota_{NT} a' + \underline{X} \, \underline{B} + U = 0$$

where

(5.3) $$X = \left[\iota_{NT} \; \underline{X}\right] \qquad \text{and} \qquad B = \begin{bmatrix} a' \\ \underline{B} \end{bmatrix}$$

Thus we have :

(5.4) $$\iota_{NT} a' + \left[Y \; \underline{X} \right] \begin{bmatrix} \Gamma \\ \underline{B} \end{bmatrix} + U = 0$$

or

(5.5) $$\iota_{NT} a' + \underline{Z} \, \underline{\Theta} + U = 0$$

where

(5.6) $$\underline{Z} = \left[Y \; \underline{X} \right] \qquad \text{and} \qquad \underline{\Theta}' = \left[\Gamma' \; \underline{B}' \right]$$

Note also that :

(5.7) $$\Gamma = L'\underline{\Theta} \qquad \text{with} \qquad L' = \left[I_M \; 0 \right]$$

We will derive one more relation before writing the likelihood function. From (3.36) we have

(5.8) $V = - U \Gamma^{-1}$

Thus :

(5.9) $\text{vec } V = - ((\Gamma^{-1})' \otimes I) \text{ vec } U$

and

$$E((\text{vec } V)(\text{vec } V)') = ((\Gamma^{-1})' \otimes I) \; E(\text{vec } U)(\text{vec } U)'(\Gamma^{-1} \otimes I)$$

$$= ((\Gamma^{-1})' \otimes I) \sum_{i=1}^{4} (\Sigma_i \otimes M_i) \; (\Gamma^{-1} \otimes I)$$

using (3.29)

(5.10) $$= \sum_{i=1}^{4} (\Gamma^{-1})' \; \Sigma_i \; \Gamma^{-1} \otimes M_i$$

But, from (3.61) we have :

(5.11) $$E((\text{vec } V)(\text{vec } V)') = \Omega = \sum_{i=1}^{4} \Omega_i \otimes M_i$$

Hence, by equating (5.10)and (5.11) we get :

(5.12) $\Omega_i = (\Gamma^{-1})' \; \Sigma_i \; \Gamma^{-1} \quad ,i=1,2,3,4$

Now, we can write the log-likelihood function of the structural form by replacing the following in the log-likelihood function of the reduced form (3.87):

(i) $$\log |\Omega_i| = \log |(\Gamma^{-1})' \; \Sigma_i \; \Gamma^{-1}|$$

$$= \log |\Sigma_i| - \frac{1}{2} \log |\Gamma|^2$$

(ii) $$\text{tr}(Y-X\Pi)'M_i(Y-X\Pi) \; \Omega_i^{-1} = \text{tr } V'M_iV \; \Omega_i^{-1}$$

$$= \text{tr}(\Gamma^{-1})'U'M_iU \; \Gamma^{-1} \; \Omega_i^{-1}$$

using (5.8)

$$= \text{tr } U'M_i \; U \; \Gamma^{-1} \; \Omega_i^{-1} \; (\Gamma^{-1})'$$

$$= \text{tr } U'M_i \; U \; \Sigma_i^{-1}$$ using (5.12)

Doing so, we get

$$(5.13)\quad \log L(Y/a,\underline{\Theta},\Sigma_\mu,\Sigma_\nu,\Sigma_\varepsilon) = -\frac{1}{2}\sum_{i=1}^{4} m_i \log|\Sigma_i|$$

$$+\frac{1}{2} NT \log|\Gamma|^2 - \frac{1}{2} \operatorname{tr} \sum_{i=1}^{4} U'M_i\, U\, \Sigma_i^{-1}$$

Now, let us use (5.5) and (5.7) in (5.13) and write :

$$\log L = -\frac{1}{2}\sum_{i=1}^{4} m_i \log|\Sigma_i| + \frac{1}{2} NT \log|L'\underline{\Theta}|^2$$

$$-\frac{1}{2} \operatorname{tr} \sum_{i=1}^{4} (\iota a' + \underline{Z}\,\underline{\Theta})'\, M_i\, (\iota a' + \underline{Z}\,\underline{\Theta})\, \Sigma_i^{-1}$$

Finally, let us expand the trace expression :

$$\sum_{i=1}^{4} (\iota\, a' + \underline{Z}\,\underline{\Theta})'\, M_i\, (\iota\, a' + \underline{Z}\,\underline{\Theta})\, \Sigma_i^{-1}$$

$$= \sum_{i=1}^{4}\Big(a\,\iota' M_i\,\iota\, a'\,\Sigma_i^{-1} + \underline{\Theta}'\underline{Z}' M_i\,\iota\, a'\,\Sigma_i^{-1}$$

$$+ a\,\iota'\, M_i\,\underline{Z}\,\underline{\Theta}\,\Sigma_i^{-1} + \underline{\Theta}'\underline{Z}'\, M_i\,\underline{Z}\,\underline{\Theta}\,\Sigma_i^{-1}\Big)$$

$$= NT\, a\, a'\,\Sigma_3^{-1} + \underline{\Theta}'\underline{Z}'\,\iota\, a'\,\Sigma_3^{-1} + a\,\iota'\underline{Z}\,\underline{\Theta}\,\Sigma_3^{-1}$$

$$+ \sum_{i=1}^{4} \underline{\Theta}'\underline{Z}'\, M_i\,\underline{Z}\,\underline{\Theta}\,\Sigma_i^{-1}$$

using (3.28) and $\iota'\iota = NT$

Thus, we obtain

$$(5.14)\quad \log L = -\frac{1}{2}\sum_{i=1}^{4} m_i \log|\Sigma_i| + \frac{1}{2} NT \log|L'\underline{\Theta}|^2$$

$$-\frac{1}{2} \operatorname{tr}(NT\, a\, a' + \underline{\Theta}'\underline{Z}'\,\iota\, a' + a\,\iota'\underline{Z}\,\underline{\Theta})\,\Sigma_3^{-1}$$

$$-\frac{1}{2} \operatorname{tr} \sum_{i=1}^{4} \underline{\Theta}'\,\underline{Z}'\, M_i\,\underline{Z}\,\underline{\Theta}\,\Sigma_i^{-1}$$

We are interested in maximising this function with respect to a, $\underline{\Theta}$, Σ_μ, Σ_ν and Σ_ε subject to two types of restrictions :

(i) a priori restrictions on the coefficients. These must include the normalisation rule for each equation (typically $\gamma^*_{ii} = -1$), the exclusion restrictions (zero a priori coefficients) and eventually other linear restrictions on the parameters. They can be expressed conveniently by

$$(5.15)\quad \begin{cases} S_o\, a = s_o \\ \underline{S}\ \text{vec}\ \underline{\Theta} = \underline{s} \end{cases}$$

where S_o is a known p_o x M matrix of full row rank, $\underline{S}$ is a known $\underline{p}$ x M(M+K-1) matrix of full row rank, s_o is a known p_o x 1 vector of constants and $\underline{s}$ is a $\underline{p}$ x 1 vector of known constants. Let $p_o + \underline{p} = p$.

(ii) symmetry conditions:

$$(5.16)\quad C\ \text{vec}\ \Sigma_j = 0 \qquad j=\mu,\nu,\varepsilon$$

where C is the matrix defined in (3.89).

Again, it can be verified that the first order conditions for maximisation automatically satisfy the symmetry conditions. These can therefore be neglected for this purpose, but they will be important for the computation of the information matrix.

We therefore write the following Lagrangian function :

$$(5.17)\quad L^* = \log L - \lambda_o' (s_o - S_o a) - \underline{\lambda}' (\underline{s} - \underline{S}\ \text{vec}\ \underline{\Theta})$$

where λ_o and $\underline{\lambda}$ are vectors of Lagrangian multipliers.

The first-order condition of maximisation is given by $dL^*=0$ for all $da \neq 0$, $d\ \text{vec}\,\underline{\Theta} \neq 0$, $d\lambda_o \neq 0$, $d\,\underline{\lambda} \neq 0$ and $d\ \text{vec}\Sigma_j \neq 0, j=\mu,\nu,\varepsilon$. Let us therefore write down the first-order differential of L* :

$$(5.18)\quad dL^* = d\log L - d\lambda_o'(s_o - S_o a) - \lambda_o' S_o da - d\underline{\lambda}'(\underline{s} - \underline{S}\ \text{vec}\ \underline{\Theta}) - \underline{\lambda}'\ \underline{S}\ d\ \text{vec}\ \underline{\Theta}$$

Now, differentiating (5.14) by applying the rules (3.92), we get :

$$(5.19)\quad d\log L = -\frac{1}{2}\sum_i m_i \operatorname{tr} \Sigma_i^{-1} d\Sigma_i + NT \operatorname{tr}(L'\underline{\Theta})^{-1}L'd\underline{\Theta}$$

$$-\frac{1}{2}\operatorname{tr}(NT(da)a' + NT\, a\, da' + (d\underline{\Theta}')\underline{Z}'\iota a' + \underline{\Theta}'\underline{Z}'\iota da'$$

$$+ da\, \iota'\underline{Z}\,\underline{\Theta} + a\,\iota'\underline{Z}\, d\underline{\Theta})\,\Sigma_3^{-1} + \frac{1}{2}\operatorname{tr}(NT\, aa' + \underline{\Theta}'\underline{Z}'\iota a' +$$

$$a\,\iota'\underline{Z}\,\underline{\Theta})\,\Sigma_3^{-1} d\Sigma_3\, \Sigma_3^{-1} + \frac{1}{2}\operatorname{tr}\sum_i \underline{\Theta}'\underline{Z}' M_i \underline{Z}\,\underline{\Theta}\,\Sigma_i^{-1} d\Sigma_i\, \Sigma_i^{-1}$$

$$-\frac{1}{2}\operatorname{tr}\sum_i (d\underline{\Theta}')\underline{Z}'M_i\underline{Z}\,\underline{\Theta}\,\Sigma_i^{-1} - \frac{1}{2}\operatorname{tr}\sum_i \underline{\Theta}'\underline{Z}'M_i\underline{Z}\, d\underline{\Theta}\,\Sigma_i^{-1}$$

Substituting (5.19) in (5.18) and using the following relationships :

$$(5.20)\quad \begin{cases} d\Sigma_1 = d\Sigma_\varepsilon + T\, d\Sigma_\mu \\ d\Sigma_2 = d\Sigma_\varepsilon + N\, d\Sigma_\nu \\ d\Sigma_3 = d\Sigma_\varepsilon + N\, d\Sigma_\mu + T\, d\Sigma_\nu \\ d\Sigma_4 = d\Sigma_\varepsilon \end{cases}$$

$$(5.21)\quad (L'\underline{\Theta})^{-1} = \Gamma^{-1} = \Sigma_4^{-1}\Gamma'\,\Omega_4 = \Sigma_4^{-1}\underline{\Theta}'L\,\Omega_4$$

we obtain :

$$(5.22)\quad dL^* = -\frac{1}{2}\sum_i m_i \operatorname{tr}\Sigma_i^{-1} d\Sigma_\varepsilon - \frac{T}{2}\sum_{1,3} m_i \operatorname{tr}\Sigma_i^{-1} d\Sigma_\mu - \frac{N}{2}\sum_{2,3} m_i \operatorname{tr}\Sigma_i^{-1} d\Sigma_\nu$$

$$+ NT \operatorname{tr}\Sigma_4^{-1}\underline{\Theta}'L\,\Omega_4\, L'd\underline{\Theta} - \frac{1}{2}\operatorname{tr}\Big(NT\,(da)\,a' + NT\, a\, da' + (d\underline{\Theta}')\underline{Z}'\iota a'$$

$$+ \underline{\Theta}'\underline{Z}'\iota\, da' + da\,\iota'\underline{Z}\,\underline{\Theta} + a\,\iota'\underline{Z}\, d\underline{\Theta}\Big)\,\Sigma_3^{-1} + \frac{1}{2}\operatorname{tr}\Sigma_3^{-1}$$

$$(NT\, a\, a' + \underline{\Theta}'\underline{Z}'\iota\, a' + a\,\iota'\underline{Z}\,\underline{\Theta})\,\Sigma_3^{-1}(d\Sigma_\varepsilon + N\, d\Sigma_\nu + T\, d\Sigma_\mu)$$

$$+ \frac{1}{2}\operatorname{tr}\sum_i \Sigma_i^{-1}\underline{\Theta}'\underline{Z}' M_i \underline{Z}\,\underline{\Theta}\,\Sigma_i^{-1} d\Sigma_\varepsilon + \frac{T}{2}\operatorname{tr}\sum_{1,3}\Sigma_i^{-1}\underline{\Theta}'\underline{Z}'M_i\underline{Z}\,\underline{\Theta}\,\Sigma_i^{-1} d\Sigma_\mu$$

$$+ \frac{N}{2}\operatorname{tr}\sum_{2,3}\Sigma_i^{-1}\underline{\Theta}'\underline{Z}'M_i\underline{Z}\,\underline{\Theta}\,\Sigma_i^{-1} d\Sigma_\nu - \frac{1}{2}\operatorname{tr}\sum_i (d\underline{\Theta})'\underline{Z}'M_i\underline{Z}\,\underline{\Theta}\,\Sigma_i^{-1}$$

$$- \frac{1}{2}\operatorname{tr}\sum_i \underline{\Theta}'\underline{Z}'M_i\underline{Z}\, d\underline{\Theta}\,\Sigma_i^{-1} - d\lambda_o'(s_o - S_o a) - \lambda_o' S_o'\, da$$

$$- d\underline{\lambda}'(\underline{s} - \underline{S}\operatorname{vec}\underline{\Theta}) - \underline{\lambda}'\underline{S}\, d\operatorname{vec}\underline{\Theta}$$

Next, using the following well-known vec-trace relationships :

(5.23) $(\text{vec } A)'(B \otimes C) \text{ vec } D = \text{tr } D\, B'A'C$

(5.24) $(\text{vec } A)'(\text{vec } B) = \text{tr } A'B$

and rearranging (5.22), we obtain :

(5.25)
$$dL^* = (-\,NT\, a'\Sigma_3^{-1} - \iota'\underline{Z}\,\underline{\Theta}\,\Sigma_3^{-1} - \lambda_o'\,S_o)\,da +$$
$$\Big[(\text{vec }\underline{\Theta})'(\Sigma_4^{-1} \otimes NT\,L\,\Omega_4\,L') + a'(\Sigma_3^{-1} \otimes \underline{Z}'\iota) - (\text{vec }\underline{\Theta})'$$
$$\sum_i (\Sigma_i^{-1} \otimes \underline{Z}'M_i\underline{Z}) + \underline{\lambda}'\underline{S}\Big]\ d\text{ vec }\underline{\Theta} +$$
$$\Big\{\text{vec}\Big[-\tfrac{1}{2}\sum_i m_i\,\Sigma_i^{-1} + \tfrac{1}{2}\Sigma_3^{-1}(NT\,aa' - \underline{\Theta}'\underline{Z}'\iota\,a' - a\iota'\underline{Z}\,\underline{\Theta})\,\Sigma_3^{-1} +$$
$$\tfrac{1}{2}\sum_i \Sigma_i^{-1}\,\underline{\Theta}'\underline{Z}'M_i\,\underline{Z}\,\underline{\Theta}\,\Sigma_i^{-1}\Big]\Big\}'d\text{ vec }\Sigma_\varepsilon +$$
$$\Big\{\text{vec}\Big[-\tfrac{T}{2}\sum_{1,3} m_i\,\Sigma_i^{-1} + \tfrac{T}{2}\Sigma_3^{-1}(NT\,aa' - \underline{\Theta}'\underline{Z}'\iota\,a' - a\iota'\underline{Z}\,\underline{\Theta})\,\Sigma_3^{-1} +$$
$$\tfrac{T}{2}\sum_{1,3} \Sigma_i^{-1}\,\underline{\Theta}'\underline{Z}'M_i\,\underline{Z}\,\underline{\Theta}\,\Sigma_i^{-1}\Big]\Big\}'d\text{ vec }\Sigma_\mu +$$
$$\Big\{\text{vec}\Big[-\tfrac{N}{2}\sum_{2,3} m_i\,\Sigma_i^{-1} + \tfrac{N}{2}\Sigma_3^{-1}(NT\,aa' - \underline{\Theta}'\underline{Z}'\iota\,a' - a\iota'\underline{Z}\,\underline{\Theta})\,\Sigma_3^{-1} +$$
$$+\tfrac{N}{2}\sum_{2,3} \Sigma_i^{-1}\,\underline{\Theta}'\underline{Z}'\,M_i\,\underline{Z}\,\underline{\Theta}\,\Sigma_i^{-1}\Big]\Big\}'d\text{ vec }\Sigma_\nu - d\lambda_o'(s_o - S_o a) - d\,\underline{\lambda}'(\underline{s} - \underline{S}\text{ vec }\underline{\Theta}$$

noting that

$$\text{tr }(da)a'\Sigma_3^{-1} = \text{tr } a\,da'\,\Sigma_3^{-1} = a'\,\Sigma_3^{-1}\,da$$

$$\text{tr } da\,\iota'\underline{Z}\,\underline{\Theta}\,\Sigma_3^{-1} = \text{tr }\underline{\Theta}'\underline{Z}'\iota\,da'\,\Sigma_3^{-1} = \iota'\underline{Z}\,\underline{\Theta}\,\Sigma_3^{-1}\,da$$

$$\text{tr }(d\underline{\Theta})'\underline{Z}'\iota\,a'\,\Sigma_3^{-1} = \text{tr } a\,\iota'\,\underline{Z}\,d\,\underline{\Theta}\,\Sigma_3^{-1} = a'(\Sigma_3^{-1} \otimes \underline{Z}'\iota)\,d\text{ vec }\underline{\Theta}$$

$$\text{tr }(d\underline{\Theta})'\underline{Z}'M_i\,\underline{Z}\,\underline{\Theta}\,\Sigma_i^{-1} = \text{tr }\underline{\Theta}'\underline{Z}'M_i\,\underline{Z}\,d\,\underline{\Theta}\,\Sigma_i^{-1} = (\text{vec }\underline{\Theta})'(\Sigma_i^{-1} \otimes \underline{Z}'M_i\underline{Z})\,d\text{ vec }\underline{\Theta}$$

Finally, denoting

(5.26)
$$W^* = \Sigma_4^{-1} \otimes NT\,L\,\Omega_4\,L' - \sum_{i=1}^{4}(\Sigma_i^{-1} \otimes \underline{Z}'M_i\underline{Z})$$

and

(5.27)
$$\tilde{W} = -\,m_3\Sigma_3^{-1} + \Sigma_3^{-1}(NT\,aa' - \underline{\Theta}'\underline{Z}'\iota\,a' - a\iota'\,\underline{Z}\,\underline{\Theta})\,\Sigma_3^{-1}$$
$$+\,\Sigma_3^{-1}\,\underline{\Theta}'\underline{Z}'M_3\,\underline{Z}\,\underline{\Theta}\,\Sigma_3^{-1}$$

we can set the first-order conditions as :

$$(5.28) \quad - NT\, \Sigma_3^{-1} a - (\Sigma_3^{-1} \otimes \underline{Z}'\iota)\, \text{vec}\, \underline{\Theta} - S_o' \lambda_o = 0$$

$$(5.29) \quad - (\Sigma_3^{-1} \otimes \underline{Z}'\iota)\, a + W^*\, \text{vec}\, \underline{\Theta} - \underline{S}'\underline{\lambda} = 0$$

$$(5.30) \quad - S_o\, a = - s_o$$

$$(5.31) \quad - \underline{S}\, \text{vec}\, \underline{\Theta} = - \underline{s}$$

$$(5.32) \quad - \frac{1}{2} \sum_{i=1,2,4} m_i \Sigma_i^{-1} + \frac{1}{2}\tilde{W} + \frac{1}{2} \sum_{i=1,2,4} \Sigma_i^{-1} \underline{\Theta}'\underline{Z}'M_i \underline{Z}\, \underline{\Theta}\, \Sigma_i^{-1} = 0$$

$$(5.33) \quad - \frac{T}{2} m_1\, \Sigma_1^{-1} + \frac{T}{2}\, \tilde{W} + \frac{T}{2}\, \Sigma_1^{-1}\, \underline{\Theta}'\underline{Z}'M_1 \underline{Z}\, \underline{\Theta}\, \Sigma_1^{-1} = 0$$

$$(5.34) \quad - \frac{N}{2} m_2\, \Sigma_2^{-1} + \frac{N}{2}\, \tilde{W} + \frac{N}{2}\, \Sigma_2^{-1}\, \underline{\Theta}'\underline{Z}'M_2 \underline{Z}\, \underline{\Theta}\, \Sigma_2^{-1} = 0$$

The last three equations (5.30), (5.31) and (5.32) can be simplified to the following :

$$(5.35) \quad \Sigma_4 = \frac{1}{m_4}\, \underline{\Theta}'\underline{Z}'M_4\, \underline{Z}\, \underline{\Theta} - \frac{1}{m_4}\, \Sigma_4\, \tilde{W}\, \Sigma_4$$

$$(5.36) \quad \Sigma_1 = \frac{1}{m_1}\, \underline{\Theta}'\underline{Z}'M_1\, \underline{Z}\, \underline{\Theta} + \frac{1}{m_1}\, \Sigma_1\, \tilde{W}\, \Sigma_1$$

$$(5.37) \quad \Sigma_2 = \frac{1}{m_2}\, \underline{\Theta}'\underline{Z}'M_2\, \underline{Z}\, \underline{\Theta} + \frac{1}{m_2}\, \Sigma_2\, \tilde{W}\, \Sigma_2$$

The maximum likelihood estimates are obtained by solving simultaneously (5.28), (5.29), (5.30), (5.31), (5.35), (5.36), (5.37) along with the two definitions :

$$(5.38) \quad \Sigma_3 = \Sigma_1 + \Sigma_2 - \Sigma_4$$

$$(5.39) \quad \Omega_4 = (\Gamma')^{-1}\, \Sigma_4\, \Gamma^{-1}$$

This system of equations is highly non linear. We notice, however, that an explicit solution can be found for a and vec $\underline{\Theta}$ in terms of the different covariance matrices. Combining (5.28), (5.29), (5.30) and (5.31) yields :

(5.40)
$$\begin{bmatrix} -NT\ \Sigma_3^{-1} & -\Sigma_3^{-1} \otimes \underline{Z}'\iota & -S_o' & 0 \\ -\Sigma_3^{-1} \otimes \iota'\underline{Z} & W^* & 0 & -\underline{S}' \\ -S_o & 0 & 0 & 0 \\ 0 & -\underline{S} & 0 & 0 \end{bmatrix} \begin{bmatrix} a \\ \text{vec}\underline{\Theta} \\ \lambda_o \\ \underline{\lambda} \end{bmatrix} = \begin{bmatrix} 0 \\ 0 \\ -s_o \\ -\underline{s} \end{bmatrix}$$

or

(5.41)
$$\begin{bmatrix} W & S' \\ S & 0 \end{bmatrix} \begin{bmatrix} \delta \\ -\lambda \end{bmatrix} = \begin{bmatrix} 0 \\ s \end{bmatrix}$$

denoting

(5.42)
$$\begin{bmatrix} -NT \otimes \Sigma_3^{-1} & -\Sigma_3^{-1} \otimes \underline{Z}'\iota \\ -\Sigma_3^{-1} \otimes \iota'\underline{Z} & W^* \end{bmatrix} = W$$

(5.43)
$$\begin{bmatrix} S_o & 0 \\ 0 & \underline{S} \end{bmatrix} = S \quad ; \quad \begin{bmatrix} s_o \\ \underline{s} \end{bmatrix} = s$$

(5.44)
$$\begin{bmatrix} a \\ \text{vec}\ \underline{\Theta} \end{bmatrix} = \delta \quad ; \quad \begin{bmatrix} \lambda_o \\ \underline{\lambda} \end{bmatrix} = \lambda$$

The first matrix on the left handside of (5.41) is non-singular iff

(i) rank $(S') = p$, which is true by hypothesis, and

(ii) rank $(I-S'(SS')^{-1}S)W(I-S'(SS')^{-1}S) = M(M+K) - p$,

which is satisfied whenever the conditions for identification are met. Its inverse (see [4]) is given by

$$\begin{bmatrix} H_1 & H_2 \\ H_2' & H_3 \end{bmatrix}$$

where

$$H_1 = F(F'WF)^{-1}F'$$

$$H_2 = - F(F'WF)^{-1}F'WS'(SS')^{-1} + S'(SS')^{-1}$$

$$H_3 = - (SS')^{-1}SWS'(SS')^{-1} + (SS')^{-1}SWF(F'WF)^{-1}F'WS'(SS')^{-1}$$

and F is a $M(K + M) \times (M^2 + MK - p)$ matrix of orthonormal vectors such that $FF' = I - S'(SS')^{-1}S$. We therefore obtain the following solution :

$$(5.45) \qquad \delta = H_2 s = - F(F'WF)^{-1}F'WS'(SS')^{-1}r + S'(SS')^{-1}s$$

It is very useful to note, from an operational point of view, that whenever only the usual restrictions are considered (normalisation and exclusion), the matrix $\underline{S}$ can be partitioned as

$$\underline{S} = \begin{bmatrix} S_1 & & & 0 \\ & S_2 & & \\ & & \ddots & \\ 0 & & & S_M \end{bmatrix}$$

and consequently S can be written as :

$$(5.46) \qquad S = \begin{bmatrix} S_o & & & & 0 \\ & S_1 & & & \\ & & \ddots & & \\ 0 & & & & S_M \end{bmatrix}$$

where the rows of S_o are elementary vectors as also those of each S_m, of dimension $p_m x(M+K-1)$ for $m=1,\ldots,M$ with $\sum_{m=1}^{M} p_m = \underline{p}$.

Hence, we have $SS'=I_p$. Also, in this case, $s_o=0$ and the subvector s_m of s, corresponding to the block S_m of $\underline{S}$, is an elementary vector (with a minus sign in front); we have

(5.47) $S_m' \, s_m = s_m \qquad m=0,1,\ldots,M$

Moreover, the matrix F' can also be partitioned in a block diagonal form :

$$(5.48) \qquad F' = \begin{bmatrix} F_o' & & & & 0 \\ & F_1' & & & \\ & & \cdot & & \\ & & & \cdot & \\ & 0 & & & F_M' \end{bmatrix}$$

where the M - p_o rows of F'_o and the M+K-1-p_m rows of F'_m, m=1,...,M are just the elementary vectors which are complementary (orthogonal) to those appearing in S_o and S_m,m=1,...,M respectively. The matrix F is therefore computed without any difficulty. In this case, the solution for δ simplifies to :

(5.49) $\delta = -F(F'WF)^{-1}F'Ws + s$

and the non-constrained coefficients are obtained by premultiplication by F' , (F'F=I,F's=0) , i.e.

(5.50) $F'\delta = -(F'WF)^{-1}F'W\,s$

We therefore suggest the following iterative procedure for the solution of the normal equations :

<u>Step 1</u> Initial conditions :

$\Sigma_i^{-1} = 0$, i=1,2,3 (their limits)

$\Sigma_4 = I$

$\Omega_4 = \frac{1}{NT}(Y - \underline{X}\hat{\Pi}_*)'Q(Y - \underline{X}\hat{\Pi}_*)$ where $\hat{\Pi}_*$ is a consistent estimation of $\Pi_* = [\pi_{*1} \ldots \pi_{*M}]$

Step 2 Use (5.45) or, in case of usual restrictions use (5.49), to estimate a and vec $\underline{\Theta}$.

Step 3 Compute Σ_i, i=1,2,4 , from (5.35), (5.36) and (5.37) using on the right hand side the current estimate for a and $\underline{\Theta}$ and the old ones (of the previous iteration) for the different Σ_i. Compute Σ_3 from (5.38) and Ω_4 from (5.39).

Step 4 Go back to Step 2 until convergence is reached.

Note that in the first iteration, one has to be careful while using (5.45) or (5.49) as, for the given initial conditions, the matrix W becomes

$$W^{(o)} = \begin{bmatrix} 0 & 0 \\ & \\ 0 & W^{*(o)} \end{bmatrix}$$

where $W^{*(o)}$ is obtained by replacing the initial conditions in (5.26); and the system (5.41) does not have a unique solution for a . Hence, instead of the full system (5.41), the following subsystem concerning vec $\underline{\Theta}$ only, should be solved for getting the first estimate for vec $\underline{\Theta}$, denoted as vec $\underline{\Theta}^{(1)}$:

$$(5.51) \qquad \begin{bmatrix} W^{*(o)} & \underline{S}' \\ & \\ \underline{S} & 0 \end{bmatrix} \begin{bmatrix} \text{vec } \underline{\Theta} \\ \\ -\underline{\lambda} \end{bmatrix} = \begin{bmatrix} 0 \\ \\ \underline{s} \end{bmatrix}$$

and the estimate for a should be derived as :

$$a_m^{(1)} = \frac{1}{NT}\, \iota'(y_m - Z_m^*\, \underline{\Theta}_m^{*(1)}) \;, \; m=1,\ldots,M$$

where $\underline{\Theta}^*_m$ is the subvector of $\underline{\Theta}_m$ (the m-th column of $\underline{\Theta}$) containing the non-zero elements only.

Also, note that in the case of usual restrictions and with the initial conditions stated in Step 1, the matrix W* becomes

$$W^{*(o)} = I \otimes NT\, L\, \Omega_4\, L' - I \otimes \underline{Z}'M_i\underline{Z}$$

$$= I \otimes [\, L(Y-\underline{X}\hat{\Pi}_*)'Q(Y-\underline{X}\hat{\Pi}_*)L' - \underline{Z}'Q\underline{Z}\,]$$

$$= I \otimes [\, L(Y-\underline{X}(\underline{X}'Q\underline{X})^{-1}\underline{X}'QY)'Q(Y-\underline{X}(\underline{X}'Q\underline{X})^{-1}\underline{X}'QY)L'-\underline{Z}'Q\underline{Z}\,]$$

$$= I \otimes [\, LY'(I-Q\underline{X}(\underline{X}'Q\underline{X})^{-1}\underline{X}')\; Q(I-\underline{X}(\underline{X}'Q\underline{X})^{-1}\underline{X}'Q)YL'-\underline{Z}'Q\underline{Z}\,]$$

$$= I \otimes [\, L(Y'QY-Y'Q\underline{X}(\underline{X}'Q\underline{X})^{-1}\underline{X}'QY)L'-\underline{Z}'Q\underline{Z}\,]$$

$$= I \otimes \left\{ \begin{bmatrix} Y'QY-Y'Q\underline{X}(\underline{X}'Q\underline{X})^{-1}\,\underline{X}'QY & 0 \\ 0 & 0 \end{bmatrix} - \begin{bmatrix} Y'QY & Y'Q\underline{X} \\ \underline{X}'QY & \underline{X}'Q\underline{X} \end{bmatrix} \right\}$$

$$= I \otimes - G$$

where

$$(5.52) \qquad G = \underline{Z}'Q\underline{X}(\underline{X}'Q\underline{X})^{-1}\,\underline{X}'Q\underline{Z}$$

In view of the block diagonal form of F' we can obtain directly, for Step 2, omitting F_o in F and substituting $I \otimes -G$ for W in (5.49) :

$$(5.53) \qquad \text{vec}\ \underline{\Theta}^{(1)} = - F\left[\text{diag} - (F_m'GF_m)^{-1}\right]\left[\text{diag} - (F_m'G)\right]\underline{s} + \underline{s}$$

which, for the coefficients of the m-th equation, gives

$$\underline{\Theta}_m^{(1)} = - F_m(F_m'\,G\,F_m)^{-1}F_m'\,G\,s_m + s_m$$

Now, $F_m'\,\underline{Z} = Z_m^*$ where Z_m^* contains all the explanatory variables of the m-th equation (both endogenous and exogenous) excluding the constant term and $\underline{Z}\,s_m = -\,y_m$, the explained variable. Therefore, we get :

$$(5.54) \qquad \underline{\Theta}_m^{(1)} = F_m\left[\underline{Z}_m^{*\prime}Q\underline{X}(\underline{X}'Q\underline{X})^{-1}\underline{X}'Q\underline{Z}_m^*\right]^{-1}\underline{Z}_m^{*\prime}Q\underline{X}(\underline{X}'Q\underline{X})^{-1}\underline{X}'Qy_m+s_m$$

which is seen to be identical (for the non-constrained coefficients) to the 2SLS covariance estimator (Method 2), see formula (4.31), page 126. Hence, the first iteration gives the 2SLS covariance estimator.

At the second iteration, let us suppose that we keep the initial values for Σ_1^{-1}, Σ_2^{-1}, Σ_3^{-1} and Ω_4 and let us compute the new Σ_4 according to (5.35), calling it $\hat{\Sigma}_4$. The new W* becomes

$$W^{*(1)} = \hat{\Sigma}_4^{-1} \otimes - G \tag{5.55}$$

where G is defined in (5.52). Once again, as the (initial) value $\Sigma_3^{-1} = 0$ leads to a singular $W^{(1)}$, we will solve the sub-system given in (5.51) with $W^{*(o)}$ replaced by $W^{*(1)}$, to obtain vec $\underline{\Theta}^{(2)}$. Thus, we get for the non-constrained coefficients of $\underline{\Theta}$:

$$F'\text{vec }\underline{\Theta}^{(2)} = - \left\{ \begin{bmatrix} F_1' & & 0 \\ & \ddots & \\ 0 & & F_M' \end{bmatrix} (\hat{\Sigma}_4^{-1} \otimes - G) \begin{bmatrix} F_1 & & 0 \\ & \ddots & \\ 0 & & F_M \end{bmatrix} \right\}^{-1} \begin{bmatrix} F_1' & & 0 \\ & \ddots & \\ 0 & & F_M' \end{bmatrix} (\hat{\Sigma}_4^{-1} \otimes - G)\, \underline{s}$$

$$= - \begin{bmatrix} \hat{\sigma}_4^{11} F_1'GF_1 & \cdots & \hat{\sigma}_4^{1M} F_1'GF_M \\ \vdots & & \vdots \\ \hat{\sigma}_4^{M1} F_M'GF_1 & \cdots & \hat{\sigma}_4^{MM} F_M'GF_M \end{bmatrix}^{-1} \begin{bmatrix} \hat{\sigma}_4^{11} F_1'G & \cdots & \hat{\sigma}_4^{1M} F_1'G \\ \vdots & & \vdots \\ \hat{\sigma}_4^{M1} F_M'G & \cdots & \hat{\sigma}_4^{MM} F_M'G \end{bmatrix} \underline{s}$$

where $\hat{\Sigma}_4^{-1} = [\, \hat{\sigma}_4^{mm'}]$, m,m'=1,...,M

$$= - \begin{bmatrix} \left[\hat{\sigma}_4^{mm'} F_m' \, G \, F_{m'}\right]_{\substack{m=1,\ldots,M \\ m'=1,\ldots,M}} \end{bmatrix}^{-1} \begin{bmatrix} \left[\sum_{m'} \hat{\sigma}_4^{mm'} F_m' \, G \, s_{m'}\right]_{m=1,\ldots,M} \end{bmatrix}$$

Now,

$$F_m' \, G \, F_{m'}' = F_m' \, \underline{Z}'Q \, \underline{X} \, (\underline{X}'Q \, \underline{X})^{-1} \, \underline{X}'Q \, \underline{Z} \, F_{m'}'$$

$$= Z_m^{*\prime}Q \, \underline{X}(\underline{X}'Q \, \underline{X})^{-1} \, \underline{X}'Q \, Z_m^*$$

and

$$- F_m' \, G \, s_{m'} = - F_m' \, \underline{Z}'Q \, \underline{X}(\underline{X}'Q \, \underline{X})^{-1} \, \underline{X}'Q \, \underline{Z} \, s_{m'}$$

$$= Z_m^{*\prime} \, Q \, \underline{X}(\underline{X}'Q \, \underline{X})^{-1}\underline{X}'Q \, y_{m'}$$

Therefore,

$$(5.56)\quad F'\text{vec}\,\underline{\Theta}^{(2)} = \left[\left[\hat{\sigma}_4^{mm'}\, Z_m^{*\prime} Q\underline{X}(X'Q\underline{X})^{-1}\underline{X}'QZ_m^{*}\right]_{\substack{m=1,\dots,M\\ m'=1,\dots,M}}\right]^{-1} \left[\left[\sum_{m'} \hat{\sigma}_4^{mm'}\, Z_m^{*\prime} Q\underline{X}(\underline{X}'Q\underline{X})^{-1}\underline{X}'Q\, y_{m'}\right]_{m=1,\dots,M}\right]$$

The above expression for vec $\underline{\Theta}$ can be shown to be that of a <u>3SLS covariance</u> estimator of vec $\underline{\Theta}$. This is done as follows :

Let us recall that our "generalised" 3SLS estimator was defined as the GLS estimator of the following system of transformed structural equations :

$$X'\,\Sigma_{mm}^{-1}\, y_m = X'\,\Sigma_{mm}^{-1}\, Z_m\, \alpha_m + X'\Sigma_{mm}^{-1}\, u_m \;,\; m=1,\dots,M$$

Now, instead of premultiplying each structural equation by the corresponding $X'\,\Sigma_{mm}^{-1}$, suppose we premultiply it by X'Q . Then, we get the following system :

$$X'Q\, y_m = X'Q\, Z_m\, \alpha_m + X'Q\, u_m \;,\; m=1,\dots,M$$

which can be simplified as

$$(5.57)\quad \underline{X}'Q\, y_m = \underline{X}'Q\, Z_m^*\, \alpha_m^* + \underline{X}'Q\, u_m \;,\; m=1,\dots,M$$

as

$$X'Q = \begin{bmatrix} \iota' \\ \underline{X}' \end{bmatrix} Q = \begin{bmatrix} \iota'Q \\ \underline{X}'Q \end{bmatrix} = \begin{bmatrix} 0 \\ \underline{X}'Q \end{bmatrix}$$

and

$$Q\, Z_m = Q\, \begin{bmatrix} \iota & Z_m^* \end{bmatrix} = \begin{bmatrix} 0 & Q\, Z_m^* \end{bmatrix}$$

We can write the system (5.57) compactly as :

$$(5.58)\quad (I\otimes\underline{X})'(I\otimes Q)\, y = (I\otimes\underline{X}')(I\otimes Q)Z^*\,\alpha^* + (I\otimes\underline{X}')(I\otimes Q)u$$

where

$$Z^* = \begin{bmatrix} Z_1^* & & \\ & \ddots & \\ & & Z_M^* \end{bmatrix} \quad\text{and}\quad \alpha^* = \begin{bmatrix} \alpha_1^* \\ \vdots \\ \alpha_M^* \end{bmatrix}$$

and $\quad E((I\otimes\underline{X}'Q)uu'(I\otimes Q\underline{X})) = (I\otimes\underline{X}'Q)\Sigma\,(I\otimes Q\underline{X}) = (\Sigma_4\otimes\underline{X}'Q\underline{X})$

Applying (feasible) GLS on (5.58), we get

$$\hat{\alpha}^*_{cov} = \left[Z^{*\prime}(I \otimes Q\underline{X})(\hat{\Sigma}_4^{-1} \otimes (\underline{X}'Q\underline{X})^{-1})(I \otimes \underline{X}'Q)Z^*\right]^{-1} Z^{*\prime}(I \otimes Q\underline{X})(\hat{\Sigma}_4^{-1} \otimes (\underline{X}'Q\underline{X})^{-1})(I \otimes \underline{X}'Q)y$$

$$= \left\{\begin{bmatrix} Z_1^{*\prime}Q\underline{X} & & 0 \\ & \ddots & \\ 0 & & Z_M^{*\prime}Q\underline{X} \end{bmatrix} \begin{bmatrix} \left[\hat{\sigma}_4^{mm'}(\underline{X}'Q\underline{X})^{-1}\right]_{\substack{m=1,\dots,M \\ m'=1,\dots,M}} \end{bmatrix} \begin{bmatrix} \underline{X}'QZ_1^* & & 0 \\ & \ddots & \\ 0 & & \underline{X}'QZ_M^* \end{bmatrix}\right\}^{-1}$$

$$\begin{bmatrix} Z_1^{*\prime}Q\underline{X} & & 0 \\ & \ddots & \\ 0 & & Z_M^{*\prime}Q\underline{X} \end{bmatrix} \begin{bmatrix} \left[\hat{\sigma}_4^{mm'}(\underline{X}'Q\underline{X})^{-1}\right]_{\substack{m=1,\dots,M \\ m'=1,\dots,M}} \end{bmatrix} \begin{bmatrix} \underline{X}'Qy_1 \\ \vdots \\ \underline{X}'Qy_M \end{bmatrix}$$

$$(5.59) \quad = \begin{bmatrix} \left[\hat{\sigma}_4^{mm'} \; Z_m^{*\prime}Q\underline{X}(\underline{X}'Q\underline{X})^{-1}\underline{X}'QZ_{m'}^*\right]_{\substack{m=1,\dots,M \\ m'=1,\dots,M}} \end{bmatrix}^{-1} \begin{bmatrix} \left[\sum_{m'} \hat{\sigma}_4^{mm'} \; Z_m^*Q\underline{X}(\underline{X}'Q\underline{X})^{-1} \underline{X}'Q\, y_{m'}\right]_{m=1,\dots,M} \end{bmatrix}$$

which is identical to (5.56), noting that F' vec $\underline{\Theta}$ and α^* are the same coefficient vectors. Thus, under the conditions stated above, the second iteration of FIML procedure gives the 3SLS covariance estimator.

Now, we derive the limiting distributions of the FIML estimators in the following form :

$$(5.60) \quad \begin{cases} \sqrt{N}\,(\hat{a}_{ML} - a) \\ \sqrt{NT}\,\mathrm{vec}(\hat{\underline{\Theta}}_{ML} - \underline{\Theta}) \\ \sqrt{NT}\,\mathrm{vec}(\hat{\Sigma}_{\varepsilon,ML} - \Sigma_\varepsilon) \\ \sqrt{N}\,\mathrm{vec}(\hat{\Sigma}_{\mu,ML} - \Sigma_\mu) \\ \sqrt{T}\,\mathrm{vec}(\hat{\Sigma}_{\nu,ML} - \Sigma_\nu) \end{cases}$$

As in the case of the reduced form ML estimators, the moments of the limiting distribution of the FIML estimators can be calculated using the inverse of the bordered information matrix. The computation of the bordered information matrix and its inverse, being rather lengthy, is presented in the form of an appendix, namely Appendix 5.A, at the end of this chapter.

The limiting distribution finally obtained is as follows :

$$(5.61)\quad \begin{bmatrix} \sqrt{N}\, I_M & 0 \\ 0 & \sqrt{NT}\, I_{M(M+K-1)} \end{bmatrix} \begin{bmatrix} \hat{a}_{ML} - a \\ \mathrm{vec}(\hat{\underline{\Theta}}_{ML} - \underline{\Theta}) \end{bmatrix}$$

$$\sim N\left(0\, , \begin{bmatrix} \Sigma_\mu + \Sigma_\nu & 0 \\ 0 & F\left[F'(\Sigma_\varepsilon^{-1} \otimes P^*)F\right]^{-1} F \end{bmatrix}\right)$$

where

$$(5.62)\quad P^* = \begin{bmatrix} \Pi_*' \\ I \end{bmatrix} \quad R \quad \begin{bmatrix} \Pi_* & I \end{bmatrix}$$

with Π_* such that $\Pi = \begin{bmatrix} \pi_o' \\ \Pi_* \end{bmatrix}$.

Let us write the fourth block of the above variance-covariance matrix in the following form :

$$F\left[F' \left(\Sigma_\varepsilon^{-1} \otimes \begin{bmatrix} \Pi_*' \\ I \end{bmatrix} R\, [\Pi_* \;\; I]\right) F\right]^{-1} F'$$

$$= F\left[F'\left(I \otimes \begin{bmatrix} \Pi_*' \\ I \end{bmatrix}\right)(\Sigma_\varepsilon^{-1} \otimes R)(I \otimes [\Pi_* \;\; I])\, F\right]^{-1} F'$$

Now, when the a priori restrictions are just the zero restrictions and the normalisation rule, then

$$(5.63)\quad F'\left(I \otimes \begin{bmatrix} \Pi_*' \\ I \end{bmatrix}\right) = \bar{\Pi}'$$

where $\bar{\Pi}$ is defined in (4.61), page 133 . Thus, the variance-covariance matrix of the limiting distribution of the unconstrained coefficients of (5.61) is :

$$(5.64)\quad \begin{bmatrix} (\Sigma_\mu + \Sigma_\nu) & 0 \\ 0 & \left[\bar{\Pi}' (\Sigma_\varepsilon^{-1} \otimes R)\, \bar{\Pi}\right]^{-1} \end{bmatrix}$$

which is seen to be equal to the variance-covariance matrix of the limiting distribution of the (feasible) G3SLS estimator (cf. (4.59), page 132). It follows that the FIML estimator and the fG3SLS estimator are asymptotically equivalent.

For the ML estimators of the variance components, we have the following limiting distributions :

$$\sqrt{NT}\ (\text{vec}\ \hat{\Sigma}_{\varepsilon,ML} - \Sigma_\varepsilon) \sim N(0\ ,\ \tilde{H}^{33})$$

$$\sqrt{N}\ (\text{vec}\ \hat{\Sigma}_{\mu,ML} - \Sigma_\mu) \sim N(0\ ,\ \tfrac{1}{2}\ (I+P)(2\ \Sigma_\mu \otimes \Sigma_\mu)\ \tfrac{1}{2}\ (I+P))$$

$$\sqrt{T}\ (\text{vec}\ \hat{\Sigma}_{\nu,ML} - \Sigma_\nu) \sim N(0\ ,\ \tfrac{1}{2}\ (I+P)(2\ \Sigma_\nu \otimes \Sigma_\nu)\ \tfrac{1}{2}\ (I+P))$$

where

$$\tilde{H}^{33} = \tfrac{1}{2}\ (I+P)(2\ \Sigma_\varepsilon \otimes \Sigma_\varepsilon)\ \tfrac{1}{2}\ (I+P)$$
$$+ \tfrac{1}{2}\ (I+P)(2I \otimes \Sigma_\varepsilon\ (L'\underline{\Theta})^{-1}L')F\left[F'(\Sigma_\varepsilon^{-1} \otimes P^*)F\right]^{-1}$$
$$F'(2I \otimes L(\underline{\Theta}'L)^{-1}\Sigma_\varepsilon)\ \tfrac{1}{2}(I+P).$$

5.2 Limited Information Maximum Likelihood (LIML) Estimation of the Structural Form

In this section, we show that, as in the classical simultaneous equations model, the LIML estimator of the parameters of a structural equation of our model, is equal to the FIML estimator of a reduced system consisting of the structural equation in question and the reduced form equations for its explanatory endogenous variables. Thus, LIML can be viewed as a special case of FIML.

The LIML estimation method for say, the first equation written as :

$$(5.65)\qquad y_1 = Y_1\ \gamma_1 + X_1\ \beta_1 + u_1$$

consists in maximising

$$(5.66)\qquad \log L\ (\text{vec}\ Y_1^*/\Pi_1^*, \Omega_1^*) = \text{const} - \tfrac{1}{2}\log|\Omega_1^*| -$$
$$\tfrac{1}{2}\ \text{vec}\ (Y_1^* - X\ \Pi_1^*)'\ \Omega_1^{*-1}\ (Y_1^* - X\Pi_1^*)$$

where

$$(5.67)\quad \begin{cases} Y_1^* = [y_1 \; Y_1] = X\,[\pi_1 \; \Pi_1] + [v_1 \; V_1] \\ V_1^* = [v_1 V_1] = V[L_1 L_I] = VL_1^* \text{ with } L_1 = \begin{bmatrix} 1 \\ 0 \\ 0 \end{bmatrix},\; L_I = \begin{bmatrix} 0 \\ I \\ 0 \end{bmatrix} \\ \Pi_1^* = [\,\pi_1 \; \Pi_1^{\prime}\,] \\ \Omega_1^* = E\,(\text{vec } V_1^*)(\text{vec } V_1^*)' \end{cases}$$

under the constraint

$$(5.68)\quad \beta_1^* + \Pi \begin{bmatrix} -1 \\ \gamma_1^* \end{bmatrix} = 0$$

By partitioning V as $[v_1 \; V_1 \; V_{\#}]$, the variance-covariance matrice of vec V, i.e. Ω, can be correspondingly partitioned as

$$(5.69)\quad \Omega = \sum_i \Omega_i \otimes M_i = \sum_i \begin{bmatrix} \omega_{i11} & \Omega_{i1I} & \Omega_{i1\#} \\ \Omega_{iI1} & \Omega_{iII} & \Omega_{iI\#} \\ \Omega_{i\#1} & \Omega_{i\#I} & \Omega_{i\#\#} \end{bmatrix} \otimes M_i$$

with

$$(5.70)\quad \Omega_1^* = E(\text{vec } V_1^*)(\text{vec } V_1^*)' = \sum_i \begin{bmatrix} \omega_{i11} & \Omega_{i1I} \\ \Omega_{iI1} & \Omega_{iII} \end{bmatrix} \otimes M_i$$

$$= \sum_i \Omega_{i}^* \otimes M_i$$

Note that

$$(5.71)\quad \Omega_1^{*-1} = \sum_i \Omega_i^{*-1} \otimes M_i$$

and as in the case of the full matrix, we can define

$$(5.72)\quad \begin{cases} \Omega_\mu^* = \Omega_1^* - T\,\Omega_4^* \\ \Omega_\nu^* = \Omega_2^* - N\,\Omega_4^* \\ \Omega_\varepsilon^* = \Omega_4^* \end{cases}$$

Let us also partition β^*_1, Π, γ^*_1 in such a way as to explicitly take into account the zero restrictions and the normalisation rule :

$$(5.73) \qquad \beta^*_1 = \begin{bmatrix} \beta_1 \\ 0 \end{bmatrix} \; ; \; \gamma^*_1 = \begin{bmatrix} -1 \\ \gamma_1 \\ 0 \end{bmatrix} \; ; \; \Pi = \begin{bmatrix} \pi_{a1} & \Pi_{a1} & \Pi_{a\#} \\ \pi_{b1} & \Pi_{b1} & \Pi_{b\#} \end{bmatrix}$$

Then, the constraint can be put in the following form :

$$\begin{bmatrix} \pi_{a1} & \Pi_{a1} \\ \pi_{b1} & \Pi_{b1} \end{bmatrix} \begin{bmatrix} -1 \\ \gamma_1 \end{bmatrix} = \begin{bmatrix} -\beta_1 \\ 0 \end{bmatrix}$$

or

$$(5.74) \qquad \begin{bmatrix} -\pi_{a1} \\ -\pi_{b1} \end{bmatrix} + \begin{bmatrix} \Pi_{a1} \\ \Pi_{b1} \end{bmatrix} \gamma_1 = \begin{bmatrix} -\beta_1 \\ 0 \end{bmatrix}$$

or

$$(5.75) \qquad -\pi_1 + \Pi_1 \gamma_1 = \begin{bmatrix} -\beta_1 \\ 0 \end{bmatrix}$$

Let us now directly substitute the constraint in log L, instead of forming the Lagrangian function, and convert the constrained maximisation problem to a problem of maximisation without constraints. The substitution of the constraint is done in the expression $Y^*_1 - X\Pi^*_1$ appearing in log L :

$$Y^*_1 - X\Pi^*_1 = [y_1 \; Y_1] - X[\pi_1 \; \Pi_1] \text{ using (5.67)}$$

$$= [y_1 \; Y_1] - X\left[\Pi_1 \gamma_1 + \begin{pmatrix} \beta_1 \\ 0 \end{pmatrix} \quad \Pi_1\right]$$

using the constraint as per (5.75)

$$= \left[y_1 - X\Pi_1\gamma_1 - X_1\beta_1 \quad Y_1 - X\Pi_1\right]$$

$$(5.76) \qquad = \left[u_1 + V_1\gamma_1 \quad V_1\right]$$

$$= \left[u_1 \; V_1\right] \begin{bmatrix} 1 & 0 \\ \gamma_1 & I \end{bmatrix}$$

$$(5.77) \qquad = V_{u1} T_1$$

denoting

$$(5.78) \quad V_{u1} = [u_1 \; V_1] \quad \text{and} \quad T_1 = \begin{bmatrix} I & 0 \\ \gamma_1 & I \end{bmatrix}$$

Note that

$$(5.79) \quad |T_1| = 1$$

and

$$(5.80) \quad T_1^{-1} = \begin{bmatrix} I & 0 \\ -\gamma_1 & I \end{bmatrix}$$

Thus, the (constrained) log L becomes :

$$(5.81) \quad \log L\,(\text{vec } Y_1^*/\beta_1, \gamma_1, \Pi_1, \Omega_1^*)$$

$$= \text{const} - \frac{1}{2} \sum_i m_i \log|\Omega_i^*| - \frac{1}{2}(\text{vec } V_1^*)' \sum_i (\Omega_i^{*-1} \otimes M_i)(\text{vec } V_1^*)$$

and the LIML estimators (of β_1, γ_1 in particular) are given by maximising (5.81) with respect to β_1, γ_1, Π_1 and Ω_ε^*, Ω_μ^*, Ω_ν^* .

Let us leave this problem here temporarily and consider the maximum likelihood estimation of the following "reduced" model :

$$(5.82) \quad \begin{cases} y_1 = Y_1\,\gamma_1 + X_1\,\beta_1 + u_1 \\ Y_1 = X\,\Pi_1 + V_1 \end{cases}$$

The above system can also be written as :

$$[y_1 \; Y_1] = [y_1 \; Y_1] \begin{bmatrix} 0 & 0 \\ \gamma_1 & 0 \end{bmatrix} + [X_1\beta_1 \quad X\,\Pi_1] + [u_1 \; V_1]$$

or

$$[y_1 \; Y_1] \begin{bmatrix} 1 & 0 \\ -\gamma_1 & I \end{bmatrix} = [X_1\beta_1 \quad X\,\Pi_1] + V_{u1} \quad \text{using (5.78)}$$

i.e.

$$Y_1^* \, T_1^{-1} = [X_1\beta_1 \quad X\,\Pi_1] + V_{u1} \qquad \text{using (5.80)}$$

or

$$Y_1^* = \Big([X_1 \beta_1 \quad X \Pi_1] + V_{u1}\Big) T_1$$

or

(5.83) $$Y_1^* = [X_1 \beta_1 - X \Pi_1 \gamma_1 \quad X \Pi_1] + V_{u1} T_1$$

Therefore, the log-likelihood function of the system (5.82) is the same as that of Y^*_1, which is in turn equal to that of V_{u1} with only change of variable, as the Jacobian of the transformation is unitary, by virtue of (5.79). Its expression is given by :

(5.84) $$\log L\,(Y_1^*/\beta_1, \gamma_1, \Pi_1, \Omega_{u1})$$

$$= \text{const.} - \frac{1}{2} \log |\Omega_{u1}| + \frac{1}{2} (\text{vec } V_{u1})' \Omega_{u1}^{-1} (\text{vec } V_{u1})$$

where

$$\Omega_{u1} = E\,(\text{vec } V_{u1})(\text{vec } V_{u1})'$$

Now,

$$\text{vec } V_{u1} = \text{vec } [u_1 \; V_1]$$

$$= \text{vec } [-V \gamma_1^* \quad V_1] \qquad \text{using } V = -U \Gamma^{-1}$$

$$= \text{vec}[v_1 \; V_1 \; V_{\neq}] \begin{bmatrix} 1 & 0 \\ -\gamma_1 & I \\ 0 & 0 \end{bmatrix} \qquad \text{writing V in partitioned form}$$

$$= \text{vec}[v_1 \; V_1] \begin{bmatrix} 1 & 0 \\ -\gamma_1 & I \end{bmatrix}$$

$$= \text{vec } V_1^* T_1^{-1} = (T_1^{-1} \otimes I) \text{ vec } V_1^*$$

Hence,

$$\Omega_{u1} = (T_1^{-1} \otimes I)\, E(\text{vec } V_1^*)(\text{vec } V_1^*)'(T_1^{-1'} \otimes I)$$

$$= (T_1^{-1} \otimes I) \sum_i (\Omega_i^* \otimes M_i)(T_1^{-1'} \otimes I)$$

with

$$\log |\Omega_{u1}| = \sum_i m_i \log |T_1^{-1} \Omega_i^* T_1^{-1'}| = \sum_i m_i \log |\Omega_i^*| \quad \text{using (5.79)}$$

and

$$\Omega_{u1}^{-1} = (T_1' \otimes I) \sum_i (\Omega_i^{*-1} \otimes M_i) (T_1 \otimes I)$$

Thus, the log-likelihood function of the reduced model is :

$$\text{const} - \frac{1}{2} \sum_i m_i \log |\Omega_i^*|$$

$$- \frac{1}{2} (\text{vec } V_1^*)'(T_1^{-1\prime} \otimes I)(T_1' \otimes I) \sum_i (\Omega_i^{*-1} \otimes M_i)(T_1 \otimes I)(T_1^{-1} \otimes I) \text{ vec } V_1^*$$

$$\text{(5.85)} \qquad = \text{const} - \frac{1}{2} \sum_i m_i \log |\Omega_i^*|$$

$$- \frac{1}{2} (\text{vec } V_1^*)' \sum_i (\Omega_i^{*-1} \otimes M_i) \text{ vec } V_1^*$$

which can be seen to be identical to the (constrained) log likelihood function of the LIML method given in (5.81).

Thus, LIML is seen to be equivalent to the FIML of a "reduced" system of equations. Having shown this, it seems irrelevant to us to go deeper in the discussion of the former method.

APPENDIX 5.A : Limiting Distribution of the FIML Estimators

5.A.1 The Information Matrix

As explained while deriving the bordered information matrix of the reduced form maximum likelihood, we start by writing down the second-order differential of the loglikelihood function (5.14) (page 172). Its first-order differential is given in (5.19) (page 174). Differentiating (5.19), using the rules recalled in (3.92) , yields :

$$\begin{aligned} d^2\log L = {} & \frac{1}{2}\sum_i m_i \operatorname{tr} \Sigma_i^{-1} d\Sigma_i \Sigma_i^{-1} d\Sigma_i - NT \operatorname{tr}(L'\underline{\Theta})^{-1} L'd\underline{\Theta}(L'\underline{\Theta})^{-1} \\ & L'd\underline{\Theta} - \frac{1}{2} \operatorname{tr}(2\ NT\ da(da)' + 4(d\underline{\Theta})'\underline{Z}'\iota(da)')\ \Sigma_3^{-1} \\ & + 2.\frac{1}{2} \operatorname{tr}(2NT(da)a' + 2(d\underline{\Theta})'\underline{Z}\iota a' + 2(da)\iota'\underline{Z\Theta})\Sigma_3^{-1}d\Sigma_3\ \Sigma_3^{-1} \\ & - 2.\frac{1}{2} \operatorname{tr}(NTaa' + \underline{\Theta}'\underline{Z}'\iota a' + a\iota'\underline{Z}\ \underline{\Theta})\ \Sigma_3^{-1}d\Sigma_3\ \Sigma_3^{-1}\ d\Sigma_3\ \Sigma_3^{-1} \\ & + 4.\frac{1}{2} \operatorname{tr} \sum_i (d\underline{\Theta})'\underline{Z}'M_i\ \underline{Z}\ \underline{\Theta}\ \Sigma_i^{-1}\ d\Sigma_i\ \Sigma_i^{-1} \\ & - 2.\frac{1}{2} \operatorname{tr} \sum_i \underline{\Theta}'\underline{Z}'M_i\ \underline{Z}\ \underline{\Theta}\ \Sigma_i^{-1}\ d\Sigma_i\ \Sigma_i^{-1}\ d\Sigma_i\ \Sigma_i^{-1} \\ & - 2.\frac{1}{2} \operatorname{tr} \sum_i (d\underline{\Theta})'\underline{Z}'M_i\underline{Z}(d\underline{\Theta})\ \Sigma_i^{-1} \end{aligned} \tag{5.A.1}$$

noting, in particular, that :

$$\operatorname{tr}(d\underline{\Theta})'\underline{Z}'M_i\underline{Z}\ \underline{\Theta}\ \Sigma_i^{-1}\ d\Sigma_i\ \Sigma_i^{-1} = \operatorname{tr}\ \underline{\Theta}'\underline{Z}'M_i\underline{Z}(d\underline{\Theta})\ \Sigma_i^{-1}\ d\Sigma_i\ \Sigma_i^{-1}$$

$$\operatorname{tr}(d\underline{\Theta})'\underline{Z}'\iota(da)'\ \Sigma_3^{-1} = \operatorname{tr}\ (da)\ \iota'\underline{Z}(d\underline{\Theta})\ \Sigma_3^{-1}$$

and also simplifying the expression by regrouping similar or identical terms.

Before taking the negative of its expectation, let us make note of the following :

$$E(\iota'\underline{Z}) = E(\iota'[Y\ \underline{X}]) = E(\iota'\ [\iota\ \pi_o' + \underline{X}\ \Pi_* + V|\underline{X}])$$

$$= E[NT\ \pi_o' + \iota'\underline{X}\ \Pi_* + \iota'V \quad \iota'\underline{X}] = [NT\ \pi_o' + \iota'\underline{X}\ \Pi_* \quad \iota'\underline{X}]$$

(5.A.2) $$= NT\ \pi_o'\ L' + \iota'\underline{X}\ [\Pi_*\ I] = NT\ \pi_o'\ L' + r_*'$$

denoting

(5.A.3) $$r_*' = \iota'\underline{X}\ [\Pi_*\ I]$$

$$E(\iota'\underline{Z}\ \underline{\Theta}) = E(\iota'\ [Y\ \underline{X}] \begin{bmatrix} \Gamma \\ \underline{B} \end{bmatrix}) = E(\iota'(Y\ \Gamma + X\ \underline{B}))$$

(5.A.4) $$= -\ E\ (\iota'\iota\ a' + \iota'U) = -\ NT\ a'$$

$$E(\underline{Z}'M_i\underline{Z}) = E \begin{bmatrix} Y' \\ \underline{X}' \end{bmatrix} M_i\ [Y\ \underline{X}]$$

$$= E\begin{bmatrix} \pi_o\ \iota' + \Pi_*'\underline{X} + V' \\ \underline{X}' \end{bmatrix} M_i\ [\iota\ \pi_o' + \underline{X}\ \Pi_* + V \quad \underline{X}]$$

$$= E\ \Big\{ L\ \pi_o\ \iota'M_i\ \iota\ \pi_o'L' + \begin{bmatrix} \Pi_*' \\ I \end{bmatrix} \underline{X}'M_i\underline{X}\ [\Pi_*\ I]$$

$$+ L\ V'M_i\ \iota\ \pi_o'\ L' + L\ \pi_o\ \iota'\ M_i\ V\ L'$$

$$+ L\ V'M_i\underline{X}\ [\Pi_*\ I] + \begin{bmatrix} \Pi_*' \\ I \end{bmatrix} \underline{X}'M_iVL' + LV'M_iVL' \Big\}$$

(5.A.5) $$= \begin{cases} P_{*i} + m_i\ L\ \Omega_i\ L' & \text{for } i=1,2,4 \\ NT\ L\ \pi_o\ \pi_o'\ L' + P_{*3} + m_3\ L\ \Omega_3\ L' & \text{for } i=3 \end{cases}$$

denoting $P_{*i} = \begin{bmatrix} \Pi_*' \\ I \end{bmatrix} \underline{X}'M_i\ \underline{X}\ [\Pi_*\ I]$ $\qquad i=1,2,3,4$

and using, in particular, (3.C.3)

$$E(\underline{Z}'M_i\ \underline{Z}\ \underline{\Theta}) = \begin{cases} m_i\ L\ \Omega_i\ L'\underline{\Theta} & \text{for } i=1,2,4 \\ -\ NT\ L\ \pi_o\ a' + m_3 L\ \Omega_3\ L'\underline{\Theta} & \text{for } i=3 \end{cases}$$ (5.A.6)

using (5.A.5) above and the following :

$$[\Pi_*\ I]\ \underline{\Theta} = 0$$

$$\pi_o'\ L'\underline{\Theta} = a'$$

$$E(\underline{\Theta}'\underline{Z}'M_i\underline{Z}\,\underline{\Theta}) = \begin{cases} m_i\,\Sigma_i & \text{for } i=1,2,4 \\ -\,NT\;aa' + m_3\,\Sigma_3 & \text{for } i=3 \end{cases} \qquad (5.A.7)$$

using (5.A.5) above and that

$$\underline{\Theta}'L'\Omega_i L'\underline{\Theta} = \Gamma'\Omega_i\Gamma = \Sigma_i$$

$$\pi_o'\,L'\,\underline{\Theta} = a'$$

$$E(\underline{Z}'M_i\,\underline{Z}\,\underline{\Theta})\,\Sigma_i^{-1} = \begin{cases} m_i\;L(\underline{\Theta}'L)^{-1} & \text{for } i=1,2,4 \\ -\,NT\;L\,\pi_o\,a'\,\Sigma_3^{-1} + m_3L(\underline{\Theta}'L)^{-1} & \text{for } i=3 \end{cases} \qquad (5.A.8)$$

using (5.A.6) above and that

$$L\,\Omega_i L'\underline{\Theta}\,\Sigma_i^{-1} = L\,\Omega_i\Gamma\,\Sigma_i^{-1} = L\,\Gamma'^{-1} = L(\underline{\Theta}'L)^{-1}$$

Now, using all the above results, the negative of the expectation of $d^2\log L$ can be written as :

$$\begin{aligned} -\,E\,(d^2\log L) = {} & -\tfrac{1}{2}\sum_i m_i\,\mathrm{tr}\,\Sigma_i^{-1}\,d\Sigma_i\,\Sigma_i^{-1}\,d\Sigma_i + NT\,\mathrm{tr}(L'\underline{\Theta})^{-1}\,L'd\,\underline{\Theta}(L'\underline{\Theta})^{-1} \\ & L'd\,\underline{\Theta} + \mathrm{tr}\Big(NT(da)(da)' + 2(d\,\underline{\Theta})'(r_* + NT\,L\,\pi_o)(da)'\Big)\,\Sigma_3^{-1} \\ & -\,2\,\mathrm{tr}\Big(NT(da)\,a' + (d\,\underline{\Theta})'(r_* + NT\,L\,\pi_o)\,a' + \\ & +\,(da)(-\,NT\,a')\Big)\,\Sigma_3^{-1}\,d\Sigma_3\,\Sigma_3^{-1} + \mathrm{tr}(NT\,aa' - NT\,aa' - NT\,aa') \\ & \Sigma_3^{-1}\,d\Sigma_3\,\Sigma_3^{-1}\,d\Sigma_3\,\Sigma_3^{-1} - 2\,\mathrm{tr}\sum_{1,2,4}(d\,\underline{\Theta})'\,m_i\,L\,(\underline{\Theta}'L)^{-1}\,d\Sigma_i\,\Sigma_i^{-1} \\ & -\,2\,(d\,\underline{\Theta})'(-\,NT\,L\,\pi_o\,a'\,\Sigma_3^{-1} + m_3\,L\,(\underline{\Theta}'L)^{-1})\,d\Sigma_3\,\Sigma_3^{-1}\;- \\ & +\,\mathrm{tr}\sum_{1,2,4} m_i\,d\Sigma_i\,\Sigma_i^{-1}\,d\Sigma_i\,\Sigma_i^{-1} + \mathrm{tr}\,(-\,NT\,aa' + m_3\,\Sigma_3)\,\Sigma_3^{-1}\,d\Sigma_3 \\ & \Sigma_3^{-1}\,d\Sigma_3\,\Sigma_3^{-1} + \mathrm{tr}\sum_{1,2,4}(d\,\underline{\Theta})'(P_{*i} + m_i\,L\,\Omega_i\,L')\,d\,\underline{\Theta}\,\Sigma_i^{-1} \\ & +\,\mathrm{tr}(d\,\underline{\Theta})'(NT\,L\,\pi_o\,\pi_o'\,L' + P_{*3} + m_3\,L\,\Omega_3\,L')(d\,\underline{\Theta})\,\Sigma_3^{-1} \end{aligned} \qquad (5.A.9)$$

Substituting formulae (5.20) and simultanously using relationships (5.23), (5.24), the above expression of $-E(d^2 \log L)$ can be modified as follows :

$$- E (d^2 \log L) =$$

$$NT (da)' \Sigma_3^{-1} (da) + 2(da)'(\Sigma_3^{-1} \otimes (r_* + NT L\pi_o) d \text{ vec } \underline{\Theta}$$

$$- NT (\text{dvec } \underline{\Theta})' \Big[((L'\underline{\Theta})^{-1} L' \otimes L'(\underline{\Theta}'L)^{-1}) P + \sum_{1,2,4} \Sigma_i^{-1} \otimes (P_{*i} + m_i L \Omega_i L')$$

$$+ \Sigma_3^{-1} \otimes (NT L\pi_o \pi_o' L' + P_{*3} + m_3 L \Omega_3 L') \Big] d \text{ vec } \underline{\Theta}$$

$$+ 2(d \text{ vec } \underline{\Theta})' \Big[\Sigma_3^{-1} \otimes (- NT a' \Sigma_3^{-1} - NT L \pi_o a' \Sigma_3^{-1} + m_3 L (\underline{\Theta}'L)^{-1}) +$$

$$+ \sum_{1,2,4} \Sigma_i^{-1} \otimes m_i L (\underline{\Theta}'L)^{-1} \Big] d \text{ vec } \Sigma_\varepsilon$$

$$+ 2(d \text{ vec } \underline{\Theta})' \Big[T \Sigma_3^{-1} \otimes (- NT a' \Sigma_3^{-1} - NT L\pi_o a' \Sigma_3^{-1} + m_3 L (\underline{\Theta}'L)^{-1}) +$$

$$+ T \Sigma_1^{-1} \otimes m_1 L (\underline{\Theta}'L)^{-1} \Big] d \text{ vec } \Sigma_\mu + 2(d \text{ vec } \underline{\Theta})' \Big[N \Sigma_3^{-1} \otimes$$

$$(- NT a' \Sigma_3^{-1} - NT L \pi_o a' \Sigma_3^{-1} + m_3 L (\underline{\Theta}'L)^{-1}) + N \Sigma_2^{-1} \otimes m_2 L (\underline{\Theta}'L)^{-1} \Big] d \text{ vec } \Sigma_\nu +$$

$$+ \frac{1}{2} (d \text{ vec } \Sigma_\varepsilon)' \sum_i m_i (\Sigma_i^{-1} \otimes \Sigma_i^{-1}) d \text{ vec } \Sigma_\varepsilon$$

$$+ \frac{T^2}{2} (d \text{ vec } \Sigma_\mu)' \sum_{1,3} m_i (\Sigma_i^{-1} \otimes \Sigma_i^{-1}) d \text{ vec } \Sigma_\mu$$

$$+ \frac{N^2}{2} (d \text{ vec } \Sigma_\nu)' \sum_{2,3} m_i (\Sigma_i^{-1} \otimes \Sigma_i^{-1}) d \text{ vec } \Sigma_\nu$$

$$+ 2.\frac{T}{2} (d \text{ vec } \Sigma_\varepsilon)' \sum_{1,3} m_i (\Sigma_i^{-1} \otimes \Sigma_i^{-1}) d \text{ vec } \Sigma_\mu$$

$$+ 2.\frac{N}{2} (d \text{ vec } \Sigma_\varepsilon)' \sum_{2,3} m_i (\Sigma_i^{-1} \otimes \Sigma_i^{-1}) d \text{ vec } \Sigma_\nu$$

$$+ 2 \frac{NT}{2} (d \text{ vec } \Sigma_\mu)' m_3 (\Sigma_3^{-1} \otimes \Sigma_3^{-1}) d \text{ vec } \Sigma_\nu$$

(5.A.10)

noting that

$$\mathrm{tr}\ (L'\underline{\Theta})^{-1} L'\, d\, \underline{\Theta}\ (L'\underline{\Theta})^{-1} L'\, d\, \underline{\Theta}$$

$$= \mathrm{tr}\ (d\, \underline{\Theta})'L\ (\underline{\Theta}'L)^{-1}\ (d\, \underline{\Theta})'\ L(\underline{\Theta}'L)^{-1}$$

$$= (d\ \mathrm{vec}\ \underline{\Theta})'\ (L'\underline{\Theta})^{-1} L' \otimes L(\underline{\Theta}'L)^{-1}\ d\ \mathrm{vec}\ \underline{\Theta}'$$

$$= (d\ \mathrm{vec}\ \underline{\Theta})'\ [(L'\underline{\Theta})^{-1} L' \otimes L(\underline{\Theta}'L)^{-1}]P\ d\ \mathrm{vec}\ \underline{\Theta}$$

From the above expression (5.A.10), the information matrix can be derived to be the following matrix :

(5.A.11) $\Psi = [\Psi_{ij}]\quad i=1,\dots,5\ ;\ j=1,\dots,5$

with

$$\Psi_{11} = NT\ \Sigma_3^{-1}$$

$$\Psi_{12} = \Sigma_3^{-1} \otimes (r_* + NT\ L\ \pi_o)$$

$$\Psi_{13} = \Psi_{14} = \Psi_{15} = 0$$

$$\Psi_{22} = [(L'\underline{\Theta})^{-1}L' \otimes L(\underline{\Theta}'L)^{-1}]P + \sum_{1,2,4} \Sigma_i^{-1} \otimes P_{*i} + m_i L\, \Omega_i L' + \Sigma_3^{-1} \otimes (NT\ L\ \pi_o\ \pi_o'\ L' + P_{*3} + m_3\ L\ \Omega_3\ L')$$

$$\Psi_{23} = \Sigma_3^{-1} \otimes \left(-\ NT\ a'\Sigma_3^{-1} - NT\ L\ \pi_o\ a'\Sigma_3^{-1} + m_3\ L(\underline{\Theta}'L)^{-1}\right) + \sum_{1,2,4} \Sigma_i^{-1} \otimes m_i\ L(\underline{\Theta}'L)^{-1}$$

$$\Psi_{24} = T\ \Sigma_3^{-1} \otimes \left(-\ NT\ a'\Sigma_3^{-1} - NT\ L\ \pi_o\ a'\Sigma_3^{-1} + m_3\ L(\underline{\Theta}'L)^{-1}\right) + T\ \Sigma_1^{-1} \otimes m_1\ L(\underline{\Theta}'L)^{-1}$$

$$\Psi_{25} = N\ \Sigma_3^{-1} \otimes \left(-\ NT\ a'\Sigma_3^{-1} - NT\ L\ \pi_o\ a'\Sigma_3^{-1} + m_3\ L(\underline{\Theta}'L)^{-1}\right) + N\ \Sigma_2^{-1} \otimes m_2\ L(\underline{\Theta}'L)^{-1}$$

$$\Psi_{33} = \frac{1}{2} \sum_i m_i\ \Sigma_i^{-1} \otimes \Sigma_i^{-1}$$

$$\Psi_{34} = \frac{T}{2} \sum_{1,3} m_i\ \Sigma_i^{-1} \otimes \Sigma_i^{-1}$$

$$\Psi_{35} = \frac{N}{2} \sum_{2,3} m_i\ \Sigma_i^{-1} \otimes \Sigma_i^{-1}$$

$$\Psi_{44} = T\,\Psi_{34}$$

$$\Psi_{45} = \frac{NT}{2}\, m_3\, \Sigma_3^{-1} \otimes \Sigma_3^{-1}$$

$$\Psi_{55} = N\,\Psi_{35}$$

and obviously, the matrix is symmetric.

Note that, by writing

$$\begin{aligned}[(L'\underline{\Theta})^{-1}L' \otimes L(\underline{\Theta}'L)^{-1}]\, P &= [I \otimes L(\underline{\Theta}'L)^{-1}]\; [(L'\underline{\Theta})^{-1}L' \otimes I]\, P \\ &= [I \otimes L(\underline{\Theta}'L)^{-1}]\; P\, [I \otimes (L'\underline{\Theta})^{-1}L' \otimes I] \\ &= [\Sigma_4^{-1} \otimes L(\underline{\Theta}'L)^{-1}]\; [\Sigma_4 \otimes I]\, P\, [\Sigma_4 \otimes I] \\ &\qquad [\Sigma_4^{-1} \otimes (L'\underline{\Theta})^{-1}L'] \\ &= [\Sigma_4^{-1} \otimes L(\underline{\Theta}'L)^{-1}]\, [\Sigma_4 \otimes \Sigma_4]\, P \\ &\qquad [\Sigma_4^{-1} \otimes (L'\underline{\Theta})^{-1}L']\end{aligned}$$

and that

$$\begin{aligned}\Sigma_4^{-1} \otimes L\Omega_4 L' &= \Sigma_4^{-1} \otimes L(\underline{\Theta}'L)^{-1}\, \Sigma_4 (L'\underline{\Theta})^{-1}\, L' \\ &= [\Sigma_4^{-1} \otimes L(\underline{\Theta}'L)^{-1}]\; [\Sigma_4 \otimes \Sigma_4]\; [\Sigma_4^{-1} \otimes (L'\underline{\Theta})^{-1}L']\end{aligned}$$

and substituting these in Ψ_{22}, we can write :

$$\begin{aligned}(5.A.12)\quad \Psi_{22} &= m_4(\Sigma_4^{-1} \otimes L(\underline{\Theta}'L)^{-1})(2\Sigma_4 \otimes \Sigma_4)\, \frac{I+P}{2}\, (\Sigma_4^{-1} \otimes (L'\underline{\Theta})^{-1}L') \\ &\quad + \Sigma_4^{-1} \otimes P_{*4} + \sum_{1,2,3} \Sigma_i^{-1} \otimes (P_{*i} + m_i L\, \Omega_i L') + \Sigma_3^{-1} \otimes NTL\pi_o \pi_o' L'\end{aligned}$$

5.A.2 Limit of the Information Matrix

First, let us recall the following limits as both N and T tend to infinity :

$$\frac{1}{NT}\, \underline{X}'M_4\, \underline{X} \rightarrow R \text{ , positive definite}$$

$$\frac{1}{NT}\, \underline{X}'M_i\, \underline{X} \rightarrow R^{(i)} \text{ , positive definite, } i=1,2,3$$

$$\frac{1}{NT}\, \underline{X}'\iota \quad \rightarrow r \text{ , a finite vector}$$

$$\frac{1}{NT}\, \underline{X}'M_i\, U \rightarrow 0 \text{ , } i=1,2,3,4$$

$$\Sigma_i^{-1} \quad \rightarrow \quad 0 \qquad \text{, } i=1,2,3$$

$$T\, \Sigma_1^{-1} \rightarrow \Sigma_\mu^{-1}$$

$$N\, \Sigma_2^{-1} \rightarrow \Sigma_\nu^{-1}$$

$$T\, \Sigma_3^{-1} \rightarrow (\Sigma_\mu + \Sigma_\nu)^{-1} \text{ assuming } \frac{N}{T} \rightarrow 1$$

$$N\, \Sigma_3^{-1} \rightarrow (\Sigma_\mu + \Sigma_\nu)^{-1} \text{ assuming } \frac{N}{T} \rightarrow 1$$

$$\frac{1}{T}\, \Omega_1 \quad \rightarrow \quad \Omega_\mu$$

$$\frac{1}{T}\, \Omega_2 \quad \rightarrow \quad \Omega_\nu$$

$$\frac{1}{T}\, \Omega_3 \quad \rightarrow \quad \Omega_\mu + \Omega_\nu \qquad \text{assuming } \frac{N}{T} \rightarrow 1$$

$$\frac{1}{N}\, \Omega_3 \quad \rightarrow \quad \Omega_\mu + \Omega_\nu \qquad \text{assuming } \frac{N}{T} \rightarrow 1$$

Using these, we derive that

$$\frac{1}{NT}\, P_{*i} \rightarrow \begin{bmatrix} \Pi_*' \\ I \end{bmatrix} \quad R^{(i)} \quad [\, \Pi_*\ \ I\,] \equiv \bar{P}_{*i} \quad , \quad i=1,2,3$$

$$\frac{1}{NT}\, P_{*4} \rightarrow \begin{bmatrix} \Pi_*' \\ I \end{bmatrix} \quad R \quad [\, \Pi_*\ \ I\,] \equiv \bar{P}_*$$

$$\frac{1}{NT}\, r_*' \quad \rightarrow r'[\, \Pi_*\ \ I\,]$$

$$\frac{1}{NT}\, NT\, L\, \pi_o \rightarrow L\, \pi_o$$

Now, we proceed to calculate the limit of

$$\eta \Psi \eta$$

where

$$\eta = \text{diag}\left[\frac{1}{\sqrt{N}} I_M, \frac{1}{\sqrt{NT}} I_{M(M+K-1)}, \frac{1}{\sqrt{NT}} I_{M^2}, \frac{1}{\sqrt{N}} I_{M^2}, \frac{1}{\sqrt{T}} I_{M^2}\right]$$

in order to obtain the second-order moments of the limiting distribution of $\eta^{-1}\left[a'(\text{vec } \underline{\Theta})'(\text{vec } \Sigma_\varepsilon)'(\text{vec } \Sigma_\mu)'(\text{vec } \Sigma_\nu)'\right]$ as stated in (5.60).

In deriving the following limits, we make use of the preliminary results on limits given above (page 198). Thus, it can be verified that

(5.A.13) $\frac{1}{N} \Psi_{11} = T \Sigma_3^{-1} \rightarrow (\Sigma_\mu + \Sigma_\nu)^{-1} \equiv \tilde{H}_{11}$

$$\frac{1}{N\sqrt{T}} \Psi_{12} = \frac{1}{\sqrt{T}} (T \Sigma_3^{-1} \otimes \frac{1}{NT} (r_* + N T L \pi_o)$$

$$= \frac{1}{\sqrt{T}} (T \Sigma_3^{-1} \otimes \left\{\frac{1}{NT} \iota' \underline{X} \begin{bmatrix} \Pi^* \\ I \end{bmatrix} + L \pi_o\right\})$$

$$\rightarrow 0$$

$$\frac{1}{NT} \Psi_{22} = \frac{m_4}{NT} (\Sigma_4^{-1} \otimes L(\underline{\Theta}'L)^{-1})(2 \Sigma_4 \otimes \Sigma_4) \frac{I+P}{2} (\Sigma_4^{-1} \otimes (L'\underline{\Theta})^{-1} L')$$

$$+ \Sigma_4^{-1} \otimes \frac{1}{NT} P_{*4} + \sum_{1,2,3} \Sigma_i^{-1} \otimes (\frac{1}{NT} P_{*i} + \frac{1}{NT} m_i L \Omega_i L')$$

(5.A.14) $\rightarrow \tilde{H}_{23} (2 \Sigma_4 \otimes \Sigma_4) \frac{I+P}{2} \tilde{H}_{23}' + \Sigma_\varepsilon^{-1} \otimes \bar{P}_* \equiv \tilde{H}_{22}$

calling

(5.A.15) $\tilde{H}_{23} = \Sigma_4^{-1} \otimes L(\underline{\Theta}'L)^{-1} \equiv \Sigma_\varepsilon^{-1} \otimes L(\underline{\Theta}'L)^{-1}$

$$\frac{1}{NT} \Psi_{23} = \Sigma_3^{-1} \otimes (- a'\Sigma_3^{-1} - L \pi_o a'\Sigma_3^{-1} + \frac{m_3}{NT} L(\underline{\Theta}'L)^{-1}) +$$

$$\sum_{1,2,4} \Sigma_i^{-1} \otimes \frac{m_i}{NT} L(\underline{\Theta}'L)^{-1}$$

(5.A.16) $\rightarrow \Sigma_4^{-1} \otimes L(\underline{\Theta}'L)^{-1} \equiv \tilde{H}_{23}$

$$\frac{1}{N\sqrt{T}} \Psi_{24} = \frac{1}{\sqrt{T}} \Big(T\, \Sigma_3^{-1} \otimes -\Big(a'T\, \Sigma_3^{-1} - L\, \pi_o\, a'T\, \Sigma_3^{-1} + \frac{m_3}{N} L(\underline{\Theta}'L)^{-1}\Big) + T\, \Sigma_1^{-1} \otimes \frac{m_1}{N} L(\underline{\Theta}'L)^{-1}\Big)$$

$$\rightarrow 0$$

and in the same way

$$\frac{1}{\sqrt{NT}} \Psi_{25} \rightarrow 0$$

Next,

$$(5.A.17)\quad \frac{1}{NT} \Psi_{33} = \frac{1}{2} \sum_i \frac{m_i}{NT} \Sigma_i^{-1} \otimes \Sigma_i^{-1} \rightarrow \frac{1}{2} \Sigma_\varepsilon^{-1} \otimes \Sigma_\varepsilon^{-1} \equiv \tilde{H}_{33}$$

$$\frac{1}{N\sqrt{T}} \Psi_{34} = \frac{1}{\sqrt{T}} \frac{1}{2} \sum_{1,3} \frac{m_i}{N} \Sigma_i^{-1} \otimes T\, \Sigma_i^{-1} \rightarrow 0$$

$$\frac{1}{\sqrt{NT}} \Psi_{35} = \frac{1}{\sqrt{N}} \frac{1}{2} \sum_{2,3} \frac{m_i}{T} \Sigma_i^{-1} \otimes N\, \Sigma_i^{-1} \rightarrow 0$$

$$(5.A.18)\quad \frac{1}{N} \Psi_{44} = \frac{1}{N} \frac{1}{2} \sum_{1,3} m_i T\, \Sigma_i^{-1} \otimes T\, \Sigma_i^{-1} \rightarrow \frac{1}{2} \Sigma_\mu^{-1} \otimes \Sigma_\mu^{-1} \equiv \tilde{H}_{44}$$

$$\frac{1}{\sqrt{NT}} \Psi_{45} = \frac{1}{\sqrt{NT}} \frac{1}{2} m_3\, N\, \Sigma_3^{-1} \otimes T\, \Sigma_3^{-1} \rightarrow 0$$

$$(5.A.19)\quad \frac{1}{T} \Psi_{55} = \frac{1}{T} \frac{1}{2} \sum_{2,3} m_i N\, \Sigma_i^{-1} \otimes N\, \Sigma_i^{-1} \rightarrow \frac{1}{2} \Sigma_\nu^{-1} \otimes \Sigma_\nu^{-1} \equiv \tilde{H}_{55}$$

Combining all the above limits, it is straightforward that

$$\lim \eta\, \Psi\, \eta = \tilde{\Psi} = \begin{bmatrix} \tilde{H}_{11} & 0 & 0 & 0 & 0 \\ 0 & \tilde{H}_{22} & \tilde{H}_{23} & 0 & 0 \\ 0 & \tilde{H}'_{23} & \tilde{H}_{33} & 0 & 0 \\ 0 & 0 & 0 & \tilde{H}_{44} & 0 \\ 0 & 0 & 0 & 0 & \tilde{H}_{55} \end{bmatrix}$$

5.A.3 The Limiting Distribution.

The bordered information matrix (in limit) is given by :

$$H = \begin{bmatrix} \tilde{\Psi} & G' \\ G & 0 \end{bmatrix}$$

where G , the matrix of gradients of the constraints of the maximum likelihood problem, is as follows :

$$G = \begin{bmatrix} S_o & & & & 0 \\ & \underline{S} & & & \\ & & C & & \\ & 0 & & C & \\ & & & & C \end{bmatrix}$$

Hence,

$$(5.A.20)\quad \tilde{H} = \left[\begin{array}{ccccc|ccccc} \tilde{H}_{11} & 0 & 0 & 0 & 0 & S_o' & 0 & 0 & 0 & 0 \\ 0 & \tilde{H}_{22} & \tilde{H}_{23} & 0 & 0 & 0 & \underline{S}' & 0 & 0 & 0 \\ 0 & \tilde{H}_{23}' & \tilde{H}_{33} & 0 & 0 & 0 & 0 & C' & 0 & 0 \\ 0 & 0 & 0 & \tilde{H}_{44} & 0 & 0 & 0 & 0 & C' & 0 \\ 0 & 0 & 0 & 0 & \tilde{H}_{55} & 0 & 0 & 0 & 0 & C' \\ \hline S_o & 0 & 0 & 0 & 0 & & & & & \\ 0 & \underline{S} & 0 & 0 & 0 & & & & & \\ 0 & 0 & C & 0 & 0 & & & \bigcirc & & \\ 0 & 0 & 0 & C & 0 & & & & & \\ 0 & 0 & 0 & 0 & C & & & & & \end{array}\right]$$

where

$$(5.A.21)\quad \tilde{H}_{11} = (\Sigma_\mu + \Sigma_\nu)^{-1}$$

$$(5.A.22)\quad \tilde{H}_{22} = \Sigma_\varepsilon^{-1} \otimes P^* + \tilde{H}_{23}(2\ \Sigma_\varepsilon \otimes \Sigma_\varepsilon)\ \frac{I+P}{2}\ \tilde{H}_{23}'$$

$$\text{with } P^* = \begin{bmatrix} \Pi_*' \\ I \end{bmatrix} R\ [\Pi_* \quad I]$$

$$(5.A.23)\quad \tilde{H}_{23} = -\ [\Sigma_{\varepsilon}^{-1} \otimes L\ (\underline{\Theta}'L)^{-1}]$$

$$(5.A.24)\quad \tilde{H}_{33} = \frac{1}{2}\ \Sigma_{\varepsilon}^{-1} \otimes \Sigma_{\varepsilon}^{-1}$$

$$(5.A.25)\quad \tilde{H}_{44} = \frac{1}{2}\ \Sigma_{\mu}^{-1} \otimes \Sigma_{\mu}^{-1}$$

$$(5.A.26)\quad \tilde{H}_{55} = \frac{1}{2}\ \Sigma_{\nu}^{-1} \otimes \Sigma_{\nu}^{-1}$$

Let us denote by $\tilde{H}^{ij}$ the corresponding block of the inverse of $\tilde{H}$. Then, like in the case of the reduced form, we have :

$$(5.A.27)\quad \tilde{H}^{11} = (\Sigma_{\mu} + \Sigma_{\nu})$$

$$(5.A.28)\quad \tilde{H}^{44} = \frac{1}{2}(I+P)\ (2\ \Sigma_{\mu} \otimes \Sigma_{\mu})\ \frac{1}{2}(I+P)$$

$$(5.A.29)\quad \tilde{H}^{55} = \frac{1}{2}(I+P)\ (2\ \Sigma_{\nu} \otimes \Sigma_{\nu})\ \frac{1}{2}(I+P)$$

For the four blocks of $\tilde{H}^{-1}$ corresponding to

$$\begin{bmatrix} \tilde{H}_{22} & \tilde{H}_{23} \\ \tilde{H}'_{23} & \tilde{H}_{33} \end{bmatrix}$$

we notice that they are equal to the first four blocks of the inverse of the following matrix $\tilde{A}$:

$$(5.A.30)\quad \tilde{A} = \left[\begin{array}{cc|cc} \tilde{H}_{22} & \tilde{H}_{23} & \underline{S}' & 0 \\ \tilde{H}'_{23} & \tilde{H}_{33} & 0 & C' \\ \hline \underline{S} & 0 & 0 & 0 \\ 0 & C & 0 & 0 \end{array}\right] = \begin{bmatrix} A_1 & G'_1 \\ G_1 & 0 \end{bmatrix}$$

Construct the following matrix F_*

$$(5.A.31)\quad F_* = \begin{bmatrix} F & 0 \\ 0 & \tilde{F} \end{bmatrix}$$

where F is a matrix of orthonormal vectors such that

(5.A.32) $FF' = I - \underline{S}'(\underline{S}\,\underline{S}')^{-1}\,\underline{S}$

and $\tilde{F}$ is a matrix of orthonormal vectors such that

(5.A.33) $\tilde{F}\tilde{F}' = I - C'C = \frac{1}{2}\,(I+P)$

It is easily checked that

(5.A.34) $F_*F_*' = I - G_1'(G_1G_1')^{-1}G_1$

Therefore the inverse corresponding to A_1, say A^1, is given by

$$A^1 = F_*(F_*'A_1F_*)^{-1}F_*'$$

Let us now compute the above matrix. First, we write :

$$F_*'\,A_1\,F_* = \begin{bmatrix} F'\tilde{H}_{22}F & F'\tilde{H}_{23}\tilde{F} \\ \tilde{F}'\tilde{H}_{23}'F & \tilde{F}'\tilde{H}_{33}\tilde{F} \end{bmatrix}$$

and notice that the block $\tilde{F}'\tilde{H}_{33}\tilde{F}$ is non-singular. Therefore, using the inversion rule for partitioned matrices and upon pre-multiplication by F_* and post-multiplication by F'_*, we obtain :

$$\tilde{H}^{22} = F\,\tilde{S}^{-1}F'$$
$$\tilde{H}^{23} = -\,F\,\tilde{S}^{-1}F'\tilde{H}_{23}\tilde{F}\,(\tilde{F}'\tilde{H}_{33}\tilde{F})^{-1}\tilde{F}'$$

$$\tilde{H}^{33} = \tilde{F}(\tilde{F}'\tilde{H}_{33}\tilde{F})^{-1}\left[I + \tilde{F}'\tilde{H}_{23}'F\,\tilde{S}^{-1}F'\tilde{H}_{23}\tilde{F}(\tilde{F}'\tilde{H}_{33}\tilde{F})^{-1}\right]\tilde{F}'$$

where

$$\tilde{S} = F'\tilde{H}_{22}F - F'\tilde{H}_{23}\tilde{F}(\tilde{F}'\tilde{H}_{33}\tilde{F})^{-1}\tilde{F}'\tilde{H}_{23}'\,F$$

$$= F'\tilde{H}_{22}F - F'\tilde{H}_{23}\tilde{F}\,\tilde{F}'\,\tilde{H}_{33}^{-1}\,\tilde{F}\,\tilde{F}'\,\tilde{H}_{23}'\,F$$

$$= F'(\Sigma_\varepsilon^{-1}\otimes P^*)\,F$$

using (5.A.22), the relation $\tilde{F}\tilde{F}' = \frac{I+P}{2}$ and the properties of the commutation matrix P.

Hence, we can simplify the above results, to obtain :

$$(5.A.35)\quad \tilde{H}^{22} = F\left[F'(\Sigma_\varepsilon^{-1} \otimes P^*)\,F\right]^{-1} F'$$

$$(5.A.36)\quad \tilde{H}^{23} = -\,F\left[F'(\Sigma_\varepsilon^{-1} \otimes P^*)\,F\right]^{-1} F'\tilde{H}_{23}\tilde{F}\;\tilde{F}'\tilde{H}_{33}^{-1}\,\tilde{F}\,\tilde{F}'$$

$$(5.A.37)\quad \tilde{H}^{33} = \tilde{F}\,(\tilde{F}'\tilde{H}_{33}\tilde{F})^{-1}\,(\tilde{F}' + \tilde{F}'\tilde{H}_{23}'\,F\left[F'(\Sigma_\varepsilon^{-1} \otimes P^*)\,F\right]^{-1} F'\tilde{H}_{23}\tilde{F}\;\tilde{F}'\tilde{H}_{33}\;\tilde{F}\,\tilde{F}')$$

$$= \tilde{F}\;\tilde{F}'\tilde{H}_{33}^{-1}\;\tilde{F}\;\tilde{F}' + \tilde{F}\;\tilde{F}'\tilde{H}_{33}^{-1}\;\tilde{F}\;\tilde{F}'\;\tilde{H}_{23}'\;F\left[F'(\Sigma_\varepsilon^{-1} \otimes P^*)F\right]^{-1} F'\tilde{H}_{23}\tilde{F}\;\tilde{F}'\tilde{H}_{33}^{-1}\tilde{F}\;\tilde{F}'$$

$$= \tilde{F}\;\tilde{F}'\tilde{H}_{33}^{-1}\;\tilde{F}\;\tilde{F}' + \tilde{F}\;\tilde{F}'\tilde{H}_{33}^{-1}\;\tilde{H}_{23}'\;F\left[F'(\Sigma_\varepsilon^{-1} \otimes P^*)\,F\right]^{-1} F'\tilde{H}_{23}\;\tilde{H}_{33}'^{-1}\;\tilde{F}\;\tilde{F}'$$

$$= \frac{1}{2}\,(I+P)\;(2\;\Sigma_\varepsilon \otimes \Sigma_\varepsilon)\;\frac{1}{2}\,(I+P)$$

$$+\;\frac{1}{2}(I+P)(2I \otimes \Sigma_\varepsilon(L'\underline{\Theta})^{-1}L')F\left[F'(\Sigma_\varepsilon^{-1} \otimes P^*)F\right]^{-1} F'(2I \otimes L(\underline{\Theta}'L)^{-1}\Sigma_\varepsilon)\;\frac{1}{2}(I+P$$

From the above calculation of inverse of the information matrix, we conclude that

$$(5.A.38)\quad \begin{bmatrix} \sqrt{N}\;I_M & 0 \\ 0 & \sqrt{NT}\;I_{M(M+K-1)} \end{bmatrix} \begin{bmatrix} \hat{a}_{ML} - a \\ \mathrm{vec}(\underline{\hat{\Theta}}_{ML} - \underline{\Theta}) \end{bmatrix}$$

$$\sim N\left(0\;,\begin{bmatrix} \Sigma_\mu + \Sigma_\nu & 0 \\ 0 & F\left[F'(\Sigma_\varepsilon^{-1} \otimes P^*)F\right]^{-1}F \end{bmatrix}\right)$$

where

$$(5.A.39)\quad P^* = \begin{bmatrix} \Pi_*' \\ I \end{bmatrix}\;R\;\begin{bmatrix} \Pi_* & I \end{bmatrix}$$

with Π_* such that $\Pi = \begin{bmatrix} \pi_o' \\ \Pi^* \end{bmatrix}$.

CHAPTER 6

THE JUST-IDENTIFIED CASE AND INDIRECT ESTIMATION OF STRUCTURAL PARAMETERS

6.1 The Identification Problem

The problem of identification in a Simultaneous Equations Model can be phrased in the form of the following question : what information can be obtained about structural parameters from information on reduced-form parameters, or, more specifically, under what conditions is it possible to derive the values (estimates) of structural parameters from given values (estimates) of reduced-form parameters.

Obviously, the first step towards answering this question, is the relationship between the two sets of parameters. Let us examine the relationship between the two types of coefficients in our model, namely (B,Γ) and Π . It is represented by the following equations :

$$(6.1) \qquad \Pi = - B\,\Gamma^{-1}$$

Since the above equations are exactly the same as the ones relating structural and reduced form coefficients in the "classical" simultaneous equations model, the conditions for identification of B and Γ from Π in the classical model, namely the rank and order conditions, also hold for our model.

Thus, when a single structural equation, say the first one, is just-identified, it means that the following system

$$(6.2) \qquad \Pi\,\gamma_1^* = - \beta_1^*$$

can be solved (using the zero restrictions and the normalisation rule) to get unique estimates of γ_1^* and β_1^* from a given estimate of Π . Of course, when the whole system is just-identified, then the system (6.1) has a unique solution for B and Γ, given Π.

In this chapter, we will first derive the limiting distribution of these indirect estimators of structural coefficients, considering both the cases - a single just-identified equation and a just-identified system - one by one. Then we will compare these limiting distributions with those of the feasible G2SLS and G3SLS estimators of structural coefficients in the case of just-identification, and also show equality between the fG2SLS and fG3SLS estimators themselves for a just-identified system.

Before going on to develop the above points, let us devote a short paragraph to the identification of the structural variance-covariance parameters through the reduced form parameters. It will be seen that, once the structural coefficients are identified, identification of the variance-covariance parameters is straightforward and does not require any further elaboration than what follows. The relationship that permits us to derive the variance-components Σ_μ, Σ_ν and Σ_ε from Ω_μ, Ω_ν, Ω_ε and Γ was established in the beginning of Chapter 5 (cf. (5.12), page 171) :

$$\Omega_i = (\Gamma^{-1})' \Sigma_i \Gamma^{-1} \qquad \text{for } i=1,2,3,4 \text{ and for } i=\mu,\nu,\varepsilon$$

or

$$\Sigma_i = \Gamma' \Omega_i \Gamma \qquad \text{for } i=1,2,3,4 \text{ and for } i=\mu,\nu,\varepsilon$$

Hence, knowing estimates of Ω_μ, Ω_ν, Ω_ε and Γ or equivalently knowing estimates of Ω_1, Ω_2, Ω_3, Ω_4 and Γ, the above equations can be used to derive estimates of Σ_1, Σ_2, Σ_3, Σ_4 or equivalently estimates of Σ_μ, Σ_ν, Σ_ε.

6.2 Derivation of the Indirect Estimators of Structural Coefficients and their Limiting Distributions

6.2.1 The Case of a Single Just-Identified Structural Equation

6.2.1.1 The Indirect Estimation Method

Let us assume, without loss of generality, that the first structural equation is just-identified. Then, estimators of γ_1^* and β_1^* can be derived from an estimator of Π, say $\hat{\Pi}$, by solving the following equations :

(6.3) $\quad \hat{\Pi}\,\gamma_1^* = \beta_1^*$

for γ_1^* and β_1^* under the zero restrictions and the normalisation rule.

Now, as done in Section 5.2, let us partition the coefficient vectors γ_1^* and β_1^* imposing the above constraints on their elements :

(6.4) $$\gamma_1^* = \begin{bmatrix} -1 \\ \gamma_1 \\ 0 \end{bmatrix} \quad , \quad \beta_1^* = \begin{bmatrix} \beta_1 \\ \\ 0 \end{bmatrix}$$

and let us also partition $\hat{\Pi}$ correspondingly :

(6.5) $$\hat{\Pi} = \begin{bmatrix} \hat{\pi}_{a1} & \hat{\Pi}_{a1} & \hat{\Pi}_{a\#} \\ \hat{\pi}_{b1} & \hat{\Pi}_{b1} & \hat{\Pi}_{b\#} \end{bmatrix}$$

Replacing (6.4) and (6.5) in (6.3), we get :

(6.6) $$\begin{bmatrix} \hat{\pi}_{a1} & \hat{\Pi}_{a1} \\ \hat{\pi}_{b1} & \hat{\Pi}_{b1} \end{bmatrix} \begin{bmatrix} -1 \\ \gamma_1 \end{bmatrix} = \begin{bmatrix} \beta_1 \\ 0 \end{bmatrix}$$

Let us further separate the constant term a_1 from the remaining elements of β_1 i.e. rewrite (6.6) as :

(6.7) $$\begin{bmatrix} \hat{\pi}^{o}_{a1} & \hat{\Pi}^{o}_{a1} \\ \hat{\pi}^{-}_{a1} & \hat{\Pi}^{-}_{a1} \\ \hat{\pi}_{b1} & \hat{\Pi}_{b1} \end{bmatrix} \begin{bmatrix} -1 \\ \gamma_1 \end{bmatrix} = \begin{bmatrix} a_1 \\ b_1 \\ 0 \end{bmatrix}$$

or

$$-\hat{\pi}^{o}_{a1} + \hat{\Pi}^{o}_{a1}\,\gamma_1 - a_1 = 0$$

$$-\hat{\pi}^{-}_{a1} + \hat{\Pi}^{-}_{a1}\,\gamma_1 - b_1 = 0$$

$$-\hat{\pi}_{b1} + \hat{\Pi}_{b1}\,\gamma_1 = 0$$

or

$$(6.8)\qquad \begin{bmatrix} 1 & \hat{\Pi}^{o}_{a1} & 0 \\ 0 & \hat{\Pi}^{-}_{a1} & I_{(K_1-1)} \\ 0 & \hat{\Pi}_{b1} & 0 \end{bmatrix} \begin{bmatrix} a_1 \\ \gamma_1 \\ b_1 \end{bmatrix} = \begin{bmatrix} \hat{\pi}^{o}_{a1} \\ \hat{\pi}^{-}_{a1} \\ \hat{\pi}_{b1} \end{bmatrix}$$

or

$$(6.9)\qquad \left[H^*_a \; \hat{\Pi}_1 \; H^*_b \right] \; \tilde{\alpha}_1 = \hat{\pi}_1$$

where

(6.10)

$H^{*\prime}_a = [1 \; 0 \; 0]$

$H^{*\prime}_b = \left[0 \; I_{K_1-1} \; 0 \right]$

$\hat{\Pi}_1 = \begin{bmatrix} \hat{\Pi}^{o}_{a1} \\ \hat{\Pi}^{-}_{a1} \\ \hat{\Pi}_{b1} \end{bmatrix}$ is the K x $\tilde{M}_1$ vector of coefficients post-multiplying X in the reduced form equations for Y_1

$\hat{\pi}_1 = \begin{bmatrix} \hat{\pi}^{o}_{a1} \\ \hat{\pi}^{-}_{a1} \\ \hat{\pi}_{b1} \end{bmatrix}$ is the first column of $\hat{\Pi}$

Therefore, knowing that, when the equation is just-identified, the matrix $\left[H^*_a \; \hat{\Pi}_1 \; H^*_b \right]$ is inversible ($K=K_1+\tilde{M}_1$), we can estimate $\tilde{\alpha}_1$ by :

$$(6.11)\qquad \hat{\tilde{\alpha}}_{1,IE} = \left[H^*_a \; \hat{\Pi}_1 \; H^*_b \right]^{-1} \hat{\pi}_1$$

where the subscript IE stands for indirect estimation.

Now, before looking at the different indirect estimators derived using different reduced form estimators, let us first rewrite the indirect estimator of (6.11) separating the constant term from the other coefficients (we have used the system (6.8) to do this) :

$$(6.12)\quad \begin{cases} \hat{a}_{1,IE} = \hat{\pi}^{o}_{a1} - \hat{\Pi}^{o}_{a1}\,\hat{\gamma}_{1,IE} \\ \begin{bmatrix} \hat{\gamma}_{1,IE} \\ \hat{b}_{1,IE} \end{bmatrix} = \begin{bmatrix} \hat{\Pi}^{-}_{a1} & I_{K_1-1} \\ \hat{\Pi}_{b1} & 0 \end{bmatrix}^{-1} \begin{bmatrix} \hat{\pi}^{-}_{a1} \\ \hat{\pi}_{b1} \end{bmatrix} \end{cases}$$

or in the notation of Chapter 3 :

$$(6.13)\quad \begin{cases} \hat{a}_{1,IE} = \hat{\pi}_{o1} - \hat{\Pi}'_{o1}\,\hat{\gamma}_{1,IE} \\ \hat{\alpha}^{*}_{1,IE} = \begin{bmatrix} \hat{\Pi}_{*1} & H^{*}_{1} \end{bmatrix}^{-1} \hat{\pi}_{*1} \end{cases}$$

with

$$\alpha^{*}_{1} = \begin{bmatrix} \gamma_1 \\ b_1 \end{bmatrix} \quad ; \quad H^{*}_{1} = \begin{bmatrix} I_{K_1-1} \\ 0 \end{bmatrix}$$

6.2.1.2 The Indirect Covariance Estimator

Let us recall from Chapter 3 that the covariance estimator of the reduced form coefficients is defined as :

$$(6.14)\quad \begin{cases} \hat{\pi}_{*m(cov)} = (\underline{X}'Q\underline{X})^{-1}\,\underline{X}'Q\,y_m \;, \quad m=1,\ldots,M \\ \hat{\pi}_{om(cov)} = \frac{1}{NT}\,\iota'_{NT}\,(y_m - \underline{X}\,\hat{\pi}_{*m(cov)}) \;, \quad m=1,\ldots,M \end{cases}$$

where π_{*m} is the coefficient vector of the m-th reduced form equation excluding the constant term and π_{om} is the constant term of the equation. Thus we have :

$$(6.15)\quad \begin{cases} \hat{\pi}_{*1(cov)} = (\underline{X}'Q\underline{X})^{-1}\,\underline{X}'Q\,y_1 \\ \hat{\Pi}_{*1(cov)} = (\underline{X}'Q\underline{X})^{-1}\,\underline{X}'Q\,Y_1 \end{cases}$$

where Π_{*1} is the submatrix of Π_1 obtained by removing the first row Π'_{o1} representing the constant terms of the equations for Y_1. In other words,

$$(6.16)\quad \Pi_{*1} \equiv \begin{bmatrix} \Pi^{-}_{a1} \\ \Pi_{b1} \end{bmatrix} \quad ; \quad \Pi'_{o1} \equiv \Pi^{o}_{a1}$$

using the notation of (6.8). In the same way,

$$(6.17) \qquad \pi_{*1} \equiv \begin{bmatrix} \bar{\pi}_{a1} \\ \pi_{b1} \end{bmatrix} \quad ; \quad \pi_{o1} \equiv \pi^{o}_{a1}$$

Using the covariance estimates given in (6.15) in the expression of $\hat{\alpha}^*_{1,IE}$ in (6.13), we get the indirect covariance estimator of α^*_1 :

$$\hat{\alpha}^*_{1,Icov} = \left[\hat{\Pi}_{*1(cov)} \quad H^*_1\right]^{-1} \hat{\pi}_{*1(cov)}$$

where the subscript 'Icov' is used to indicate indirect covariance estimation. Replacing $\hat{\Pi}_{*1(cov)}$ and $\hat{\pi}_{*1(cov)}$ by equations (6.15), we get :

$$(6.18) \quad \hat{\alpha}^*_{1,Icov} = \left[(\underline{X}'Q\underline{X})^{-1}\underline{X}'QY_1 \quad (\underline{X}'Q\underline{X})^{-1}\underline{X}'QX^*_1\right]^{-1}(\underline{X}'Q\underline{X})^{-1}\underline{X}'Qy_1$$

noting that H^*_1 can be written as :

$$H^*_1 = (\underline{X}'Q\underline{X})^{-1}\underline{X}'Q\underline{X}H^*_1 = (\underline{X}'Q\underline{X})^{-1}\underline{X}'QX^*_1$$

Simplifying (6.18) yields

$$\hat{\alpha}^*_{1,Icov} = (\underline{X}'Q \left[Y_1 \ X^*_1\right])^{-1} \underline{X}'Q\, y_1$$

$$(6.19) \qquad = (\underline{X}'Q\, Z^*_1)^{-1} \underline{X}'Q\, y_1$$

Now, if we go back to Chapter 4 and look at the expression of the covariance 2SLS estimator of α^*_1 ((4.31), page 126), we have :

$$(6.20) \quad \hat{\alpha}^*_{1,C2SLS} = \left[Z^{*\prime}_1Q\underline{X}(\underline{X}'Q\underline{X})^{-1}\underline{X}'QZ^*_1\right]^{-1}Z^{*\prime}_1Q\underline{X}(\underline{X}'Q\underline{X})^{-1}\underline{X}'Qy_1$$

As $Z^{*\prime}_1Q\underline{X}$ is square and non-singular if the first equation is just-identified, we can simplify (6.20) and write :

$$(6.21) \quad \hat{\alpha}^*_{1,C2SLS} = (\underline{X}'Q\, Z^*_1)^{-1} \underline{X}'Q\, y_1$$

Hence, comparing (6.19) and (6.21), we prove

$$(6.22) \quad \hat{\alpha}^*_{1,C2SLS} = \hat{\alpha}^*_{1,Icov}$$

Turning to the indirect covariance estimator of the constant term, from (6.13) we have :

$$\hat{a}_{1,Icov} = \hat{\pi}_{o1(cov)} - \hat{\Pi}'_{o1(cov)} \hat{\gamma}_{1,Icov}$$

or using (6.14) :

$$\hat{a}_{1,Icov} = \frac{1}{NT} \iota' (y_1 - \underline{X}\, \hat{\pi}_{*1(cov)}) - \frac{1}{NT} \iota'(Y_1 - \underline{X}\, \hat{\Pi}_{*1(cov)}) \hat{\gamma}_{1,Icov}$$

$$= \frac{1}{NT} \left[\iota' y_1 - \iota' Y_1 \hat{\gamma}_{1,Icov} - \iota' \underline{X}\, (\hat{\pi}_{*1(cov)} - \hat{\Pi}_{*1(cov)} \hat{\gamma}_{1,Icov}) \right]$$

But, from (6.13) we also have :

$$\hat{\pi}_{*1(cov)} - \hat{\Pi}_{*1(cov)} \hat{\gamma}_{1,Icov} = H_1^* \hat{b}_{1,Icov}$$

Therefore,

$$\hat{a}_{1,Icov} = \frac{1}{NT} \iota' \left[y_1 - Y_1 \hat{\gamma}_{1,Icov} - \underline{X} H_1^* \hat{b}_{1,Icov} \right]$$

$$= \frac{1}{NT} \iota' \left[y_1 - Z_1^* \hat{\alpha}_{1,Icov}^* \right]$$

$$= \hat{a}_{1,C2SLS} \qquad \text{(cf. (4.32), page 126)}$$

Thus, we have shown that the indirect estimator of the coefficients of a just-identified structural equation derived using the covariance estimator of Π is exactly equal to the corresponding covariance 2SLS estimator.

6.2.1.3 <u>The Indirect Feasible GLS Estimator</u>

Let us now take the feasible GLS estimator of Π or equivalently that of $\tilde{\pi}$ (which allows us to treat the constant terms separately) defined in (3.73), Chapter 3, i.e.

$$\hat{\tilde{\pi}}_{fGLS} = (\tilde{X}' \hat{\Omega}^{-1} \tilde{X})^{-1} \tilde{X}' \hat{\Omega}^{-1} y$$

and call the indirect estimator of $\tilde{\alpha}_1$ obtained using this fGLS estimator of $\tilde{\pi}$, the indirect fGLS (IfGLS) estimator of $\tilde{\alpha}_1$ i.e.:

$$(6.23) \qquad \hat{\tilde{\alpha}}_{1,fIGLS} = \left[H_a^* \hat{\Pi}_{1,fGLS} H_b^* \right]^{-1} \hat{\pi}_{1,fGLS}$$

Before determining its limiting distribution, let us first derive a convenient expression for $\hat{\pi}_{1,fGLS}$ and substitute it in (6.23). We will omit the subscript fGLS in the following calculations in order to avoid cumbersome notations.

Recall that :

$$\tilde{\pi} = \begin{bmatrix} \pi_o \\ \pi_* \end{bmatrix} \quad ; \quad \Pi = \begin{bmatrix} \pi_1 & \cdots & \pi_M \end{bmatrix}$$

where π_o is the column vector representing the transpose of the first row of Π ; π_* is the column vector containing the remaining rows of Π one below the other and where π_m denotes the m-th column of Π, m=1,...,M. Then, the first column of Π can be written as :

(6.24) $\pi_1 = L_*' \tilde{\pi}$

where

(6.25) $$\underset{(MxMK)}{L_*'} = \begin{bmatrix} e_1' & 0 & 0 \\ 0 & I_{K-1} & 0 \end{bmatrix} \quad \text{with} \quad \underset{(1xM)}{e_1'} = (1\ 0\ \ldots\ 0)$$

Hence

(6.26) $\hat{\pi}_1 = L_*' (\tilde{X}' \hat{\Omega}^{-1} \tilde{X})^{-1} \tilde{X}' \hat{\Omega}^{-1} y$

By writing

$$y_m = Y \gamma_m + \iota a_m + X_m^* b_m + u_m \ , \ m=1,\ldots,M$$

or

$$y_m = Y L_m \gamma_m + \iota a_m + \underline{X} H_m^* b_m + u_m \ , \ m=1,\ldots,M$$

where H^*_m and L_m are appropriate selection matrices, we have :

$$y = \begin{bmatrix} Y L_1 \gamma_1 + \iota a_1 + \underline{X} H_1^* b_1 + u_1 \\ \vdots \\ Y L_M \gamma_M + \iota a_M + \underline{X} H_M^* b_M + u_M \end{bmatrix}$$

(6.27) $$= (I_M \otimes Y) \begin{bmatrix} L_1 \gamma_1 \\ \vdots \\ L_M \gamma_M \end{bmatrix} + (I_M \otimes \iota) \begin{bmatrix} a_1 \\ \vdots \\ a_M \end{bmatrix} + (I \otimes \underline{X}) \begin{bmatrix} H_1^* b_1 \\ \vdots \\ H_M^* b_M \end{bmatrix} + \begin{bmatrix} u_1 \\ \vdots \\ u_M \end{bmatrix}$$

Now ,

$$(I \otimes Y) = I \otimes (\iota\ \hat{\pi}_o' + \underline{X}\ \hat{\Pi}_* + \hat{V})$$

$$= (I \otimes \iota)(I \otimes \hat{\pi}_o') + (I \otimes \underline{X})(I \otimes \hat{\Pi}_*) + (I \otimes \hat{V})$$

$$= \left[(I \otimes \iota) \quad (I \otimes \underline{X})\right] \begin{bmatrix} I \otimes \hat{\pi}_o' \\ I \otimes \hat{\Pi}_* \end{bmatrix} + (I \otimes \hat{V})$$

$$(6.28) \qquad = \tilde{X} \begin{bmatrix} I \otimes \hat{\pi}_o' \\ I \otimes \hat{\Pi}_* \end{bmatrix} + I \otimes \hat{V}$$

Thus, replacing (6.28) in (6.27), we get :

$$(6.29) \quad y = \tilde{X} \begin{bmatrix} I \otimes \hat{\pi}_o' \\ I \otimes \hat{\Pi}_* \end{bmatrix} \begin{bmatrix} L_1 \gamma_1 \\ \vdots \\ L_M \gamma_M \end{bmatrix} + (I \otimes \hat{V}) \begin{bmatrix} L_1 \gamma_1 \\ \vdots \\ L_M \gamma_M \end{bmatrix}$$

$$+ (I \otimes \iota) \begin{bmatrix} a_1 \\ \vdots \\ a_M \end{bmatrix} + (I \otimes \underline{X}) \begin{bmatrix} H_1^* b_1 \\ \vdots \\ H_M^* b_M \end{bmatrix} + \begin{bmatrix} u_1 \\ \vdots \\ u_M \end{bmatrix}$$

Let us now substitute (6.29) for y in (6.26) and expand each term of the resulting expression one by one.

(i) First, we have :

$$L_*'(\tilde{X}'\ \hat{\Omega}^{-1}\ \tilde{X})^{-1}\ \tilde{X}'\ \hat{\Omega}^{-1}\ \tilde{X} \begin{bmatrix} I \otimes \hat{\pi}_o' \\ I \otimes \hat{\Pi}_* \end{bmatrix} \begin{bmatrix} L_1 \gamma_1 \\ \vdots \\ L_M \gamma_M \end{bmatrix}$$

$$= L_*' \begin{bmatrix} I \otimes \hat{\pi}_o' \\ I \otimes \hat{\Pi}_* \end{bmatrix} \begin{bmatrix} L_1 \gamma_1 \\ \vdots \\ L_M \gamma_M \end{bmatrix}$$

$$= \begin{bmatrix} e_1' & 0 & 0 \\ 0 & I_{K-1} & 0 \end{bmatrix} \begin{bmatrix} \hat{\pi}_o' L_1 \gamma_1 \\ \vdots \\ \hat{\pi}_o' L_M \gamma_M \\ \hat{\Pi}_* L_1 \gamma_1 \\ \vdots \\ \hat{\Pi}_* L_M \gamma_M \end{bmatrix}$$

$$= \begin{bmatrix} \hat{\pi}_o' \\ \hat{\Pi}_* \end{bmatrix} L_1 \gamma_1$$

$$= \hat{\Pi}\, L_1\, \gamma_1$$

$$(6.30) = \hat{\Pi}_1\, \gamma_1$$

(ii) Next, we have :

$$(6.31)\qquad L_*' (\tilde{X}' \hat{\Omega}^{-1} \tilde{X})^{-1} \tilde{X} \hat{\Omega}^{-1} (I \otimes V) \begin{bmatrix} L_1 \gamma_1 \\ \vdots \\ L_M \gamma_M \end{bmatrix}$$

Let us note that :

$$(I \otimes \hat{V}) \begin{bmatrix} L_1 \gamma_1 \\ \vdots \\ L_M \gamma_M \end{bmatrix} = (I \otimes \hat{V}) \operatorname{vec} \begin{bmatrix} L_1 \gamma_1 & \cdots & L_M \gamma_M \end{bmatrix}$$

$$= (I \otimes \hat{V}) \operatorname{vec} L^* \qquad \text{denoting } \begin{bmatrix} L_1\gamma_1 & \cdots & L_M\gamma_M \end{bmatrix} \text{ as } L^*$$

$$(6.32)\qquad = (L^{*\prime} \otimes I) \operatorname{vec} \hat{V}$$

But

$$L^{*\prime} = \Gamma' + I$$

Thus, expression (6.32) becomes :

$$((\Gamma' + I) \otimes I) \operatorname{vec} \hat{V}$$

$$= ((\Gamma' + I) \otimes I)(I - \tilde{X}(\tilde{X}' \hat{\Omega}^{-1} \tilde{X})^{-1} \tilde{X}' \hat{\Omega}^{-1}) \operatorname{vec} V$$

$$= ((\Gamma' + I) \otimes I) M_{\tilde{X}\hat{\Omega}} \operatorname{vec} V$$

$$= (\Gamma' \otimes I) M_{\tilde{X}\hat{\Omega}} \operatorname{vec} V + M_{\tilde{X}\hat{\Omega}} \operatorname{vec} V$$

Hence, expression (6.31) can be written as :

$$L_*'(\tilde{X}'\hat{\Omega}^{-1}\tilde{X})^{-1} \tilde{X}'\hat{\Omega}^{-1} \left[(\Gamma' \otimes I) M_{\tilde{X}\hat{\Omega}} \operatorname{vec} V + M_{\tilde{X}\hat{\Omega}} \operatorname{vec} V\right]$$

$$= L_*'(\tilde{X}'\hat{\Omega}^{-1}\tilde{X})^{-1} \tilde{X}'\hat{\Omega}^{-1} (\Gamma' \otimes I) M_{\tilde{X}\hat{\Omega}} \operatorname{vec} V$$

$$\text{noting that } \tilde{X}'\hat{\Omega}^{-1} M_{\tilde{X}\hat{\Omega}} = 0$$

$$(6.33)\qquad = - L_*'(\tilde{X}'\hat{\Omega}^{-1}\tilde{X})^{-1} \tilde{X}'\hat{\Omega}^{-1}(\Gamma' \otimes I) M_{\tilde{X}\hat{\Omega}} (\Gamma^{-1\prime} \otimes I) u$$

$$\text{using } V = - U \Gamma^{-1}$$

(iii) Next, let us regroup the third and fourth terms of (6.29) and substitute them in (6.26) :

$$L_*'(\tilde{X}'\hat{\Omega}^{-1}\tilde{X})^{-1}\,\tilde{X}'\hat{\Omega}^{-1}(I \otimes \iota)\begin{bmatrix} a_1 \\ \vdots \\ a_M \end{bmatrix} +$$

$$L_*'(\tilde{X}'\hat{\Omega}^{-1}\tilde{X})^{-1}\tilde{X}'\hat{\Omega}^{-1}(I \otimes \underline{X})\begin{bmatrix} H_1^* \, b_1 \\ \vdots \\ H_M^* \, b_M \end{bmatrix}$$

$$= L_*'(\tilde{X}'\hat{\Omega}^{-1}\tilde{X})^{-1}\,\tilde{X}'\hat{\Omega}^{-1}\left[(I \otimes \iota) \quad (I \otimes \underline{X})\right]\begin{bmatrix} a_1 \\ \vdots \\ a_M \\ H_1^* \, b_1 \\ \vdots \\ H_M^* \, b_M \end{bmatrix}$$

$$= L_*'\begin{bmatrix} a_1 \\ \vdots \\ a_M \\ H_1^* \, b_1 \\ \vdots \\ H_M^* \, b_M \end{bmatrix}$$ noting that $\left[(I \otimes \iota) \quad (I \otimes \underline{X})\right] = \tilde{X}$ and simplifying accordingly

$$= \begin{bmatrix} a_1 \\ H_1^* \, b_1 \end{bmatrix}$$ using (6.25) and multiplying out

$$= \begin{bmatrix} 1 & 0 \\ 0 & H_1^* \end{bmatrix}\begin{bmatrix} a_1 \\ b_1 \end{bmatrix}$$

$$(6.34) = \begin{bmatrix} H_a^* & H_b^* \end{bmatrix}\begin{bmatrix} a_1 \\ b_1 \end{bmatrix}$$

(iv) Finally, substituting the last term of (6.29) in (6.26), we have :

$$(6.35) \quad L_*'(\tilde{X}'\,\hat{\Omega}^{-1}\,\tilde{X})^{-1}\,\tilde{X}'\,\hat{\Omega}^{-1}\,u$$

Therefore, combining (6.30), (6.33), (6.34) and (6.35), we obtain :

$$\hat{\pi}_1 = \hat{\Pi}_1\gamma_1 - L_*'(\tilde{X}'\hat{\Omega}^{-1}\tilde{X})^{-1}\tilde{X}'\hat{\Omega}^{-1}(\Gamma'\otimes I)\, M_{\tilde{X}\hat{\Omega}}\, (\Gamma^{-1'}\otimes I)\, u$$

$$+\begin{bmatrix} H_a^* & H_b^* \end{bmatrix}\begin{bmatrix} a_1 \\ b_1 \end{bmatrix} + L_*'\, (\tilde{X}'\hat{\Omega}^{-1}\tilde{X})^{-1}\, \tilde{X}'\, \hat{\Omega}^{-1}\, u$$

$$(6.36)\quad = \begin{bmatrix} H_a^* & \hat{\Pi}_1 & H_b^* \end{bmatrix}\begin{bmatrix} a_1 \\ \gamma_1 \\ b_1 \end{bmatrix} + L_*'\, (\tilde{X}'\, \hat{\Omega}^{-1}\, \tilde{X})^{-1}\, \tilde{X}'\, \hat{\Omega}^{-1}\, (\Gamma'\otimes I)(I - M_{\tilde{X}\hat{\Omega}})(\Gamma^{-1'}\otimes I)\, u$$

as $\tilde{X}'\hat{\Omega}^{-1}u$ of the last term in the previous expression can be written as $\tilde{X}'\hat{\Omega}^{-1}(\Gamma'\otimes I)\, I(\Gamma^{-1'}\otimes I)u$

Inserting the above expression in place of $\hat{\pi}_1$ in (6.11) yields :

$$\hat{\alpha}_{1,IfGLS} = \begin{bmatrix} H_a^* & \hat{\Pi}_1 & H_b^* \end{bmatrix}^{-1} \left\{ \begin{bmatrix} H_a^* & \hat{\Pi}_1 & H_b^* \end{bmatrix}\begin{bmatrix} a_1 \\ \gamma_1 \\ b_1 \end{bmatrix} \right.$$

$$\left. + L_*'(\tilde{X}'\, \hat{\Omega}^{-1}\, \tilde{X})^{-1}\, \tilde{X}'\, \hat{\Omega}^{-1}\, (\Gamma'\otimes I)(I - M_{\tilde{X}\hat{\Omega}})(\Gamma^{-1'}\otimes I)u \right\}$$

$$(6.37)\quad = \tilde{\alpha}_1 + \begin{bmatrix} H_a^* & \hat{\Pi}_1 & H_b^* \end{bmatrix}^{-1} L_*'(\tilde{X}'\, \hat{\Omega}^{-1}\, \tilde{X})^{-1}\, \tilde{X}'\hat{\Omega}^{-1}\, (\Gamma'\otimes I)(I - M_{\tilde{X}\hat{\Omega}})\, w$$

writing $(\Gamma^{-1'}\otimes I)\, u = \text{vec}\, V = w$

Now, let us slightly modify the following expression :

$$\tilde{X}'\, \hat{\Omega}^{-1}\, (\Gamma'\otimes I)\, (I - M_{\tilde{X}\hat{\Omega}})\, w$$

Replacing the definition of $M_{\tilde{X}\hat{\Omega}}$ in it, it becomes :

$$\tilde{X}'\, \hat{\Omega}^{-1}\, (\Gamma'\otimes I)\, \tilde{X}\, (\tilde{X}'\, \hat{\Omega}^{-1}\, \tilde{X})^{-1}\, \tilde{X}'\, \hat{\Omega}^{-1}\, w$$

Writing

$$\tilde{X} = \begin{bmatrix} I\otimes\iota & I\otimes\underline{X} \end{bmatrix}$$

we have

$$(\Gamma'\otimes I)\, \begin{bmatrix} (I\otimes\iota) & (I\otimes\underline{X}) \end{bmatrix}$$

$$= \begin{bmatrix} (I\otimes\iota)(\Gamma'\otimes 1) & (I\otimes\underline{X})(\Gamma'\otimes I_{K-1}) \end{bmatrix}$$

$$= \left[(I \otimes \iota)\ (I \otimes \underline{X})\right] \begin{bmatrix} \Gamma' \otimes 1 & 0 \\ 0 & \Gamma' \otimes I_{K-1} \end{bmatrix}$$

$$= \tilde{X} \begin{bmatrix} \Gamma' \otimes 1 & 0 \\ 0 & \Gamma' \otimes I_{K-1} \end{bmatrix}$$

Thus, the expression

$$L_*'(\tilde{X}'\hat{\Omega}^{-1}\tilde{X})^{-1}\ \tilde{X}'\hat{\Omega}^{-1}(\Gamma' \otimes I)\ (I - M_{\tilde{X}\hat{\Omega}})\ w$$

becomes

$$L_*'(\tilde{X}'\hat{\Omega}^{-1}\tilde{X})^{-1}\tilde{X}'\hat{\Omega}^{-1}\tilde{X} \begin{bmatrix} \Gamma' & 0 \\ 0 & \Gamma'\otimes I_{K-1} \end{bmatrix} (\tilde{X}'\hat{\Omega}^{-1}\tilde{X})^{-1}\ \tilde{X}'\hat{\Omega}^{-1}w$$

$$(6.38) = L_*' \begin{bmatrix} \Gamma' & 0 \\ 0 & \Gamma'\otimes I_{K-1} \end{bmatrix} (\tilde{X}'\ \hat{\Omega}^{-1}\ \tilde{X})^{-1}\ \tilde{X}'\ \hat{\Omega}^{-1}\ w$$

Substituting (6.38) in (6.37) we obtain :

$$\hat{\tilde{\alpha}}_{1,\text{fIGLS}} - \tilde{\alpha}_1 = \left[H_a^*\hat{\Pi}_1 H_b^*\right]^{-1} L_*' \begin{bmatrix} \Gamma' & 0 \\ 0 & \Gamma'\otimes I_{K-1} \end{bmatrix} (\tilde{X}'\hat{\Omega}^{-1}\tilde{X})^{-1}\ \tilde{X}'\hat{\Omega}^{-1}w$$

or

$$(6.39) \qquad = \left[H_a^*\ \hat{\Pi}_1\ H_b^*\right]^{-1} L_*'\ \Gamma_*'\ (\tilde{X}'\hat{\Omega}^{-1}\tilde{X})^{-1}\ \tilde{X}'\hat{\Omega}^{-1}\ w$$

$$\text{with } \Gamma_*' = \begin{bmatrix} \Gamma' & 0 \\ 0 & \Gamma'\otimes I_{K-1} \end{bmatrix}$$

Now that we have expressed the indirect fGLS estimator in a more convenient form to manipulate, we go on to derive its limiting distribution.

The derivation of the limiting distribution is given in detail in Appendix 6.A. We conclude that

$$(6.40) \qquad \bar{D}_2^{-\frac{1}{2}}(\hat{\tilde{\alpha}}_{1,\text{IfGLS}} - \tilde{\alpha}_1) \sim N\ (\ 0,\ \tilde{P}_1^{-1}\ \tilde{R}_1^{-1}\ \tilde{P}_1'^{-1}\ )$$

where $\bar{D}_2$ is the same matrix as defined in (4.C.7) with the index "m" changed to "1" in $\tilde{M}_m$ and K_m; similarly for $\tilde{P}_1$ and $\tilde{R}_1$ as in (4.41) and (4.42) respectively with "m" changed to "1".

6.2.2 The Case of a Just-Identified System

When the whole system is just-identified, the indirect estimators of all the structural coefficients are obtained by solving

$$(6.41)\qquad \hat{\Pi}\,\Gamma = B$$

for Γ , B given an estimator of Π .

In other words, we have to solve :

$$\hat{\Pi}\begin{bmatrix}\gamma_1^* & \cdots & \gamma_M^*\end{bmatrix} = \begin{bmatrix}\beta_1^* & \cdots & \beta_M^*\end{bmatrix}$$

or

$$\Pi\,\gamma_m^* = \beta_m^* \qquad \text{for } m=1,\ldots,M .$$

By proceeding in the same way as in the case of a single just-identified equation (Section 6.2.1) for each of the M just-identified structural equations, in can be shown that, in general, the limiting distribution of

$$\bar{D}_2^{-\frac{1}{2}}\,(\hat{\tilde{\alpha}}_{m,\text{IfGLS}} - \tilde{\alpha}_m) = \begin{bmatrix}\sqrt{N}\ (\hat{a}_{m,\text{fIGLS}} - a_m)\\ \sqrt{NT}\ (\hat{\alpha}^*_{m,\text{fIGLS}} - \alpha^*_m)\end{bmatrix}$$

is

$$(6.42)\qquad N\,(\,0,\ \tilde{P}_m^{-1}\tilde{R}_m^{-1}\tilde{P}_m^{-1'}\,)$$

for m=1,...,M .

6.3 Comparison of the IfGLS Estimator with the fG2SLS and fG3SLS Estimators

6.3.1 The Case of a Single Just-Identified Equation

It was established at the end of Section 4.2 (cf. (4.39), p. 128) that the limiting distribution of

$$\bar{D}_2^{-\frac{1}{2}}\,(\hat{\tilde{\alpha}}_{m,\text{fG2SLS}} - \tilde{\alpha}_m)$$

is normal with zero mean and variance-covariance matrix equal to

$$\left[\tilde{P}_m' \tilde{R}_m \tilde{P}_m \right]^{-1}$$

Now, when say the first equation is just-identified, the rank conditions ensure that the matrix

$$\tilde{P}_1 = \begin{bmatrix} 1 & 0 & 0 \\ 0 & \Pi_{*1} & H_1^* \end{bmatrix}$$

is square and non-singular. Hence,

$$\left[\tilde{P}_1' \tilde{R}_1 \tilde{P}_1 \right]^{-1} = \tilde{P}_1^{-1} \tilde{R}_1^{-1} \tilde{P}_1^{-1'}$$

Therefore, in this case, the limiting distribution of

$$\bar{D}_2^{-\frac{1}{2}} (\hat{\tilde{\alpha}}_{1,fG2SLS} - \tilde{\alpha}_1)$$

is normal with zero mean and variance-covariance matrix equal to $\tilde{P}_1^{-1} \tilde{R}_1^{-1} \tilde{P}_1^{-1'}$, which can be seen to be exactly the same as the limiting distibution of

$$\bar{D}_2^{-\frac{1}{2}} (\hat{\tilde{\alpha}}_{1,IfGLS} - \tilde{\alpha}_1)$$

arrived at on page 217 (cf. (6.40)).

Thus, we have proved the asymptotic equivalence between the feasible generalised two stage least squares estimator and the indirect feasible generalised least squares estimator of the coefficients of a just-identified structural equation, a result similar to the one we already have for a classical simultaneous equation model.

6.3.2 The Case of a Just-Identified System

6.3.2.1 Comparison between fG2SLS and fG3SLS Estimators

In Chapter 4, we defined the G3SLS estimator as :

$$\hat{\alpha}_{G3SLS} = \left[Z_*' \Sigma_*^{-1} \mathcal{X} (\mathcal{X}' \Sigma_*^{-1} \Sigma \Sigma_*^{-1} \mathcal{X})^{-1} \mathcal{X}' \Sigma_*^{-1} Z_* \right]^{-1}$$

$$Z_*' \Sigma_*^{-1} \mathcal{X} (\mathcal{X}' \Sigma_*^{-1} \Sigma \Sigma_*^{-1} \mathcal{X})^{-1} \mathcal{X}' \Sigma_*^{-1} y$$

Now, if the whole system is just-identified, then $Z_*' \Sigma_*^{-1} \mathcal{X}$ is square and non singular as

$$Z_*' \Sigma_*^{-1} \mathcal{X} = \begin{bmatrix} Z_1' \Sigma_{11}^{-1} X & & 0 \\ & \ddots & \\ 0 & & Z_M' \Sigma_{MM}^{-1} X \end{bmatrix}$$

and the just-identification of each of the M structural equations ensures the non-singularity of each of the blocks $Z_m' \Sigma_{mm}^{-1} X$, $m=1,\ldots,M$.

Thus we can write :

$$\begin{aligned} \hat{\alpha}_{G3SLS} &= (\mathcal{X}' \Sigma_*^{-1} Z_*)^{-1} (\mathcal{X}' \Sigma_*^{-1} \Sigma \Sigma_*^{-1} \mathcal{X})(Z_*' \Sigma_*^{-1} \mathcal{X})^{-1} \\ &\quad (Z_*' \Sigma_*^{-1} \mathcal{X})(\mathcal{X}' \Sigma_*^{-1} \Sigma \Sigma_*^{-1} \mathcal{X})^{-1} \mathcal{X}' \Sigma_*^{-1} y \\ &= (\mathcal{X}' \Sigma_*^{-1} Z_*)^{-1} \mathcal{X}' \Sigma_*^{-1} y \end{aligned}$$

or

$$(6.43) \quad \begin{bmatrix} \hat{\alpha}_{1,G3SLS} \\ \vdots \\ \hat{\alpha}_{M,G3SLS} \end{bmatrix} = \begin{bmatrix} (X' \Sigma_{11}^{-1} Z_1)^{-1} X' \Sigma_{11}^{-1} y_1 \\ \vdots \\ (X' \Sigma_{MM}^{-1} Z_M)^{-1} X' \Sigma_{MM}^{-1} y_M \end{bmatrix}$$

Now, if we look at the expression of the G2SLS estimator of the coefficients of any structural equation, say the m-th one, we have :

$$\begin{aligned} \hat{\alpha}_{m,G2SLS} &= \left[Z_m' \Sigma_{mm}^{-1} X (X' \Sigma_{mm}^{-1} X)^{-1} X' \Sigma_{mm}^{-1} Z_m \right]^{-1} \\ &\quad Z_m' \Sigma_{mm}^{-1} X (X' \Sigma_{mm}^{-1} X)^{-1} X' \Sigma_{mm}^{-1} y_m \end{aligned}$$

Using the fact that $X' \Sigma_{mm}^{-1} Z_m$ is non-singular when the equation is just-identified, the above expression can be simplified to the following :

$$(6.44) \quad \hat{\alpha}_{m,G2SLS} = (X' \Sigma_{mm}^{-1} Z_m)^{-1} X' \Sigma_{mm}^{-1} y_m$$

By comparing the two expressions (6.43) and (6.44), we conclude that the G3SLS estimator is exactly equal to the G2SLS estimator when the whole system is just identified. Naturally, both these estimators will also have the same limiting distributions.

Equality between the feasible G3SLS estimator and the feasible G2SLS estimator is immediate provided we use the same estimators for the variance components in both cases.

6.3.2.2 Comparison between fG2SLS, fG3SLS and IfGLS estimators

Since, in the case of a just-identified system, the fG3SLS and the fG2SLS estimators are identical and that the fG2SLS estimator of each equation is in turn asymptotically equivalent to the corresponding IfGLS estimator (result proved in Section 6.3.1), we arrive at the following interesting conclusion that when a system of simultaneous equations is just-identified, the (feasible) G3SLS estimator is exactly equal to the (feasible) G2SLS estimator and they are both asymptotically equivalent to the (feasible) IGLS estimator.

APPENDIX 6.A : Limiting Distribution of the Indirect Feasible GLS Estimator

As in the case of the (feasible) generalised 2SLS estimator, we will determine the limiting distribution of

$$\begin{bmatrix} \sqrt{N} & 0 \\ 0 & \sqrt{NT}\, I_{\tilde{M}_1+K_1-1} \end{bmatrix} (\hat{\tilde{\alpha}}_{1,fIGLS} - \tilde{\alpha}_1) = \bar{D}_2^{-\frac{1}{2}}(\hat{\tilde{\alpha}}_{1,fIGLS} - \tilde{\alpha}_1)$$

Using expression (6.39), we can write :

$$(6.A.1) \quad \bar{D}_2^{-\frac{1}{2}}(\hat{\tilde{\alpha}}_{1,fIGLS} - \tilde{\alpha}_1) = \bar{D}_2^{-\frac{1}{2}} \left[H_a^* \hat{\Pi}_1 \; H_b^*\right]^{-1} L_*' \Gamma_*' D^{\frac{1}{2}} (D^{\frac{1}{2}} \tilde{X}'\hat{\Omega}^{-1} \tilde{X} D^{\frac{1}{2}})^{-1} D^{\frac{1}{2}} \tilde{X}' \hat{\Omega}^{-1} w$$

where

$$D = \begin{bmatrix} \frac{1}{N} I & 0 \\ 0 & \frac{1}{NT} I_{M(K-1)} \end{bmatrix}$$

In order to find the limiting distribution of (6.A.1), we need to derive the following limits :

I. $\operatorname{plim} \bar{D}_2^{-\frac{1}{2}} \left[H_a^* \hat{\Pi}_1 \; H_b^*\right]^{-1} L_*' \Gamma_*' D^{\frac{1}{2}}$:

First, we note that

$$\left[H_a^* \hat{\Pi}_1 \; H_b^*\right]^{-1} L_*' = \begin{bmatrix} 1 & \hat{\Pi}_{a1}^{o} & 0 \\ 0 & \hat{\Pi}_{a1}^{-} & I \\ 0 & \hat{\Pi}_{b1} & 0 \end{bmatrix}^{-1} \begin{bmatrix} e_1' & 0 & 0 \\ 0 & I_{K-1} & 0 \end{bmatrix}$$

$$= \begin{bmatrix} 1 & \hat{\Pi}_{ao} \\ 0 & \hat{\Pi}_{ab} \end{bmatrix}^{-1} \begin{bmatrix} e_1' & 0 \\ 0 & L_{**} \end{bmatrix}$$

denoting $\hat{\Pi}_{ao} = [\hat{\Pi}_{a1}^{o} \quad 0]$

$$\hat{\Pi}_{ab} = \begin{bmatrix} \hat{\Pi}_{a1}^{-} & I \\ \hat{\Pi}_{b1} & 0 \end{bmatrix}$$

$$L_{**} = [I_{K-1} \quad 0]$$

$$\equiv \begin{bmatrix} \bar{M}_{11} & \bar{M}_{12} \\ 0 & \bar{M}_{22} \end{bmatrix} \begin{bmatrix} \bar{N}_{11} & 0 \\ 0 & \bar{N}_{22} \end{bmatrix} = \begin{bmatrix} \bar{M}_{11}\bar{N}_{11} & \bar{N}_{12}\bar{N}_{22} \\ 0 & \bar{M}_{22}\bar{N}_{22} \end{bmatrix}$$

defining $\bar{M}_{11}$, $\bar{M}_{12}$, $\bar{M}_{22}$, $\bar{N}_{11}$ and $\bar{N}_{22}$ appropriately.

Then we write :

$$\bar{D}_2^{-\frac{1}{2}} \quad [H_a^* \hat{\Pi}_1 H_b^*]^{-1} L_*' \Gamma_*' D^{\frac{1}{2}}$$

$$= \begin{bmatrix} \sqrt{N} & \bar{M}_{11} \bar{N}_{11} & \sqrt{N} & \bar{M}_{12} \bar{N}_{22} \\ & 0 & \sqrt{NT} & \bar{M}_{22} \bar{N}_{22} \end{bmatrix} \begin{bmatrix} \frac{1}{\sqrt{N}} \Gamma' & 0 \\ 0 & \frac{1}{\sqrt{NT}} (\Gamma' \otimes I_{K-1}) \end{bmatrix}$$

$$= \begin{bmatrix} \bar{M}_{11} \bar{N}_{11} \Gamma' & \frac{1}{\sqrt{T}} \bar{M}_{12} \bar{N}_{22} (\Gamma' \otimes I_{K-1}) \\ 0 & \bar{M}_{22} \bar{N}_{22} (\Gamma' \otimes I_{K-1}) \end{bmatrix}$$

Finally, we conclude that the plim of the above expression is given by :

$$\begin{bmatrix} (\text{plim } \bar{M}_{11}) \bar{N}_{11} & 0 \\ 0 & (\text{plim } \bar{M}_{22}) \bar{N}_{22} \end{bmatrix} \begin{bmatrix} \Gamma' & 0 \\ 0 & \Gamma' \otimes I_{K-1} \end{bmatrix}$$

$$= \begin{bmatrix} (\text{plim } \bar{M}_{11}) & 0 \\ 0 & (\text{plim } \bar{M}_{22}) \end{bmatrix} \begin{bmatrix} \bar{N}_{11} & 0 \\ 0 & \bar{N}_{22} \end{bmatrix} \Gamma_*'$$

$$= \begin{bmatrix} 1 & 0 \\ 0 & \Pi_{ab}^{-1} \end{bmatrix} \begin{bmatrix} e_1' & 0 \\ 0 & L_{**} \end{bmatrix} \Gamma_*'$$

(6.A.2)

replacing plim $\bar{M}_{11}$ and plim $\bar{M}_{22}$ by their respective values in the inverse of the limit of $\begin{bmatrix} 1 & \hat{\Pi}_{ao} \\ 0 & \hat{\Pi}_{ab} \end{bmatrix}$ and also replacing $\bar{N}_{11}$ and $\bar{N}_{22}$ by their respective values. Now, the expression (6.A.2) can be further rewritten as :

$$= \begin{bmatrix} 1 & 0 \\ 0 & \Pi_{ab}^{-1} \end{bmatrix} L_*' \Gamma_*'$$

$$= \begin{bmatrix} 1 & 0 \\ 0 & \begin{bmatrix} \Pi_{a1}^{-} & I \\ \Pi_{b1} & 0 \end{bmatrix}^{-1} \end{bmatrix} L_*' \Gamma_*'$$

$$= \begin{bmatrix} 1 & 0 \\ 0 & [\Pi_{*1} \; H_1^*] \end{bmatrix}^{-1} L_*' \Gamma_*'$$

(in the notations of Chapter 3)

Hence, the required plim is given by

(6.A.3) $\tilde{P}_1^{-1} L_*' \Gamma_*'$

using the notation of Chapter 4 for $\tilde{P}_m$ (see (4.41)).

II. $\text{plim}\,(D^{\frac{1}{2}} \tilde{X}' \hat{\Omega}^{-1} \tilde{X} D^{\frac{1}{2}})^{-1}$:

First,

$$\tilde{X}'\hat{\Omega}^{-1}\tilde{X} = \begin{bmatrix} \sum_i \hat{\Omega}_i^{-1} \otimes \iota' M_i \iota & \sum_i \hat{\Omega}_i^{-1} \otimes \iota' M_i \underline{X} \\ \sum_i \hat{\Omega}_i^{-1} \otimes \underline{X}' M_i \iota & \sum_i \hat{\Omega}_i^{-1} \otimes \underline{X}' M_i \underline{X} \end{bmatrix}$$

$$= \begin{bmatrix} \hat{\Omega}_3^{-1} \cdot NT & \hat{\Omega}_3^{-1} \otimes \iota' \underline{X} \\ \hat{\Omega}_3^{-1} \otimes \underline{X}'\iota & \sum_i \hat{\Omega}_i^{-1} \otimes \underline{X}' M_i \underline{X} \end{bmatrix} \quad \text{using (3.28)}$$

$= D_1$ using the notation of Chapter 3, see (3.B.3).

Hence, the required plim becomes :

$$\text{plim}\,(D^{\frac{1}{2}} D_1 D^{\frac{1}{2}})^{-1} = \begin{bmatrix} (\Omega_\mu + \Omega_\nu) & 0 \\ 0 & \Omega_\varepsilon \otimes R^{-1} \end{bmatrix}$$

(cf. (3.B.6) , page 102)

(6.A.4) $= \bar{D}$ as denoted in (3.B.6)

Knowing the two limits (6.A.3) and (6.A.4), we can say that the limiting distribution of $\bar{D}_2^{-\frac{1}{2}} (\hat{\tilde{\alpha}}_{1,IfGLS} - \tilde{\alpha}_1)$ is the same as that of

$$\tilde{P}_1^{-1} L_*' \Gamma_*' \bar{D} D^{\frac{1}{2}} \tilde{X}' \hat{\Omega}^{-1} w$$

Now,

$$\tilde{X}' \hat{\Omega}^{-1} = \begin{bmatrix} \sum_i \hat{\Omega}_i^{-1} \otimes \iota' M_i \\ \sum_i \hat{\Omega}_i^{-1} \otimes \underline{X}' M_i \end{bmatrix}$$

$$\tilde{X}'\hat{\Omega}^{-1} = \begin{bmatrix} \hat{\Omega}_3^{-1} \otimes \iota' \\ \sum_i \hat{\Omega}_i^{-1} \otimes \underline{X}' M_i \end{bmatrix}$$

$= D_2$ using the notation of Chapter 3, cf. (3.B.4), page 101.

Thus,

$$D^{\frac{1}{2}}\ \tilde{X}'\ \hat{\Omega}^{-1}\ w = D^{\frac{1}{2}} D_2\ w$$

whose limiting distribution was derived to be normal with zero mean and variance-covariance matrix equal to $\bar{D}^{-1}$ (see page 105 and expression (3.B.35)).

It follows that the limiting distribution of $\bar{D}_2^{-\frac{1}{2}}\ (\hat{\tilde{\alpha}}_{1,IfGLS} - \tilde{\alpha}_1)$ is normal with zero mean and variance-covariance matrix equal to

$$\tilde{P}_1^{-1}\ L_*'\ \Gamma_*'\ \bar{D}\ \bar{D}^{-1}\ \bar{D}'\ \Gamma_*\ L_*\ \tilde{P}_1^{-1'}$$

$$= \tilde{P}_1^{-1}\ L_*'\ \Gamma_*'\ \bar{D}\ \Gamma_*\ L_*\ \tilde{P}_1'^{-1}$$

$$= \tilde{P}_1^{-1}\ \tilde{R}_1^{-1}\ \tilde{P}_1'^{-1}$$

since

$$L_*'\ \Gamma_*'\ \bar{D}\ \Gamma_*\ L_* =$$

$$\begin{bmatrix} e_1' & 0 & 0 \\ 0 & I_{K-1} & 0 \end{bmatrix} \begin{bmatrix} \Gamma' & 0 \\ 0 & \Gamma' \otimes I_{K-1} \end{bmatrix} \begin{bmatrix} (\Omega_\mu + \Omega_\nu) & 0 \\ 0 & (\Omega_\varepsilon \otimes R^{-1}) \end{bmatrix} \begin{bmatrix} \Gamma & 0 \\ 0 & \Gamma \otimes I \end{bmatrix} \begin{bmatrix} e_1 & 0 \\ 0 & I_{K-1} \\ 0 & 0 \end{bmatrix}$$

$$= \begin{bmatrix} e_1' & 0 & 0 \\ 0 & I_{K-1} & 0 \end{bmatrix} \begin{bmatrix} \Gamma'(\Omega_\mu + \Omega_\nu)\Gamma & 0 \\ 0 & \Gamma'\Omega_\varepsilon\Gamma \otimes R^{-1} \end{bmatrix} \begin{bmatrix} e_1 & 0 \\ 0 & I_{K-1} \\ 0 & 0 \end{bmatrix}$$

$$= \begin{bmatrix} e_1' & 0 & 0 \\ 0 & I_{K-1} & 0 \end{bmatrix} \begin{bmatrix} (\Sigma_\mu + \Sigma_\nu) & 0 \\ 0 & \Sigma_\varepsilon \otimes R^{-1} \end{bmatrix} \begin{bmatrix} e_1 & 0 \\ 0 & I_{K-1} \\ 0 & 0 \end{bmatrix}$$

$$= \begin{bmatrix} e_1'(\Sigma_\mu + \Sigma_\nu)'e_1 & 0 \\ 0 & \begin{bmatrix} I_{K-1} & 0 \end{bmatrix} (\Sigma_\varepsilon \otimes R^{-1}) \begin{bmatrix} I_{K-1} \\ 0 \end{bmatrix} \end{bmatrix}$$

$$= \begin{bmatrix} \sigma_{\mu 11} + \sigma_{\nu 11} & 0 \\ 0 & \sigma_{\varepsilon 11}\ R^{-1} \end{bmatrix}$$

$$= \tilde{R}_1^{-1}$$ using the definition of $\tilde{R}_m$ given in (4.42)

Thus

$$(6.A.5)\quad \bar{D}_2^{-\frac{1}{2}}(\hat{\tilde{\alpha}}_{1,IfGLS} - \tilde{\alpha}_1) \sim N\left(0,\ \tilde{P}_1^{-1}\ \tilde{R}_1^{-1}\ \tilde{P}_1'^{-1}\right)$$

CHAPTER 7

BIAS OF THE FEASIBLE ESTIMATORS OF REDUCED FORM AND STRUCTURAL VARIANCE COMPONENTS AND COEFFICIENTS

7.1 The Unbiasedness of the Feasible AOV Estimators of Reduced Form Variance Components

Let us first briefly recall the structure of the reduced form variance-covariance matrix and the AOV formulae used for estimating its variance-components. The variance-covariance matrix is given in (3.61) :

$$(7.1) \qquad E\,(\text{vec } V)(\text{vec } V)' = \Omega = \sum_{i=1}^{4} \Omega_i \otimes M_i$$

where

$$(7.2) \qquad \Omega_i = [\omega_{imm'}] \ , \ m,m'=1,\ldots,M$$

$$(7.3) \qquad \begin{cases} \omega_{1mm'} = \omega_{\varepsilon mm'} + T\,\omega_{\mu mm'} \\ \omega_{2mm'} = \omega_{\varepsilon mm'} + N\,\omega_{\nu mm'} \\ \omega_{3mm'} = \omega_{\varepsilon mm'} + T\,\omega_{\mu mm'} + N\,\omega_{\nu mm'} = \omega_{1mm'} + \omega_{2mm'} - \omega_{\varepsilon mm'} \\ \omega_{4mm'} = \omega_{\varepsilon mm'} \end{cases}$$

The AOV estimators of these components are :

$$(7.4) \qquad \begin{cases} \hat{\omega}_{1mm'} = \frac{1}{N-1}\,\hat{v}'_m\,M_1\,\hat{v}_{m'} \\ \hat{\omega}_{2mm'} = \frac{1}{T-1}\,\hat{v}'_m\,M_2\,\hat{v}_{m'} \\ \hat{\omega}_{\varepsilon mm'} = \frac{1}{(N-1)(T-1)}\,\hat{v}'_m\,Q\,\hat{v}_{m'} \\ \hat{\omega}_{3mm'} = \hat{\omega}_{1mm'} + \hat{\omega}_{2mm'} - \hat{\omega}_{\varepsilon mm'} \end{cases}$$

where

$$\hat{v}_m = y_m - \iota\,\hat{\pi}_{om,cov} - \underline{X}\,\hat{\pi}_{*m,cov}$$

with

$$\hat{\pi}_{*m,cov} = (\underline{X}'Q\,\underline{X})^{-1}\,\underline{X}'Q\,y_m$$

$$\hat{\pi}_{om,cov} = \frac{1}{NT}\,\iota'\left[y_m - \underline{X}\,\hat{\pi}_{*m,cov}\right]$$

Thus, we can write :

$$\hat{v}_m = y_m - \frac{1}{NT}\,\iota\left[\iota' y_m - \iota'\underline{X}\,\hat{\pi}_{*m,cov}\right] - \underline{X}\,\hat{\pi}_{*m,cov}$$

$$= \left[I - \frac{1}{NT}\,\iota\iota'\right] y_m - \left[I - \frac{1}{NT}\,\iota\iota'\right]\underline{X}\,\hat{\pi}_{*m,cov}$$

$$= (I - M^*)\,(y_m - \underline{X}\,\hat{\pi}_{*m,cov})$$

denoting

(7.5) $\quad M^* = \frac{1}{NT}\,\iota\iota'$

Now,

$$y_m - \underline{X}\,\hat{\pi}_{*m,cov} = \iota\,\pi_{om} + \underline{X}\,\pi_{*m} + v_m - \underline{X}(\underline{X}'Q\underline{X})^{-1}\underline{X}'Q(\iota\,\pi_{om} + \underline{X}\,\pi_{*m} + v_m)$$

$$= \iota\,\pi_{om} + v_m - \underline{X}(\underline{X}'Q\,\underline{X})^{-1}\,\underline{X}'Q\,v_m$$

$$= \iota\,\pi_{om} + (I - \underline{X}(\underline{X}'Q\,\underline{X})^{-1}\,\underline{X}'Q)\,v_m$$

$$= \iota\,\pi_{om} + M_{QX}\,v_m$$

denoting

(7.6) $\quad M_{QX} = I - \underline{X}(\underline{X}'Q\,\underline{X})^{-1}\,\underline{X}'Q$

Hence, we have

$$\hat{v}_m = (I - M^*)\,(\iota\,\pi_{om} + M_{QX}\,v_m)$$

or

(7.7) $\quad \hat{v}_m = (I - M^*)\,M_{QX}\,v_m \qquad$ as $(I - M^*)\,\iota = 0$

Substituting (7.7) for v_m in (7.4) we get

(7.8)
$$\begin{cases} \hat{\omega}_{1mm'} = \frac{1}{N-1}\,v_m'\,M_{QX}'\,(I - M^*)'\,M_1\,(I - M^*)\,M_{QX}\,v_{m'} \\ \hat{\omega}_{2mm'} = \frac{1}{T-1}\,v_m'\,M_{QX}'\,(I - M^*)'\,M_2\,(I - M^*)\,M_{QX}\,v_{m'} \\ \hat{\omega}_{\varepsilon mm'} = \frac{1}{(N-1)(T-1)}\,v_m'M_{QX}'(I - M^*)'Q\,(I - M^*)\,M_{QX}\,v_{m'} \end{cases}$$

By noting that

$$(7.9) \qquad M^{*'} M_1 = M^{*'} M_2 = M^{*'} Q = 0$$

the above expressions are simplified as :

$$(7.10) \qquad \begin{cases} \hat{\omega}_{1mm'} = \frac{1}{N-1} v_m' M_{QX}' M_1 M_{QX} v_{m'} \\ \hat{\omega}_{2mm'} = \frac{1}{T-1} v_m' M_{QX}' M_2 M_{QX} v_{m'} \\ \hat{\omega}_{\varepsilon mm'} = \frac{1}{(N-1)(T-1)} v_m' M_{QX}' Q M_{QX} v_{m'} \end{cases}$$

In what follows, we will prove the unbiasedness of only one of these estimators, say $\hat{\omega}_{1mm'}$, the unbiasedness of the other two being proved in the same way. Obviously, if $\hat{\omega}_{1mm'}$, $\hat{\omega}_{2mm'}$ and $\hat{\omega}_{\varepsilon mm'}$ are unbiased, then $\hat{\omega}_{3mm'}$ is also unbiased by virtue of its definition in (7.4).

Let us therefore calculate the expectation of $\hat{\omega}_{1mm'}$:

$$E(\hat{\omega}_{1mm'}) = \frac{1}{N-1} E(\mathrm{tr}\ v_m' M_{QX}' M_1 M_{QX} v_m)$$

$$= \frac{1}{N-1} E(\mathrm{tr}\ M_{QX}' M_1 M_{QX} v_m v_m')$$

$$(7.11) \qquad = \frac{1}{N-1} \mathrm{tr}\ M_{QX}' M_1 M_{QX} \Omega_{mm'} \quad \text{as } M_{QX} \text{ and } M_1 \text{ are non-stochastic}$$

Now,

$$\mathrm{tr}\ M_{QX}' M_1 M_{QX} \Omega_{mm'} = \mathrm{tr}\ M_1 M_{QX} \Omega_{mm'} \quad \text{as}$$

$$M_{QX}' M_1 = (I-\underline{X}(\underline{X}'Q\underline{X})^{-1} \underline{X}'Q) M_1$$

$$= M_1 \text{ since } Q M_1 = 0$$

$$= \mathrm{tr}\ M_1 (I-\underline{X}(\underline{X}'Q\underline{X})^{-1}\underline{X}'Q)(\sum_i \omega_{imm'} M_i)$$

$$= \mathrm{tr}\ M_1(\sum_i \omega_{imm'} M_i - \omega_{\varepsilon mm'} \underline{X}(\underline{X}'Q\underline{X})^{-1}\underline{X}'Q)$$

as $Q M_i = 0$, $i=1,2,3$

and $Q M_4 = Q Q = Q$

$$= \text{tr}\left[\omega_{1mm'} M_1 - \omega_{\varepsilon mm'} M_1 \underline{X}(\underline{X}'Q\underline{X})^{-1}\underline{X}'Q\right]$$

$$= \omega_{1mm'} \text{ tr } M_1 - \omega_{\varepsilon mm'} \text{ tr } \underline{X}(\underline{X}'Q\underline{X})^{-1}\underline{X}'Q M_1$$

(7.12) $$= \omega_{1mm'} (N-1) \qquad \text{as } Q M_1 = 0$$

Thus, combining (7.11) and (7.12), we obtain :

$$E(\hat{\omega}_{1mm'}) = \frac{1}{N-1} \omega_{1mm'} (N-1)$$

(7.13) $$= \omega_{1mm'}$$

which proves that $\hat{\omega}_{1mm'}$ is unbiased. Similarly, as mentioned earlier, $\hat{\omega}_{2mm'}$, $\hat{\omega}_{\varepsilon mm'}$ and $\hat{\omega}_{3mm'}$ can be verified to be unbiased too.

7.2 The Unbiasedness of the Feasible GLS Estimator of the Reduced Form Coefficients

Once again, let us recall that the feasible GLS estimator of vec Π is given by :

(7.14) $$\text{vec } \hat{\Pi} = \left[\sum_i \hat{\Omega}_i^{-1} \otimes X'M_iX\right]^{-1} \left[\sum_i \hat{\Omega}_i^{-1} \otimes X'M_i\right] y$$

Substituting

(7.15) $$y = (I \otimes X) \text{ vec } \Pi + \text{vec } V$$

in (7.14) and simplifying, we get

(7.16) $$\text{vec}\hat{\Pi} = \text{vec}\Pi + (\sum_i \hat{\Omega}_i^{-1} \otimes X'M_iX)^{-1} (\sum_i \hat{\Omega}_i^{-1} \otimes X'M_i) \text{ vec}V$$

or

(7.17) $$\text{vec}\hat{\Pi} - \text{vec } \Pi = \Psi(V) \text{ vec } V$$

where $\Psi(V)$ is obtained by expressing $\hat{\Omega}_i$, $i=1,2,3,4$ in terms of V in the expression premultiplying vecV in (7.16). More explicitly, we have :

$$(7.18)\quad \begin{cases} \hat{\Omega}_1 = \frac{1}{N-1}\hat{V}' M_1 \hat{V} \\ \hat{\Omega}_2 = \frac{1}{T-1}\hat{V}' M_2 \hat{V} \\ \hat{\Omega}_\varepsilon = \frac{1}{(N-1)(T-1)}\hat{V}' Q \hat{V} \\ \hat{\Omega}_3 = \hat{\Omega}_1 + \hat{\Omega}_2 - \hat{\Omega}_\varepsilon \end{cases}$$

denoting $\hat{V} = \left[\hat{v}_1 \ \ldots \ \hat{v}_M\right]$ and writing equations (7.4) in matrix form. Now, using (7.7) we can write

$$(7.19)\quad \hat{V} = (I - M^*)\, M_{QX}\left[v_1 \ \ldots \ v_M\right] = (I - M^*)\, M_{QX}\, V$$

Substituting (7.19) for $\hat{V}$ in (7.18), simplifying using (7.9) and then replacing the resulting expressions $\hat{\Omega}_i(V)$, $i=1,2,3,4$ in (7.16), we can identify the function $\Psi(V)$.

Now, we will establish the unbiasedness of vec $\hat{\Pi}$ using a result, due to Kakwani [23] , which states that "any estimator which has a sampling error of the form H(u).u can be proved to be unbiased, if its mean exists, provided H is even and u is symmetrically distributed around its mean". For a proof of this, see Kakwani [23].

First we note, from (7.19), that changing the sign of V changes that of $\hat{V}$ but leaves unchanged the signs of $\hat{\Omega}_1$, $\hat{\Omega}_2$, $\hat{\Omega}_3$, $\hat{\Omega}_\varepsilon$ as these are quadratic forms in $\hat{V}$ (cf. (7.18)). Hence,

$$\hat{\Omega}_i(V) = \hat{\Omega}_i(-V)$$

and

$$\hat{\Omega}_i^{-1}(V) = \hat{\Omega}_i^{-1}(-V)$$

Therefore,

$$\Psi(V) = \Big(\sum_i \hat{\Omega}_i^{-1}(V) \otimes \underline{X}'M_i\underline{X}\Big)^{-1} \Big(\sum_i \hat{\Omega}_i^{-1}(V) \otimes \underline{X}'M_i\Big)$$

is an even function of V.

Thus, we have shown that

$$(7.20)\quad \text{vec}\,\hat{\Pi} - \text{vec}\,\Pi = \Psi(V)\,.\,\text{vec}\,V$$

with

(7.21) $\psi(-V) = \psi(V)$

Hence, substituting - V for V in (7.20) merely changes the sign of (vec $\hat{\Pi}$ - vec Π). So, assuming that vec V is distributed symmetrically around its mean i.e. vec V and vec (- V) have the same probability law, it implies that (vec $\hat{\Pi}$ - vec Π) and (vec Π - vec $\hat{\Pi}$) have the same probability density function. In other words, vec $\hat{\Pi}$ is symmetrically distributed about the value vec Π thus proving that vec $\hat{\Pi}$ is unbiased if its mean exists.

7.3 Bias of Structural Variance Components Estimators

7.3.1 A General Note

In this section, we will assume that there are no time effects in the model i.e. each structural equation error term is composed of only two parts, namely a unit effect and an overall effect. This assumption has been made mainly to make the presentation of our approach as clear as possible. However, we believe that our approach can be extended to the general case with a few additional assumptions.

In the simpler case, the variance-covariance matrix of the error vector of, say, the first structural equation, reduces to the following :

(7.22) $\Sigma_{11} = E((\text{vec } u_1)(\text{vec } u_1')) = \sigma_{111} N_1 + \sigma_{411} N_4$

where

(7.23) $\sigma_{111} = \sigma_{\varepsilon 11} + T\,\sigma_{\mu 11}\ ,\ \sigma_{411} = \sigma_{\varepsilon 11}$

(7.24) $N_1 = \frac{A}{T}\ ,\ N_4 = I - \frac{A}{T}$

and the AOV estimators of the two variance components are :

(7.25) $\hat{\sigma}_{411} = \hat{\sigma}_{\varepsilon 11} = \frac{1}{N(T-1)}\, \hat{u}_1'\, N_4\, \hat{u}_1$

and

(7.26) $\hat{\sigma}_{111} = \frac{1}{N}\, \hat{u}_1'\, N_1\, \hat{u}_1$

whereby

$$(7.27) \qquad \hat{\sigma}_{\mu 11} = \frac{1}{T}\left[\frac{1}{N} u_1' N_1 u_1 - \hat{\sigma}_{\varepsilon 11}\right]$$

In order to find the bias of these estimators, we have to calculate their expectations. As the exact moments are impossible to determine, we will derive an asymptotic approximation to it, by expanding it in series and retaining only terms of order up to $\frac{1}{N}$, keeping T fixed. The validity of this procedure is proved in a paper by Bhattacharya and Ghosh [11] in which they show that, under certain moment conditions (cf. Theorem 2(b) and Remark 1.1., [11]), the moment approximation obtained by the above method is in fact the moment of the asymptotic approximation to the exact distribution of the estimator. We keep T fixed as it is typically the number of units that can become large in panel data compared to the number of time periods.

7.3.2 Bias of $\hat{\sigma}_{\varepsilon 11}$

Let us now proceed to calculate the bias of each of the three estimators given above, one by one. First, let us examine the estimator of $\hat{\sigma}_{\varepsilon 11}$ which is given by :

$$(7.28) \qquad \hat{\sigma}_{\varepsilon 11} = \frac{1}{N(T-1)} \quad \hat{u}_1' N_4 \hat{u}_1$$

Recall that

$$(7.29) \qquad \hat{u}_1 = y_1 - Z_1 \hat{\alpha}_{1,C2SLS}$$

where

$$\hat{\alpha}_{1,C2SLS} = \begin{bmatrix} \hat{a}_{1,C2SLS} \\ \hat{\alpha}^*_{1,C2SLS} \end{bmatrix}, \quad Z_1 = \begin{bmatrix} \iota & Z_1^* \end{bmatrix}$$

with the covariance 2SLS estimators in our particular case being given by :

$$(7.30) \begin{cases} \hat{\alpha}^*_{1,C2SLS} = \left[Z_1^{*\prime} N_4 \underline{X}(\underline{X}' N_4 \underline{X})^{-1} \underline{X}' N_4 Z_1^*\right]^{-1} Z_1^{*\prime} N_4 \underline{X}(\underline{X}' N_4 \underline{X})^{-1} \underline{X}' N_4 y_1 \\ \hat{a}_{1,C2SLS} = \frac{1}{NT} \iota'(y_1 - Z_1^* \hat{\alpha}^*_{1,C2SLS}) \end{cases}$$

Thus,

$$\hat{u}_1 = y_1 - Z_1^* \hat{\alpha}^*_{1,C2SLS} - \frac{1}{NT} \iota\iota'(y_1 - Z_1^* \hat{\alpha}^*_{1,C2SLS})$$

$$= (I - \frac{1}{NT} \iota\iota')(y_1 - Z_1^* \hat{\alpha}^*_{1,C2SLS})$$

(7.31) $$= (I - M^*)(y_1 - Z_1^*(Z_1^{*\prime} N_4 \underline{X}(\underline{X}'N_4\underline{X})^{-1}\underline{X}'N_4 Z_1^*)^{-1} Z_1^{*\prime} N_4 \underline{X}(\underline{X}'N_4\underline{X})^{-1}\underline{X}'N_4' y_1)$$

using (7.5) and substituting (7.30) for $\hat{\alpha}^*_{1,C2SLS}$

(7.32) $$= (I - M^*)(I - Z_1^*(Z_1^{*\prime}\, C\, Z_1^*)^{-1}\, Z_1^{*\prime}\, C)\, u_1$$

by substituting

$$y_1 = Z_1\, \alpha_1 + u_1$$

in (7.31), simplifying it and denoting

(7.33) $$N_4\, \underline{X}(\underline{X}'N_4\underline{X})^{-1}\underline{X}'N_4 \equiv C$$ [1]

Hence

$$\hat{\sigma}_{\varepsilon 11} = \frac{1}{(N-1)(T-1)}\, u_1'(I - Z_1^*(Z_1^{*\prime}CZ_1^*)^{-1}Z_1^{*\prime}C)'(I-M^*)'N_4(I-M^*)(I-Z_1^*(Z_1^{*\prime}CZ_1^*)^{-1}Z_1^{*\prime}C)u_1$$

$$= \frac{1}{(N-1)(T-1)}\, u_1'(I - Z_1^*(Z_1^{*\prime}CZ_1^*)^{-1}Z_1^{*\prime}C)'N_4(I-Z_1^*(Z_1^{*\prime}CZ_1^*)^{-1}Z_1^{*\prime}C)u_1 \quad \text{as } N_4M^* = 0$$

(7.34) $$= \frac{1}{(N-1)(T-1)}\Big[u_1'N_4u_1 - u_1'C\, Z_1^*(Z_1^{*\prime}C\, Z_1^*)^{-1}Z_1^{*\prime}N_4u_1 - u_1'N_4Z_1^*(Z_1^{*\prime}C\, Z_1^*)^{-1}Z_1^{*\prime}C\, u_1 + u_1'C\, Z_1^*(Z_1^{*\prime}C\, Z_1^*)^{-1}Z_1^{*\prime}N_4Z_1^*(Z_1^{*\prime}C\, Z_1^*)^{-1}Z_1^{*\prime}C\, u_1\Big]$$

Let us consider these four terms one by one, expanding each one in series and retaining terms of order greater than or equal to $\frac{1}{N}$ in each of the series. After doing this, we will calculate the expectations of all the retained terms, which will allow us to obtain the bias.

---

1) This C matrix, which appears throughout this chapter, should not be confused with the earlier C matrix of (3.121) used till the previous chapter.

(i) First term of (7.34) :

The first term of (7.34) is

$$(7.35) \qquad \frac{1}{(N-1)(T-1)}\, u_1'\, N_4\, u_1$$

This term is O(1) as

$$E(\frac{1}{N(T-1)}\, u_1'\, N_4\, u_1) = \frac{1}{N(T-1)} \operatorname{tr} N_4\, \Sigma_{11}$$

$$(7.36) \qquad = \frac{\sigma_{\varepsilon 11} \operatorname{tr} N_4}{N(T-1)} = \sigma_{\varepsilon 11} = O(1)$$

(ii) Second Term of (7.34) :

$$\frac{1}{N(T-1)}\, u_1' C\, Z_1^*(Z_1^{*\prime} C\, Z_1^*)^{-1}\, Z_1^{*\prime} N_4\, u_1$$

Let us write, following Nagar's approach (see [32]),

$$(7.37) \qquad Z_1^* = \bar{Z}_1 + V_z = \begin{bmatrix}\bar{Y}_1 & X_1^*\end{bmatrix} + \begin{bmatrix}V_1 & 0\end{bmatrix} = \begin{bmatrix}X\Pi_1 & X_1^*\end{bmatrix} + \begin{bmatrix}V_1 & 0\end{bmatrix}$$

where Π_1 is the submatrix of Π and V_1 the submatrix of V appearing in the reduced form equations explaining Y_1 i.e.

$$Y_1 = X\,\Pi_1 + V_1$$

Then,

$$u_1' C\, Z_1^*(Z_1^{*\prime} C\, Z_1^*)^{-1}\, Z_1^{*\prime} N_4\, u_1$$

$$= u_1' C(\bar{Z}_1 + V_z)\left[(\bar{Z}_1' + V_z')C(\bar{Z}_1 + V_z)\right]^{-1}(\bar{Z}_1' + V_z')N_4\, u_1$$

$$= (u_1' C\,\bar{Z}_1 + u_1' C\, V_z)\left[\bar{Z}_1' C\,\bar{Z}_1 + V_z' C\,\bar{Z}_1 + \bar{Z}_1' C\, V_z + V_z' C\, V_z\right]^{-1}(\bar{Z}_1' N_4 u_1 + V_z' N_4 u_1)$$

$$= (u_1' C\,\bar{Z}_1 + u_1' C\, V_z)\left[I + (\bar{Z}_1' C\,\bar{Z}_1)^{-1}(V_z' C\,\bar{Z}_1 + \bar{Z}_1' C\, V_z + V_z' C\, V_z)\right]^{-1}(\bar{Z}_1' C\,\bar{Z}_1)^{-1}(\bar{Z}_1' N_4 u_1 + V_z' N_4 u_1)$$

$$= (u_1' C\,\bar{Z}_1 + u_1' C\, V_z)\Big\{I - (\bar{Z}_1' C\,\bar{Z}_1)^{-1}(V_z' C\,\bar{Z}_1 + \bar{Z}_1' C\, V_z + V_z' C\, V_z) + (\bar{Z}_1' C\,\bar{Z}_1)^{-1}(V_z' C\,\bar{Z}_1 + \bar{Z}_1' C\, V_z +$$

$$(7.38) \qquad + V_z' C\, V_z)(\bar{Z}_1' C\,\bar{Z}_1)^{-1}(V_z' C\,\bar{Z}_1 + \bar{Z}_1' C\, V_z + V_z' C\, V_z) - \ldots\Big\}(\bar{Z}_1' C\,\bar{Z}_1)^{-1}(\bar{Z}_1' N_4 u_1 + V_z' N_4 u_1)$$

by expanding $\left[I + (\bar{Z}_1' C\,\bar{Z}_1)^{-1}(V_z' C\,\bar{Z}_1 + \bar{Z}_1' C\, V_z + V_z' C\, V_z)\right]^{-1}$ in series after verifying that the order of the relevant matrix is $\frac{1}{\sqrt{N}}$ using the order results of Appendix 7.A (page 268).

From the orders calculated in Appendix 7.A, the above expression can be notationally simplified as :

(7.39) $$(A_{\frac{1}{2}} + A_o)(B_o + B_{-\frac{1}{2}} + B_{-1} + B_{-\frac{3}{2}} + B_{-2})(D_{-\frac{1}{2}} + D_o)$$

where

(7.40) $$\begin{cases} A_{\frac{1}{2}} = u_1' C \bar{Z}_1 \quad ; \quad A_o = u_1' C V_z \\ B_o = I \quad ; \quad B_{-\frac{1}{2}} = -(\bar{Z}_1' C \bar{Z}_1)^{-1} (V_z' C \bar{Z}_1 + \bar{Z}_1' C V_z) \end{cases}$$

(7.41) $$B_{-1} = -(\bar{Z}_1'C\,\bar{Z}_1)^{-1} V_z'C\,V_z + (\bar{Z}_1'C\,\bar{Z}_1)^{-1}(V_z'C\,\bar{Z}_1 + \bar{Z}_1'C\,V_z)(\bar{Z}_1'C\,\bar{Z}_1)^{-1}(V_z'C\,\bar{Z}_1 + \bar{Z}_1'C\,V_z)$$

(7.42) $$B_{-\frac{3}{2}} = -(\bar{Z}_1'C\,\bar{Z}_1)^{-1} V_z'C\,V_z\,(\bar{Z}_1'C\,\bar{Z}_1)^{-1}(V_z'C\,\bar{Z}_1 + \bar{Z}_1'C\,V_z) + (\bar{Z}_1'C\,\bar{Z}_1)^{-1}(V_z'C\,\bar{Z}_1 + \bar{Z}_1'C\,V_z)(\bar{Z}_1'C\,\bar{Z}_1)^{-1} V_z'C\,V_z$$

(7.43) $$B_{-2} = -(\bar{Z}_1'C\,\bar{Z}_1)^{-1} V_z'C\,V_z\,(\bar{Z}_1'C\,\bar{Z}_1)^{-1}\,V_z'C\,V_z$$

(7.44) $$D_{-\frac{1}{2}} = (\bar{Z}_1'C\,\bar{Z}_1)^{-1}\,\bar{Z}_1'\,N_4\,u_1 \;;\; D_o = (\bar{Z}_1'C\,\bar{Z}_1)^{-1}\,V_z'\,N_4\,u_1$$

and where the indices indicate the power of N in the order of the term. Expanding the brackets in (7.39), we get :

$$(A_{\frac{1}{2}}B_o + A_oB_o + A_{\frac{1}{2}}B_{-\frac{1}{2}} + A_oB_{-\frac{1}{2}} + A_{\frac{1}{2}}B_{-1} + A_oB_{-1} + A_oB_{-\frac{3}{2}} + A_{\frac{1}{2}}B_{-\frac{3}{2}} + A_{\frac{1}{2}}B_{-2} + A_oB_{-2})(D_{-\frac{1}{2}} + D_o)$$

(7.45) $$\begin{aligned} = {} & A_{\frac{1}{2}}B_oD_{-\frac{1}{2}} + A_oB_oD_{-\frac{1}{2}} + A_{\frac{1}{2}}B_{-\frac{1}{2}}D_{-\frac{1}{2}} + A_oB_{-\frac{1}{2}}D_{-\frac{1}{2}} + A_{\frac{1}{2}}B_{-1}D_{-\frac{1}{2}} \\ & + A_oB_{-1}D_{-\frac{1}{2}} + A_{\frac{1}{2}}B_{-\frac{3}{2}}D_{-\frac{1}{2}} + A_oB_{-\frac{3}{2}}D_{-\frac{1}{2}} + A_{\frac{1}{2}}B_{-2}D_{-\frac{1}{2}} + A_oB_{-2}D_{-\frac{1}{2}} \\ & + A_{\frac{1}{2}}B_oD_o + A_oB_oD_o + A_{\frac{1}{2}}B_{-\frac{1}{2}}D_o + A_oB_{-\frac{1}{2}}D_o + A_{\frac{1}{2}}B_{-1}D_o + \\ & + A_oB_{-1}D_o + A_{\frac{1}{2}}B_{-\frac{3}{2}}D_o + A_{\frac{1}{2}}B_{-2}D_o + A_oB_{-2}D_o + A_oB_{-\frac{3}{2}}D_o \end{aligned}$$

Using the result that the order of a product is the sum of the orders of its components and neglecting terms of order less than $\frac{1}{N}$ after division by N(T-1), we can write :

$$\frac{1}{N(T-1)} u_1'C\, z_1^*(z_1^{*\prime}Cz_1^*)^{-1}\, z_1^{*\prime}N_4u_1 \overset{1)}{\cong} \frac{1}{N(T-1)}(A_{\frac{1}{2}}B_oD_{-\frac{1}{2}} + A_{\frac{1}{2}}B_oD_o + A_oB_oD_o + A_{\frac{1}{2}}B_{-\frac{1}{2}}D_o)$$

$$= \frac{1}{N(T-1)}\Big[u_1'C\,\bar{z}_1(\bar{z}_1'C\,\bar{z}_1)^{-1}(\bar{z}_1'N_4u_1 + V_z'N_4u_1) - u_1'C\,\bar{z}_1(\bar{z}_1'C\,\bar{z}_1)^{-1}(V_z'C\,\bar{z}_1 + \bar{z}_1'C\,V_z)$$

$$\cdot(\bar{z}_1'C\,\bar{z}_1)^{-1}V_z'N_4u_1 + u_1'C\,V_z(\bar{z}_1'C\,\bar{z}_1)^{-1}V_z'N_4u_1\Big] \tag{7.46}$$

(iii) Third Term of (7.34) :

$$\frac{1}{N(T-1)} u_1'\,N_4\,z_1^*\,(z_1^{*\prime}C\,z_1^*)^{-1}\,z_1^{*\prime}\,C\,u_1$$

This term is the transpose of the second term and hence the terms of order upto $\frac{1}{N}$ in its expansion are the transpose of those in the expansion of the previous term i.e. we will have the following truncated expansion for this term :

$$\frac{1}{N(T-1)}\Big[(u_1'N_4\bar{z}_1 + u_1'N_4V_z)(\bar{z}_1'C\,\bar{z}_1)^{-1}\,\bar{z}_1'C\,u_1 + u_1'N_4V_z(\bar{z}_1'C\,\bar{z}_1)^{-1}(\bar{z}_1'C\,V_z + V_1'C\,\bar{z}_1)$$

$$\cdot(\bar{z}_1'C\,\bar{z}_1)^{-1}\,\bar{z}_1'C\,u_1 + u_1'N_4V_z(\bar{z}_1'C\,\bar{z}_1)^{-1}V_z'C\,u_1\Big] \tag{7.47}$$

(iv) Fourth Term of (7.34) :

$$\frac{1}{N(T-1)} u_1'C\,z_1^*\,(z_1^{*\prime}C\,z_1^*)^{-1}\,z_1^{*\prime}N_4z_1^*\,(z_1^{*\prime}C\,z_1^*)^{-1}\,z_1^{*\prime}C\,u_1$$

As before, by replacing z_1^* by $\bar{z}_1 + V_z$ (see (7.37)) and expanding the inverse in series, we obtain :

$$\frac{1}{N(T-1)}(u_1'C\,z_1 + u_1'C\,V_z)\Big\{I - (\bar{z}_1'C\,\bar{z}_1)^{-1}(V_z'C\,\bar{z}_1 + \bar{z}_1'C\,V_z + V_1'C\,V_z) - \ldots\Big\}$$

$$\cdot(\bar{z}_1'C\,\bar{z}_1)^{-1}(\bar{z}_1'N_4\bar{z}_1 + \bar{z}_1'N_4V_z + V_z'N_4\bar{z}_1 + V_z'N_4V_z)\Big\{I + (\bar{z}_1'C\,\bar{z}_1)^{-1}$$

$$\cdot(V_z'C\,\bar{z}_1 + \bar{z}_1'C\,V_z + V_z'C\,V_z) - \ldots\Big\}(\bar{z}_1'C\,\bar{z}_1)^{-1}(\bar{z}_1'C\,u_1 + V_z'C\,u_1) \tag{7.48}$$

Once again, simplifying the notation using (7.40) to (7.44), we can write (7.48) as :

---

1) an equality sign with ~ above indicates that only terms upto a certain order are retained in the expansion on the RHS.

$$(7.49) \quad \frac{1}{N(T-1)} (A_{\frac{1}{2}} + A_o)(B_o + B_{-\frac{1}{2}} + B_{-1} + B_{-\frac{3}{2}} + B_{-2})(C_{-\frac{1}{2}} + C_o)$$
$$(B_o + B_{-\frac{1}{2}} + B_{-1} + B_{-\frac{3}{2}} + B_{-2})(E_{-\frac{1}{2}} + E_{-1})$$

where

$$(7.50) \quad \begin{cases} C_{-\frac{1}{2}} = (\bar{Z}_1' \, C \, \bar{Z}_1)^{-1} \, (\bar{Z}_1' \, N_4 \, V_z + V_z' \, N_4 \, \bar{Z}_1) \\ C_o = (\bar{Z}_1' \, C \, \bar{Z}_1)^{-1} \, (\bar{Z}_1' \, N_4 \, \bar{Z}_1 + V_z' \, N_4 \, V_z) \end{cases}$$

$$(7.51) \quad \begin{cases} E_{-\frac{1}{2}} = (\bar{Z}_1' \, C \, \bar{Z}_1)^{-1} \, \bar{Z}_1' \, C \, u_1 \\ E_{-1} = (\bar{Z}_1' \, C \, \bar{Z}_1)^{-1} \, V_z' \, C \, u_1 \end{cases}$$

We will straight away remove $B_{-\frac{3}{2}}$ and B_{-2} from expression (7.49) as they are of order less than $\frac{1}{N}$ and are further multiplied by terms of order less than 1 all the time.

Now, multiplying the brackets in (7.49) and neglecting terms of order less than $\frac{1}{N}$ (in the same way as done on pages 236 - 237) we are left with only one "significant"1) term, which is :

$$\frac{1}{N(T-1)} A_{\frac{1}{2}} \, B_o \, C_o \, B_o \, E_{-\frac{1}{2}}$$

$$(7.52) = \frac{1}{N(T-1)} u_1'C\bar{Z}_1(\bar{Z}_1'C\bar{Z}_1)^{-1}(\bar{Z}_1'N_4\bar{Z}_1 + V_z'N_4V_z)(\bar{Z}_1'C\bar{Z}_1)^{-1}\bar{Z}_1'Cu_1$$

Now that we have obtained all the "significant" terms in the expansion of $\hat{\sigma}_{\varepsilon 11}$, the next step is to find their expectations. Let us consider these terms one by one, after multiplying all of them by N(T-1) in order to avoid repeating this factor each time.

---

1) The term "significant", wherever used in this chapter, will be taken to mean "of order greater than or equal to $\frac{1}{N}$". Equivalently the term " non-negligible" is also used.

Let us recall that the "significant" terms are (7.35), (7.46), (7.47) and (7.52). The paragraphs dealing with expectations of these four terms will be numbered I, II, III & IV respectively.

I.

$$E\,(u_1'N_4u_1) = N(T-1)\,\sigma_{\varepsilon 11} \qquad \text{(see (7.36) above)} \tag{7.53}$$

II.

$$E\Big[u_1'C\,\bar{Z}_1(\bar{Z}_1'C\,\bar{Z}_1)^{-1}(\bar{Z}_1'N_4u_1 + V_z'N_4u_1) - u_1'C\,\bar{Z}_1(\bar{Z}_1'C\,\bar{Z}_1)^{-1} (V_z'C\,\bar{Z}_1 + \bar{Z}_1'C\,V_z)(\bar{Z}_1'C\,\bar{Z}_1)^{-1}\,V_z'N_4u_1 + u_1'C\,V_z(\bar{Z}_1'C\,\bar{Z}_1)^{-1}V_z'N_4u_1\Big] \tag{7.54}$$

Let us take each of the above five terms one by one and find its expectation.

II.1

$$E\,u_1'\,C\,\bar{Z}_1\,(\bar{Z}_1'\,C\,\bar{Z}_1)^{-1}\,\bar{Z}_1'\,N_4\,u_1 \tag{7.55}$$

$$= E\,\mathrm{tr}\,C\,\bar{Z}_1\,(\bar{Z}_1'C\,\bar{Z}_1)^{-1}\,\bar{Z}_1'\,N_4\,u_1\,u_1'$$

$$= \mathrm{tr}\,C\,\bar{Z}_1\,(\bar{Z}_1'C\,\bar{Z}_1)^{-1}\,\bar{Z}_1'\,N_4\,\Sigma_{11}$$

$$= \sigma_{\varepsilon 11}\,\mathrm{tr}\,(\bar{Z}_1'C\,\bar{Z}_1)^{-1}\,\bar{Z}_1'\,N_4C\,\bar{Z}_1$$

$$= \sigma_{\varepsilon 11}\,\mathrm{tr}\,I_{(K_1+\tilde{M}_1-1)} \qquad \text{as } N_4C = C$$

$$= \sigma_{\varepsilon 11}(K_1+\tilde{M}_1-1) \tag{7.56}$$

II.2

$$E\,u_1'C\,\bar{Z}_1(\bar{Z}_1'C\,\bar{Z}_1)^{-1}\,V_z'N_4u_1 \tag{7.57}$$

$$= E\,u_1'\bar{P}\,V_z'N_4u_1$$

by writing

$$\bar{P} = C\,\bar{Z}_1\,(\bar{Z}_1'C\,\bar{Z}_1)^{-1} \tag{7.58}$$

$$= E - u_1'\bar{P}_1L_1'\,\Gamma^{-1'}\,U'N_4u_1$$

by writing

$$\bar{P} = \left[\bar{P}_1\;\bar{P}_2\right] \tag{7.59}$$

(7.60) $V_1 = V\, L_1 = -\, U\, \Gamma^{-1}\, L_1$

where $\bar{P}_1$ and $\bar{P}_2$ are of appropriate dimensions and L_1 is a selection matrix.

Denoting

(7.61) $\underset{(NTxM)}{\bar{F}} = -\, \bar{P}_1\, L_1\, \Gamma^{-1'}$

we can also write (7.57) as

(7.62) $E\, u_1'\bar{F}\, U'N_4 u_1 = 0$ (See Section 7.B.1 of Appendix 7.B for the derivation of the the expectation value.)

II.3

(7.63) $-\, E\, u_1'C\, \bar{Z}_1(\bar{Z}_1'C\, \bar{Z}_1)^{-1}\, V_z'C\, \bar{Z}_1(\bar{Z}_1'C\, \bar{Z}_1)^{-1}\, V_z'N_4 u_1$

$= -\, E\, u_1'\bar{P}\, V_z'\bar{P}\, V_z'N_4 u_1$ using (7.58)

$= -\, E\, u_1'\begin{bmatrix}\bar{P}_1 \bar{P}_2\end{bmatrix} \begin{bmatrix} V_1' \\ 0 \end{bmatrix} \begin{bmatrix}\bar{P}_1 \bar{P}_2\end{bmatrix} \begin{bmatrix} V_1' \\ 0 \end{bmatrix} N_4 u_1$

using (7.59) and the definition of V_z in (7.37)

$= -\, E\, u_1'\bar{P}_1 V_1'\bar{P}_1 V_1' N_4 u_1$

$= -\, E\, u_1'\bar{P}_1 L_1'\Gamma^{-1'} U'\bar{P}_1 L_1 \Gamma^{-1'} U'N_4 u_1$ using (7.60)

(7.64) $= -\, E\, u_1'\bar{F}\, U'\bar{F}\, U'\, N_4\, u_1$ using (7.61)

(7.65) $$= -\sum_k \sum_\ell \Big[\sigma_{\varepsilon 1k}\, \sigma_{\varepsilon \ell 1}\, \bar{R}_{k\ell}\, (N(T-1) - \frac{T-1}{T}) + \sigma_{\varepsilon 1\ell}\, \sigma_{\varepsilon k1}\, \bar{R}_{k\ell}\, \frac{1}{T} + \sigma_{\varepsilon 11}\, \sigma_{\varepsilon k\ell}\, \bar{R}_{k\ell}\, \frac{1}{T} + m^4_{\varepsilon 11k}\, \bar{R}_{k\ell}\, (\frac{T-1}{T})\Big]$$

(See section 7.B.2 of Appendix 7.B for the derivation of the above expectation value as well as for the explanations of notations $\bar{R}_{k\ell}$ and $m^4_{\varepsilon 11k\ell}$.)

II.4

(7.66) $$- E\ u_1'C\ \bar{Z}_1(\bar{Z}_1'C\ \bar{Z}_1)^{-1}\ \bar{Z}_1'C\ V_z(\bar{Z}_1'C\ \bar{Z}_1)^{-1}\ V_z'N_4u_1$$

$$= - E\ u_1'\bar{P}\ \bar{Z}_1'C\ V_z\ \bar{W}\ V_z'N_4u_1$$ using (7.58) and writing

(7.67) $$\bar{W} = (\bar{Z}_1'C\ \bar{Z}_1)^{-1}$$

$$= - E\ u_1'\bar{H}\ V_1\ \bar{W}_{11}\ V_1'N_4u_1$$ where

(7.68) $\bar{H} = \bar{P}\ \bar{Z}_1'C$ and $\bar{W}$ is partitioned appropriately as

$$\bar{W} = \begin{bmatrix} \bar{W}_{11} & \bar{W}_{12} \\ \bar{W}_{21} & \bar{W}_{22} \end{bmatrix}$$

$$= - E\ u_1'\bar{H}\ U\ \Gamma^{-1}\ L_1\bar{W}_{11}L_1'\ \Gamma^{-1'}U'N_4u_1$$ using (7.60)

(7.69) $$= - E\ u_1'\bar{H}\ U\ \bar{R}\ U'N_4u_1$$ where

(7.70) $$\bar{R} = \Gamma^{-1}\ L_1\bar{W}_{11}L_1'\Gamma^{-1'}$$

(7.71) $$= - \sum_k \sum_\ell \bar{R}_{k\ell} \Big[\sigma_{\varepsilon 1k}\ \sigma_{\varepsilon \ell 1}\ (K_1+\tilde{M}_1-1)\left(N(T-1) - \frac{T-1}{T}\right) + \sigma_{\varepsilon 1\ell}\ \sigma_{\varepsilon k1}\ (K_1+\tilde{M}_1-1)\ \frac{1}{T} + \sigma_{\varepsilon 11}\ \sigma_{\varepsilon k\ell}\ (K_1+\tilde{M}_1-1)\ \frac{1}{T} + m^4_{\varepsilon 11k\ell}\ (K_1+\tilde{M}_1-1)\ \frac{(T-1)}{T} \Big]$$

(This step is derived in section 7.B.3 of Appendix 7.B.)

II.5

(7.72) $$E\ u_1'C\ V_z(\bar{Z}_1'C\ \bar{Z}_1)^{-1}\ V_z'N_4u_1$$

$$= E\ u_1'C\ V_1\bar{W}_{11}V_1'N_4u_1$$ using (7.67) and (7.68)

$$= E\ u_1'C\ U\ \Gamma^{-1}\ L_1\bar{W}_{11}L_1'\ \Gamma^{-1'}\ U'N_4u_1$$ using (7.60)

$$(7.73) \quad = E\ u_1'C\ U\ \bar{R}\ U'N_4u_1 \qquad \text{using (7.70)}$$

$$(7.74) \quad = \sum_k \sum_\ell \bar{R}_{k\ell} \Big[\sigma_{\varepsilon 1k}\ \sigma_{\varepsilon \ell 1}\ (K-1)\Big(N(T-1) - \frac{T-1}{T}\Big) + \sigma_{\varepsilon 1\ell}\ \sigma_{\varepsilon k1}\ (K-1)\ \frac{1}{T} + \sigma_{\varepsilon 11}\ \sigma_{\varepsilon k\ell}\ (K-1)\ \frac{1}{T} + m^4_{\varepsilon 11k\ell}\ (K-1)\ \frac{(T-1)}{T}\Big]$$

(See section 7.B.4 of Appendix 7.B for a proof of this last step.)

III.

$$(7.75) \quad \Big[E\ (u_1'N_4\bar{Z}_1 + u_1'N_4V_z)(\bar{Z}_1'C\ \bar{Z}_1)^{-1}\bar{Z}_1'C\ u_1 + u_1'N_4V_z(\bar{Z}_1'C\ \bar{Z}_1)^{-1} (\bar{Z}_1'C\ V_z + V_z'C\ \bar{Z}_1)(\bar{Z}_1'C\ \bar{Z}_1)^{-1}\bar{Z}_1'C\ u_1 + u_1'N_4V_z(\bar{Z}_1'C\ \bar{Z}_1)^{-1}V_z'C\ u_1\Big]$$

Let us calculate the expectation of each term separately and add them to get the total expectation.

III.1

$$(7.76) \quad E\ u_1'N_4\bar{Z}_1(\bar{Z}_1'C\ \bar{Z}_1)^{-1}\ \bar{Z}_1'C\ u_1$$

$$= \operatorname{tr}\ N_4\bar{Z}_1(\bar{Z}_1'C\ \bar{Z}_1)^{-1}\ \bar{Z}_1'C\ \Sigma_{11}$$

$$= \operatorname{tr}\ (\bar{Z}_1'C\ \bar{Z}_1)^{-1}\ \bar{Z}_1'C\ \Sigma_{11}N_4\bar{Z}_1$$

$$= \sigma_{\varepsilon 11}\ \operatorname{tr}\ I_{(K_1+\tilde{M}_1-1)} \qquad \text{as } \Sigma_{11}N_4 = \sigma_{\varepsilon 11}N_4 \text{ and } C\ N_4 = C$$

$$(7.77) \quad = (K_1+\tilde{M}_1-1)\ \sigma_{\varepsilon 11}$$

III.2

$$(7.78) \quad E\ u_1'N_4V_z(\bar{Z}_1'C\ \bar{Z}_1)^{-1}\ \bar{Z}_1'C\ u_1$$

$$= E\ u_1'N_4V_z\ \bar{P}'u_1 \qquad \text{using (7.58)}$$

$$= E\ u_1'N_4V_1\ \bar{P}_1'u_1 \quad \text{using the definition of } V_z \text{ in (7.37)}$$

$$= E\ u_1'N_4U\ \bar{F}'U_1$$

$$(7.79) \quad = 0 \qquad \text{(as this is the transpose of (7.62) which equals zero)}$$

III.3

$$(7.80) \qquad - E\ u_1'N_4V_z(\bar{Z}_1'C\ \bar{Z}_1)^{-1}\ \bar{Z}_1'C\ V_z(\bar{Z}_1'C\ \bar{Z}_1)^{-1}\ \bar{Z}_1'C\ u_1$$

$$(7.81) \qquad = - E\ u_1'N_4U\ \bar{F}'U\ \bar{F}'u_1$$

using the definitions of $\bar{F}(7.61)$ and $\bar{P}_1(7.59)$

$$(7.82) \qquad = - \sum_k \sum_\ell \Big[\sigma_{\varepsilon 1k}\ \sigma_{\varepsilon \ell 1}\ \bar{R}_{k\ell}(N(T-1) - \frac{(T-1)}{T}) + \sigma_{\varepsilon 11}\ \sigma_{\varepsilon k\ell}\ \bar{R}_{k\ell}\ \frac{1}{T} + \sigma_{\varepsilon 1k}\ \sigma_{\varepsilon \ell 1}\ \bar{R}_{k\ell}\ \frac{1}{T} + m^4_{\varepsilon 11k\ell}\ \bar{R}_{k\ell}\ \frac{(T-1)}{T} \Big]$$

(as (7.81) is the transpose of (7.64) which is in turn equal to the above expression.)

III.4

$$(7.83) \qquad - E\ u_1'N_4V_z(\bar{Z}_1'C\ \bar{Z}_1)^{-1}\ \bar{V}_z'C\ Z_1(\bar{Z}_1'C\ \bar{Z}_1)^{-1}\ \bar{Z}_1'C\ u_1$$

$$= - E\ u_1'N_4U\ \bar{R}\ U'\ \bar{H}\ u_1$$

using the definitions of $\bar{R}(7.70)$ and $\bar{H}(7.68)$

The above expectation is exactly equal to (7.69) and hence its value is given by (7.71).

III.5

$$(7.84) \qquad E\ u_1'N_4V_z(\bar{Z}_1'C\ \bar{Z}_1)^{-1}\ V_z'C\ u_1$$

$$= E\ u_1'N_4U\ \bar{R}\ U'\ C\ u_1$$

using the definition of $\bar{R}(7.70)$

Once again it can be easily verified that the above expectation is exactly equal to (7.73) and hence its value is given by (7.74).

IV.

$$(7.85) \qquad E\ u_1'C\ \bar{Z}_1(\bar{Z}_1'C\ \bar{Z}_1)^{-1}(\bar{Z}_1'N_4\bar{Z}_1+V_z'N_4V_z)(\bar{Z}_1'C\ \bar{Z}_1)^{-1}\ \bar{Z}_1'C\ u_1$$

IV.1

$$(7.86) \qquad E\ u_1'C\ \bar{Z}_1(\bar{Z}_1'C\ \bar{Z}_1)^{-1}(\bar{Z}_1'N_4\bar{Z}_1)(\bar{Z}_1'C\ \bar{Z}_1)^{-1}\ \bar{Z}_1'C\ u_1$$

$$= E\, u_1' \Psi\, u_1$$

by writing $\Psi = C\, \bar{Z}_1(\bar{Z}_1'C\, \bar{Z}_1)^{-1}\bar{Z}_1'N_4\bar{Z}_1(\bar{Z}_1'C\, \bar{Z}_1)^{-1}\bar{Z}_1'C$

$$= \text{tr}\, \Psi\, \Sigma_{11}$$

$$= \text{tr}\, (\bar{Z}_1'C\, \bar{Z}_1)^{-1}\, \bar{Z}_1'N_4\bar{Z}_1(\bar{Z}_1'C\, \bar{Z}_1)^{-1}(\bar{Z}_1'C\, \Sigma_{11}\, C\, \bar{Z}_1)$$

$$= \sigma_{\varepsilon 11}\, \text{tr}\, (\bar{Z}_1'C\, \bar{Z}_1)^{-1}\, \bar{Z}_1'N_4\bar{Z}_1(\bar{Z}_1'C\, \bar{Z}_1)^{-1}(\bar{Z}_1'C\, \bar{Z}_1)$$

as $C\, \Sigma_{11}\, C = \sigma_{\varepsilon 11}\, C$

(7.87) $$= \sigma_{\varepsilon 11}\, \text{tr}\, (\bar{Z}_1'C\, \bar{Z}_1)^{-1}\, \bar{Z}_1'\, N_4\, \bar{Z}_1$$

IV.2

(7.88) $$E\, u_1'C\, \bar{Z}_1(\bar{Z}_1'C\, \bar{Z}_1)^{-1}\, V_z'N_4V_z(\bar{Z}_1'C\, \bar{Z}_1)^{-1}\, \bar{Z}_1'C\, u_1$$

$$= E\, u_1'\bar{P}\, V_z'N_4V_z\bar{P}'u_1$$ using the definition of $\bar{P}$ given in (7.58)

$$= E\, u_1'\bar{F}\, U'N_4U\, \bar{F}'u_1$$

(7.89) $$= \sum_k \sum_\ell \Big[\sigma_{\varepsilon 11}\, \sigma_{\varepsilon k\ell}\, \bar{R}_{k\ell}(N(T-1) - \frac{T-1}{T}) + \sigma_{\varepsilon 1k}\, \sigma_{\varepsilon 1\ell}\, \bar{R}_{k\ell}\, (\frac{1}{T}) + \sigma_{\varepsilon 1\ell}\, \sigma_{\varepsilon 1k}\, \bar{R}_{k\ell}\, (\frac{1}{T}) + m^4_{\varepsilon 11k\ell}\, \bar{R}_{k\ell}\, \frac{T-1}{T} \Big]$$

(See section 7.B.5 of Appendix 7.B for a proof of (7.89).)

Thus, the expectation of $\hat{\sigma}_{\varepsilon 11}$ (to the order of $\frac{1}{N}$), which is the sum of all the expectation values of I, II, III and IV, is as follows :

(7.90) $$E(\hat{\sigma}_{\varepsilon 11}) \cong \frac{1}{N(T-1)} \Big[N(T-1)\sigma_{\varepsilon 11} + (7.56)+(7.65)+(7.71)+(7.74)+(7.77)+(7.82)+(7.71)+(7.74)+(7.87)+(7.89) \Big]$$

Now, as explained in Appendix 7.B (see footnote 1) on page 287), $\bar{R}_{k\ell}$ denotes a typical element of the matrix :

$$\bar{R} = \Gamma^{-1} L_1 \bar{W}_{11} L_1' \Gamma^{-1'}$$

Thus

$$\text{(7.91)} \qquad \bar{R}_{k\ell} = \gamma^{k'} L_1 \bar{W}_{11} L_1' \gamma^{\ell}$$

where γ^k and γ^ℓ denote respectively the k-th and ℓ-th columns of $\Gamma^{-1'}$. (Let us also recall that $\bar{W}_{11}$ is the $(\tilde{M}_1 x \tilde{M}_1)$ matrix appearing in $\bar{W} = (\bar{Z}_1' C \bar{Z}_1)^{-1} = \begin{bmatrix} \bar{W}_{11} & \bar{W}_{12} \\ \bar{W}_{21} & \bar{W}_{22} \end{bmatrix}$ and L_1 is a $(Mx\tilde{M}_1)$ selection matrix.)

So we can write :

$$\bar{R}_{k\ell} = \gamma^{k*'} \bar{W}_{11} \gamma^{\ell *}$$

$$\text{where } \gamma^{k*'} = \gamma^{k'} L_1 \quad (1x\tilde{M}_1)$$

$$\gamma^{\ell *} = L_1' \gamma^{\ell} \quad (\tilde{M}_1 x1)$$

$$= \begin{bmatrix} \gamma_1^{k*} & \cdots & \gamma_{\tilde{M}_1}^{k*} \end{bmatrix} \begin{bmatrix} \bar{W}_{11}^{11} & \cdots & \bar{W}_{1\tilde{M}_1}^{11} \\ \bar{W}_{\tilde{M}_1 1}^{11} & \cdots & \bar{W}_{\tilde{M}_1\tilde{M}_1}^{11} \end{bmatrix} \begin{bmatrix} \gamma_1^{\ell *} \\ \vdots \\ \gamma_{\tilde{M}_1}^{\ell *} \end{bmatrix}$$

where γ_m^{k*}, $\gamma_m^{\ell *}$ denote the m-th components of γ^{k*} and $\gamma^{\ell *}$ respectively, $m=1,\ldots,\tilde{M}_1$;

and $\bar{W}_{mm'}^{11}$ is the (m,m')-th element of $\bar{W}_{11}$, $m,m'=1,\ldots,\tilde{M}_1$.

Therefore,

$$\text{(7.92)} \qquad \bar{R}_{k\ell} = \sum_{m=1}^{\tilde{M}_1} \sum_{m'=1}^{\tilde{M}_1} \gamma_m^{k*} \bar{W}_{mm'}^{11} \gamma_{m'}^{\ell *}$$

Now, since $(\bar{Z}_1' C \bar{Z}_1)^{-1}$ is of order $\frac{1}{N}$, each element of $\bar{W}_{11}$ is also of order $\frac{1}{N}$ and since γ_m^{k*}, $\gamma_{m'}^{\ell *}$ are constants, each term in the above sum (7.92) is of order $\frac{1}{N}$ and hence the sum , i.e. $\bar{R}_{k\ell}$, is of order $\frac{1}{N}$.

Thus we find that there are still certain terms of order less than $\frac{1}{N}$ in the expression (7.90), namely, all the terms that are products of $\bar{R}_{k\ell}$ with terms of order less than 1. Neglecting such terms in each of the expressions appearing in (7.90) leaves us finally with the following "non-negligible" terms :

$$E(\hat{\sigma}_{\varepsilon 11}) \cong \sigma_{\varepsilon 11} + 2\sigma_{\varepsilon 11}\frac{(K_1+\tilde{M}_1-1)}{N(T-1)} + \frac{2}{N(T-1)}\sum_k\sum_\ell \bar{R}_{k\ell}\,\sigma_{\varepsilon 1k}\sigma_{\varepsilon\ell 1}$$

$$\Big(-N(T-1) - N(T-1)(K_1+\tilde{M}_1-1) + N(T-1)(K-1)\Big) + \sigma_{\varepsilon 11}\cdot$$

$$\frac{tr(\bar{Z}_1'C\bar{Z}_1)^{-1}\bar{Z}_1'N_4\bar{Z}_1}{N(T-1)} + \frac{1}{N(T-1)}\sum_k\sum_\ell \sigma_{\varepsilon 11}\sigma_{\varepsilon k\ell}\,\bar{R}_{k\ell}\,N(T-1)$$

$$(7.93)\qquad = \sigma_{\varepsilon 11} + 2\sigma_{\varepsilon 11}\frac{(K_1+\tilde{M}_1-1)}{N(T-1)} + 2\sum_k\sum_\ell \bar{R}_{k\ell}\,\sigma_{\varepsilon 1k}\,\sigma_{\varepsilon\ell 1}\,(K-1-(K_1+\tilde{M}_1))$$

$$+\ \sigma_{\varepsilon 11}\frac{tr(\bar{Z}_1'C\bar{Z}_1)^{-1}\bar{Z}_1'N_4\bar{Z}_1}{N(T-1)} + \sum_k\sum_\ell \sigma_{\varepsilon 11}\,\sigma_{\varepsilon k\ell}\,\bar{R}_{k\ell}$$

or

$$(7.94)\qquad E(\hat{\sigma}_{\varepsilon 11}) = \sigma_{\varepsilon 11} + b_{\varepsilon 11}$$

where $b_{\varepsilon 11}$, the bias of $\hat{\sigma}_{\varepsilon 11}$ upto order $\frac{1}{N}$, is formed of all the terms in the expression (7.93) excluding $\sigma_{\varepsilon 11}$. Note that the first and the second terms of the bias are very similar to the last term and the first term respectively, of Nagar's expression of the bias of the residual variance estimator, (cf. (2.12), [33]).

7.3.2 Bias of $\hat{\sigma}_{\mu 11}$

The AOV estimator of $\sigma_{\mu 11}$, the variance of the unit effect in the first structural equation, is given by :

$$(7.95)\qquad \hat{\sigma}_{\mu 11} = \frac{1}{T}\left[\hat{\sigma}_{111} - \hat{\sigma}_{\varepsilon 11}\right]$$

Taking expectations, we get

(7.96) $$E(\hat{\sigma}_{\mu 11}) = E(\frac{1}{T}\hat{\sigma}_{111}) - E(\frac{1}{T}\hat{\sigma}_{\varepsilon 11})$$

The previous section tells us that

(7.97) $$E(\frac{1}{T}\hat{\sigma}_{\varepsilon 11}) \cong \frac{1}{T}\sigma_{\varepsilon 11} + \frac{1}{T}b_{\varepsilon 11} \qquad \text{(cf. (7.94))}$$

Thus, in order to determine the bias (upto $O(\frac{1}{N})$) of $\hat{\sigma}_{\mu 11}$, we need to calculate the expectation of $\frac{1}{T}\hat{\sigma}_{111}$ i.e. that of $\frac{1}{T}\frac{1}{N}\hat{u}_1'N_1\hat{u}_1$. Using the expression (7.32) of $\hat{u}_1$, we obtain :

$$\frac{1}{T}\frac{1}{N}\hat{u}_1'N_1\hat{u}_1 = \frac{1}{TN}u_1'\left[I-Z_1^*(Z_1^{*\prime}CZ_1^*)^{-1}Z_1^{*\prime}C\right]'(I-M^*)N_1(I-M^*)$$
$$\left[I-Z_1^*(Z_1^{*\prime}CZ_1^*)^{-1}Z_1^{*\prime}C\right]u_1$$
$$= \frac{1}{TN}u_1'\left[I-Z_1^*(Z_1^{*\prime}CZ_1^*)^{-1}Z_1^{*\prime}C\right]'N_1$$
$$\left[I-Z_1^*(Z_1^{*\prime}CZ_1^*)^{-1}Z_1^{*\prime}C\right]u_1$$

as $N_1M^* = 0$

(7.98) $$= \frac{1}{TN}\left[u_1'N_1u_1 - u_1'CZ_1^*(Z_1^{*\prime}CZ_1^*)^{-1}Z_1^{*\prime}N_1u_1 - u_1'N_1Z_1^*(Z_1^{*\prime}CZ_1^*)^{-1}Z_1^{*\prime}Cu_1 + u_1'CZ_1^*(Z_1^{*\prime}CZ_1^*)^{-1}Z_1^{*\prime}N_1Z_1^*(Z_1^{*\prime}CZ_1^*)^{-1}Z_1^{*\prime}Cu_1\right]$$

Note that the above expression (7.98) is very similar to that of $\hat{\sigma}_{\varepsilon 11}$ in (7.34), the only difference between the two being the replacement of N_4 in (7.34) by N_1 in (7.98) and of (T-1) in the denominator of (7.34) by T in that of (7.98). Therefore, the expectation of (7.98) can be found by following the same procedure as in the previous section i.e. by writing $Z_1^* = \bar{Z}_1 + V_z$ and expanding each term in series, truncating the series upto $O(\frac{1}{N})$ and then computing the expectation of the terms retained.

From the results on the computation of orders given in Appendix 7.A, it is obvious that the "non-negligible" or "significant" terms of $\frac{1}{TN} u_1'N_1u_1$ are the same as those of $\hat{\sigma}_{\varepsilon 11}$ with N_4 replaced by N_1 in the numerator and (T-1) by T in the denominator. Thus, they are as follows :

(i) $\frac{1}{T(N-1)} u_1'N_1u_1$

(ii) $$\frac{1}{T(N-1)} \Big[u_1'C\bar{Z}_1(\bar{Z}_1'C\bar{Z}_1)^{-1}(\bar{Z}_1'N_1u_1+V_z'N_1u_1)-u_1'C\bar{Z}_1(\bar{Z}_1'C\bar{Z}_1)^{-1} (V_z'C\bar{Z}_1 + \bar{Z}_1'CV_z)(\bar{Z}_1'C\bar{Z}_1)^{-1}V_z'N_1u_1 + u_1'CV_z(\bar{Z}_1'C\bar{Z}_1)^{-1}V_z'N_1u_1 \Big]$$

(iii) the transpose of (ii)

(iv) $\frac{1}{T(N-1)} u_1'C\bar{Z}_1(\bar{Z}_1'C\bar{Z}_1)^{-1}(\bar{Z}_1'N_1\bar{Z}_1 + V_z'N_1V_z)(\bar{Z}_1'C\bar{Z}_1)^{-1}\bar{Z}_1'Cu_1$

Let us now calculate the expectations of the above terms one by one. As done in the previous case, we will give the Roman numerals I, II, III and IV to the paragraphs dealing with the expectations of (i), (ii), (iii) and (iv) respectively, subdividing each paragraph in the same way as before, if necessary. Further, we will not give all the details of the calculations, as they are similar to those of the previous section, but state only the final results. Lastly, we will write the value of the expectations, omitting the factor $\frac{1}{TN}$ in order to avoid repeating it each time. This factor will, obviously, be included in the calculations at a later stage.

I.

(7.99) $$E(u_1'N_1u_1) = N\,\sigma_{111}$$

II.1

(7.100) $$E(u_1'C\,\bar{Z}_1(\bar{Z}_1'C\,\bar{Z}_1)^{-1}\,Z_1'N_1u_1) = 0$$

II.2

(7.101) $$E\,u_1'C\,\bar{Z}_1(\bar{Z}_1'C\,\bar{Z}_1)^{-1}\,V_z'N_1u_1 = 0$$

II.3

$$- E\, u_1'C\, \bar{Z}_1(\bar{Z}_1'C\, \bar{Z}_1)^{-1}\, V_z'C\, \bar{Z}_1(\bar{Z}_1'C\bar{Z}_1)^{-1}V_z'N_1u_1$$

$$(7.102) \quad = - \sum_k \sum_\ell \bar{R}_{k\ell} \Big[\sigma_{\varepsilon 1k}\, \sigma_{\varepsilon \ell 1}\, (N - \tfrac{1}{T}) + \sigma_{\varepsilon 1\ell}\, \sigma_{\varepsilon k1}\, (- \tfrac{1}{T}) + \sigma_{\varepsilon 11}\, \sigma_{\varepsilon k\ell}\, (- \tfrac{1}{T}) + m^4_{\varepsilon 11k\ell}\, \tfrac{1}{T} \Big]$$

II.4

$$- E\, u_1'C\, \bar{Z}_1(\bar{Z}_1'C\, \bar{Z}_1)^{-1}\, \bar{Z}_1'C\, V_z(\bar{Z}_1'C\bar{Z}_1)^{-1}V_z'N_1u_1$$

$$(7.103) \quad = - \sum_k \sum_\ell \bar{R}_{k\ell} \Big[\sigma_{\varepsilon 1k}\sigma_{\varepsilon \ell 1}\, (K_1+\tilde{M}_1-1)(N- \tfrac{1}{T}) + \sigma_{\varepsilon 1\ell}\sigma_{\varepsilon k1}\, (- \tfrac{1}{T}) + \sigma_{\varepsilon 11}\sigma_{\varepsilon k\ell}\, (K_1+\tilde{M}_1-1)(- \tfrac{1}{T}) + m^4_{\varepsilon 11k\ell}\, (K_1+\tilde{M}_1-1)\, \tfrac{1}{T} \Big]$$

II.5

$$E\, u_1'C\, V_z(\bar{Z}_1'C\, \bar{Z}_1)^{-1}\, V_z'N_1u_1$$

$$(7.104) \quad = \sum_k \sum_\ell \bar{R}_{k\ell} \Big[\sigma_{\varepsilon 1k}\, \sigma_{\varepsilon 1\ell}(K-1)(N- \tfrac{1}{T}) + \sigma_{\varepsilon 1\ell}\sigma_{\varepsilon k1}\, (K-1)(- \tfrac{1}{T}) + \sigma_{\varepsilon 11}\, \sigma_{\varepsilon k\ell}\, (K-1)\, (- \tfrac{1}{T}) + m^4_{\varepsilon 11k\ell}\, (K-1)(\tfrac{1}{T}) \Big]$$

III.1

$$(7.105) \qquad E\, u_1'N_1\bar{Z}_1(\bar{Z}_1'C\, \bar{Z}_1)^{-1}\, \bar{Z}_1'C\, u_1 = 0$$

III.2

$$(7.106) \qquad E\, u_1'N_1V_z(\bar{Z}_1'C\, \bar{Z}_1)^{-1}\, \bar{Z}_1'C\, u_1 = 0$$

III.3

$$- E\, u_1'N_1V_z(\bar{Z}_1'C\, \bar{Z}_1)^{-1}\, \bar{Z}_1'C\, V_z(\bar{Z}_1'C\bar{Z}_1)^{-1}\, \bar{Z}_1'C\, u_1$$

= (7.102) as this is the transpose of II.3

III.4

Same as the result of II.4 i.e. (7.103)

III.5

Same as the result of II.5 i.e. (7.104)

IV.1

$$E\, u_1'C\, \bar{Z}_1(\bar{Z}_1'C\, \bar{Z}_1)^{-1}\, (\bar{Z}_1'N_1\bar{Z}_1)(\bar{Z}_1'C\bar{Z}_1)^{-1}\, \bar{Z}_1'C\, u_1$$

$$(7.107) \quad = \sigma_{\varepsilon 11}\, \mathrm{tr}\, (\bar{Z}_1'C\, \bar{Z}_1)^{-1}\, \bar{Z}_1'N_1\bar{Z}_1$$

IV.2

$$-\, E\, u_1'C\, \bar{Z}_1(\bar{Z}_1'C\, \bar{Z}_1)^{-1}\, V_z'N_1V_z(\bar{Z}_1'C\bar{Z}_1)^{-1}\, \bar{Z}_1'C\, u_1$$

$$(7.108) \quad = \sum_k \sum_\ell \bar{R}_{k\ell} \Big[\sigma_{\varepsilon 11}\, \sigma_{\varepsilon k\ell}\, (N - \tfrac{1}{T}) + \sigma_{\varepsilon 1k}\, \sigma_{\varepsilon \ell 1}\, (-\tfrac{1}{T}) + \sigma_{\varepsilon 1\ell}\, \sigma_{\varepsilon 1k}\, (-\tfrac{1}{T}) + m^4_{\varepsilon 11k\ell}\, \tfrac{1}{T} \Big]$$

Thus, the expectation of $\hat{\sigma}_{111}$ (upto $O(\frac{1}{N})$) is given by :

$$E(\hat{\sigma}_{111}) \cong \frac{1}{N} \Big[N\sigma_{111} + 2x(7.102) + 2x(7.103) + 2x(7.104) + (7.105) + (7.106) + (7.107) + (7.108) \Big]$$

$$(7.109) \quad = \sigma_{111} + \frac{1}{N} \Big[2x(7.102) + 2x(7.103) + 2x(7.104) + (7.105) + (7.106) + (7.107) + (7.108) \Big]$$

Once again, noting that $\bar{R}_{k\ell}$ is $O(\frac{1}{N})$, we can still neglect certain terms of the above expression (7.109), i.e. terms which are products of $\bar{R}_{k\ell}$ with terms of order less than 1. Doing so, we are left with the following expression :

$$E(\hat{\sigma}_{111}) \cong \sigma_{111} + \frac{2}{N} \sum_k \sum_\ell \bar{R}_{k\ell}\, \sigma_{\varepsilon 1k}\, \sigma_{\varepsilon \ell 1}\, \big(-N - N(K_1+\tilde{M}_1-1) + N(K-1)\big)$$

$$(7.110) \qquad + \sigma_{\varepsilon 11} \frac{\mathrm{tr}(\bar{Z}_1'C\bar{Z}_1)^{-1}(\bar{Z}_1'N_1\bar{Z}_1)}{N} + \frac{1}{N} \sum\sum_{k\ell} \bar{R}_{k\ell}\sigma_{\varepsilon 11}\sigma_{\varepsilon k\ell}\, N$$

$$(7.111) \quad = \sigma_{111} + 2 \sum_k \sum_\ell \bar{R}_{k\ell} \Big[\sigma_{\varepsilon 1k}\sigma_{\varepsilon \ell 1}((K-1)-(K_1+\tilde{M}_1)+\sigma_{\varepsilon 11}\sigma_{\varepsilon k\ell} \Big] + \sigma_{\varepsilon 11} \frac{\mathrm{tr}(\bar{Z}_1'C\bar{Z}_1)^{-1}\bar{Z}_1'N_1\bar{Z}_1}{N}$$

or

$$(7.112) \quad E(\hat{\sigma}_{111}) \cong \sigma_{111} + b_{111}$$

where b_{111} denotes all the terms on the RHS of (7.111) except σ_{111} and is the bias of $\hat{\sigma}_{111}$ upto order $\frac{1}{N}$.

Now, replacing (7.112) in the expression of $E(\hat{\sigma}_{\mu 11})$ given in (7.96), we obtain :

$$E(\hat{\sigma}_{\mu 11}) \cong \frac{1}{T}(\sigma_{111} + b_{111}) - \frac{1}{T}\sigma_{\varepsilon 11} - \frac{1}{T}b_{\varepsilon 11}$$

$$= \frac{1}{T}(\sigma_{111} - \sigma_{\varepsilon 11}) + \frac{1}{T}(b_{111} - b_{\varepsilon 11})$$

or

$$(7.113) \quad E(\hat{\sigma}_{\mu 11}) \cong \sigma_{\mu 11} + b_{\mu 11}$$

where the bias upto order $\frac{1}{N}$ of $\hat{\sigma}_{\mu 11}$, denoted as $b_{\mu 11}$, is :

$$(7.114) \quad b_{\mu 11} = \frac{1}{T}(b_{111} - b_{\varepsilon 11})$$

7.4 Bias of Structural Coefficients Estimators

7.4.1 A General Note

Before beginning the long and tedious calculations involved in the derivation of the bias of coefficient estimators, we would like to make three remarks. Firstly, as in the case of structural variance components estimators, only the bias of order upto $\frac{1}{N}$ will be derived using series expansions.

Secondly, we still consider a model with only unit effect and overall effect (i.e. omitting time effects) in order to simplify the presentation.

Thirdly, we examine the bias of only the two stage estimators, namely that of the covariance 2SLS and the feasible G2SLS estimators of the coefficients of any single structural equation. The reason for this is the fact that the two stage estimators are the most frequently used in practical applications and also are the ones for which results on small sample moments are available in the classical simultaneous equation case. Of course, it is possible to extend our technique to the three stage estimator but it is beyond the scope of our present work to do so.

7.4.2 Bias of the Covariance 2SLS Estimator

Let us recall that the covariance 2SLS estimator of any structural equation, say the first one, is given by :

$$(7.115)\quad \hat{\alpha}^*_{1,C2SLS} = \left[Z_1^{*\prime} N_4 \underline{X}(\underline{X}' N_4 \underline{X})^{-1} \underline{X}' N_4 Z_1\right]^{-1} Z_1^{*\prime} N_4 \underline{X}(\underline{X}' N_4 \underline{X})^{-1} \underline{X}' N_4 y_1$$

where α_1^* is the vector of all the coefficients of the equation except for the constant term a_1. Substituting

$$y_1 = \iota_{NT}\, a_1 + Z_1^* \alpha_1^* + u_1$$

for y_1 in (7.115) and simplifying, we obtain

$$(7.116)\quad \hat{\alpha}^*_{1,C2SLS} = \alpha_1^* + \left[Z_1^{*\prime} N_4 \underline{X}(\underline{X}' N_4 \underline{X})^{-1} \underline{X}' N_4 Z_1^*\right]^{-1}$$

$$Z_1^{*\prime} N_4 \underline{X}(\underline{X}' N_4 \underline{X})^{-1} \underline{X}' N_4 u_1$$

Using the following notation, introduced in Section 7.3.2,

$$C = N_4\, \underline{X}(\underline{X}' N_4 \underline{X})^{-1}\, \underline{X}' N_4$$

equation (7.116) can be rewritten as :

$$(7.117)\quad \hat{\alpha}^*_{1,C2SLS} = \alpha_1^* + (Z_1^{*\prime} C\, Z_1^*)^{-1}\, Z_1^{*\prime} C\, u_1$$

Thus the bias of the covariance 2SLS estimator is given by the expectation of

$$(7.118)\quad \hat{\alpha}^*_{1,C2SLS} - \alpha_1^* = (Z_1^{*\prime} C\, Z_1^*)^{-1}\, Z_1^{*\prime} C\, u_1$$

Once again, since the exact expectation of the above expression cannot be determined, we will calculate an asymptotic approximation to it, upto order $\frac{1}{N}$, using the same technique as before. That is, we start by writing

$$(7.119)\quad Z_1^* = \bar{Z}_1 + V_z \qquad \text{(cf. (7.37), page 235)}$$

and

$$(7.120)\quad (Z_1^{*}{'}C\,Z_1^{*})^{-1} = \left[I + (\bar{Z}_1'C\,\bar{Z}_1)^{-1}(V_z'C\,\bar{Z}_1 + \bar{Z}_1'C\,V_z + V_z'C\,V_z)\right]^{-1}(\bar{Z}_1'C\,\bar{Z}_1)^{-1}$$

and then expand in series the first inverse appearing in the RHS of (7.120) above. Doing so, we obtain :

$$(7.121)\quad \begin{aligned} \hat{\alpha}^{*}_{1,C2SLS} - \alpha_1^{*} = \Big\{ & I - (\bar{Z}_1'C\bar{Z}_1)^{-1}(V_z'C\bar{Z}_1 + \bar{Z}_1'CV_z + V_z'CV_z) \\ & + (\bar{Z}_1'C\bar{Z}_1)^{-1}(V_z'C\bar{Z}_1 + \bar{Z}_1'CV_z + V_z'CV_z)(\bar{Z}_1'C\bar{Z}_1)^{-1} \\ & (V_z'C\bar{Z}_1 + \bar{Z}_1'CV_z + V_z'CV_z) - \ldots\Big\}(\bar{Z}_1'C\bar{Z}_1)^{-1} \\ & (\bar{Z}_1'Cu_1 + V_z'Cu_1) \end{aligned}$$

Now, using the notation (7.40) to (7.44) along with

$$(7.122)\quad \begin{cases} F_{-\frac{1}{2}} = (\bar{Z}_1'C\,\bar{Z}_1)^{-1}\,\bar{Z}_1'C\,u_1 \\ F_{-1} = (\bar{Z}_1'C\,\bar{Z}_1)^{-1}\,V_z'C\,u_1 \end{cases}$$

we can write (7.121) as[1] :

$$(B_o + B_{-\frac{1}{2}} + B_{-1} + B_{-\frac{3}{2}} + B_{-2})\;(F_{-\frac{1}{2}} + F_{-1})$$

Next, we neglect terms of order less than $\frac{1}{N}$ in the above product to obtain :

$$(7.123)\quad \begin{aligned} \hat{\alpha}^{*}_{1,C2SLS} - \alpha_1^{*} & \cong B_o\,F_{-\frac{1}{2}} + B_{-\frac{1}{2}}\,F_{-\frac{1}{2}} + B_o\,F_{-1} \\ & = (\bar{Z}_1'C\bar{Z}_1)^{-1}\bar{Z}_1'Cu_1 - (\bar{Z}_1'C\bar{Z}_1)^{-1}(V_z'C\bar{Z}_1 + \\ & \quad \bar{Z}_1'CV_z)(\bar{Z}_1'C\bar{Z}_1)^{-1}\bar{Z}_1'Cu_1 + (\bar{Z}_1'C\bar{Z}_1)^{-1}V_z'Cu_1 \end{aligned}$$

---

1) Recall that the indices indicate the power of N in the order of the terms.

Finally, we calculate the expectation of each of the four terms above, one by one and add them up to get the required bias.

I.

(7.124) $E\,(\bar{Z}_1'C\,\bar{Z}_1)^{-1}\,\bar{Z}_1'C\,u_1 = 0$

II. $E - (\bar{Z}_1'C\,\bar{Z}_1)^{-1}\,V_z'C\,\bar{Z}_1(\bar{Z}_1'C\,\bar{Z}_1)^{-1}\,Z_1'C\,u_1$

(7.125) $= -\,E\,\bar{W}_1 L_1'\Gamma^{-1}{}'U'\bar{H}\,u_1$

using the definitions of $\bar{W}$ (7.67), $\bar{H}$ (7.68) and V_z (7.37) and partitioning $\bar{W}$ as $\left[\bar{W}_1\bar{W}_2\right]$ appropriately

$= -\left[\bar{W}_1 L_1'\Gamma^{-1}{}'\ \mathrm{tr}\,\bar{H}\,\Sigma_{1m}\right]$, $m=1,\ldots,M$

(7.126) $= -\,(K_1+\tilde{M}_1-1)\,\bar{W}_1 L_1'\Gamma^{-1}{}' \begin{bmatrix} \sigma_{\varepsilon 11} \\ \vdots \\ \sigma_{\varepsilon M1} \end{bmatrix}$

as $\bar{H}\,\Sigma_{1m} = \sigma_{\varepsilon 1m}\,\bar{H}\,N_4 = \sigma_{\varepsilon 1m}\,\bar{H}$

and $\mathrm{tr}\,\bar{H} = K_1+\tilde{M}_1-1$ (cf.(7.B.22))

III.

$E - (\bar{Z}_1'C\,\bar{Z}_1)^{-1}\,\bar{Z}_1'C\,V_z(\bar{Z}_1'C\,\bar{Z}_1)^{-1}\,Z_1'C\,u_1$

$= -\,E\,\bar{P}'U\,\Gamma^{-1}\,L_1\bar{P}_1'u_1$

using the definition of $\bar{P}$ given in (7.58) and the partition (7.59) of $\bar{P}$ as $\left[\bar{P}_1\bar{P}_2\right]$

$= -\,E\,\bar{P}'U\,\bar{F}'\,u_1$

using the definition of $\bar{F}$ given in (7.61)

$= -\,E\,\bar{P}'\left[\sum_m u_m\,\bar{F}_m'\,u_1\right]$

$$= - E\ \bar{P}'\left[\sum_m u_m\ u_1'\ \bar{F}_m\right]$$

$$= - \bar{P}'\sum_m \Sigma_{1m}\ \bar{F}_m$$

$$= - \sum_m \bar{P}'\ \Sigma_{1m}\ \bar{P}_1 L_1 \gamma^m$$

$$= - \sum_m \bar{P}'\ \Sigma_{1m}\ \bar{P}\ L_1 L_1 \gamma^m \qquad \text{writing } \bar{P}_1 = \bar{P}\ L_1$$

$$= - \sum_m \sigma_{\varepsilon m1}\ \bar{P}'\ \bar{P}\ L_1 L_1 \gamma^m \qquad \text{as } \Sigma_{1m}\ \bar{P} = \sigma_{\varepsilon m1}\ \bar{P}$$

$$= - \sum_m \sigma_{\varepsilon m1}\ \bar{W}\ L_1 L_1 \gamma^m \qquad \text{as } \bar{P}'\ \bar{P} = (\bar{Z}_1'C\ \bar{Z}_1)^{-1} = \bar{W}$$

$$= - \bar{W}_1 L_1 \left[\gamma^1 \dots \gamma^M\right]\begin{bmatrix}\sigma_{\varepsilon 11}\\ \vdots \\ \sigma_{\varepsilon M1}\end{bmatrix}$$

noting that $\bar{W}\ L_1 = \bar{W}_1$ of (7.125)

$$(7.127) \qquad = - \bar{W}_1 L_1 \Gamma^{-1}\begin{bmatrix}\sigma_{\varepsilon 11}\\ \vdots \\ \sigma_{\varepsilon M1}\end{bmatrix}$$

IV. $\quad E\ (\bar{Z}_1'C\ \bar{Z}_1)^{-1}\ V_z'C\ u_1$

$$= E\ \bar{W}_1 L_1' \Gamma^{-1'} U'C\ u_1$$

$$= \bar{W}_1 L_1' \Gamma^{-1}\left[\text{tr } C\ \Sigma_{m1}\right],\ m=1,\dots,M$$

$$= \bar{W}_1 L_1' \Gamma^{-1}\left[\text{tr}(\sigma_{\varepsilon m1}\ C)\right]\ m=1,\dots,M \qquad \text{as } C\ \Sigma_{m1} = \sigma_{\varepsilon m1}\ C$$

$$(7.128) \qquad = (K-1)\ \bar{W}_1 L_1' \Gamma^{-1}\begin{bmatrix}\sigma_{\varepsilon 11}\\ \vdots \\ \sigma_{\varepsilon M1}\end{bmatrix}$$

Thus, the bias, upto order $\frac{1}{N}$, of the covariance 2SLS estimator is given by :

$$E(\hat{\alpha}^*_{1,C2SLS} - \alpha^*_1) = (-(K_1+\tilde{M}_1-1) - 1+K-1)\ \bar{W}_1 L_1' \Gamma^{-1\prime} \begin{bmatrix} \sigma_{\varepsilon 11} \\ \vdots \\ \sigma_{\varepsilon M1} \end{bmatrix}$$

$$(7.129) \qquad = \left[(K-1) - (K_1+\tilde{M}_1)\right]\ \bar{W}_1 L_1' \Gamma^{-1\prime} \begin{bmatrix} \sigma_{\varepsilon 11} \\ \vdots \\ \sigma_{\varepsilon M1} \end{bmatrix}$$

Note that the above expression of the bias has a striking similarity to the one obtained by Nagar (cf. (2.14), [32]) for the bias of the 2SLS estimator, in that both contain the same factor $(K-1)-(K_1+\ \tilde{M}_1)$ and the expression multiplying it in our (7.129) is in fact the covariance between the error term and the explanatory variables of the covariance equation (with $\bar{W}_1$ in front), the analogue of Nagar's Qq.

7.4.3 Bias of the Feasible G2SLS Estimator

The feasible G2SLS estimator of the coefficients of any structural equation, say the first one, is given by (cf. (4.37), page 127) :

$$(7.130) \quad \hat{\alpha}_{1,fG2SLS} = \left[Z_1'\hat{\Sigma}^{-1}_{11}X\ (X'\hat{\Sigma}^{-1}_{11}X)^{-1}X'\hat{\Sigma}^{-1}_{11}Z_1\right]^{-1} Z_1'\hat{\Sigma}^{-1}_{11}X\ (X'\hat{\Sigma}^{-1}_{11}X)^{-1}X'\hat{\Sigma}^{-1}_{11}y_1$$

Replacing

$$y_1 = Z_1\ \alpha_1 + u_1$$

in (7.130) and simplifying, we get

$$(7.131) \quad \hat{\alpha}_{1,fG2SLS} - \alpha_1 = \left[Z_1'\hat{\Sigma}^{-1}_{11}X\ (X'\hat{\Sigma}^{-1}_{11}X)^{-1}X'\hat{\Sigma}^{-1}_{11}Z_1\right]^{-1} Z_1'\hat{\Sigma}^{-1}_{11}X\ (X'\hat{\Sigma}^{-1}_{11}X)^{-1}X'\hat{\Sigma}^{-1}_{11}u_1$$

Since we assume absence of time effects, we have :

$$(7.132) \quad \hat{\Sigma}^{-1}_{11} = \sum_{i=1,4} \frac{1}{\hat{\sigma}_{i11}}\ N_i$$

where N_1 and N_4 are defined in (7.24) and $\hat{\sigma}_{111}$ and $\hat{\sigma}_{411}$ are given in (7.26) and (7.25) respectively[1]. Substituting (7.132) in (7.131), we can write :

$$(7.133)\quad \hat{\alpha}_{1,fG2SLS} - \alpha_1 = \left[(\sum_i \frac{1}{\hat{\sigma}_{i11}} Z_1'N_iX)(\sum_i \frac{1}{\hat{\sigma}_{i11}} X'N_iX)^{-1}(\sum_i \frac{1}{\hat{\sigma}_{i11}} X'N_iZ_1)\right]^{-1} (\sum_i \frac{1}{\hat{\sigma}_{i11}} Z_1'N_iX)(\sum_i \frac{1}{\hat{\sigma}_{i11}} X'N_iX)^{-1}(\sum_i \frac{1}{\hat{\sigma}_{i11}} Z_1'N_iu_1)$$

Hence, the bias of the feasible G2SLS estimator is the expectation of the expression on the RHS of (7.133).

As done in Sections 7.3.2 and 7.4.2, we approximate the bias by developing it in series. In this case, we propose the following technique, to arrive at a series expansion.

Let

$$(7.134)\quad \eta_{i11} = \hat{\sigma}_{i11} - \sigma_{i11} \qquad\qquad i=1,4$$

Then, we can write

$$(7.135)\quad \hat{\sigma}_{i11} = \sigma_{i11} + (\hat{\sigma}_{i11} - \sigma_{i11}) = \sigma_{i11} + \eta_{i11}$$

Using this split-up, we can develop the inverse of $\hat{\sigma}_{i11}$ as follows :

$$\frac{1}{\hat{\sigma}_{i11}} = \frac{1}{\sigma_{i11}+\eta_{i11}} = \frac{1}{\sigma_{i11}(1+\frac{\eta_{i11}}{\sigma_{i11}})}$$

$$= \frac{1}{\sigma_{i11}} (1 - \frac{\eta_{i11}}{\sigma_{i11}} + \frac{\eta^2_{i11}}{\sigma^2_{i11}} - \dots)$$

For the above expansion to be valid, we should verify that successive powers of $\frac{\eta_{i11}}{\sigma_{i11}}$ are of decreasing orders in N . In what follows we will show that $\frac{\eta_{i11}}{\sigma_{i11}}$ is of order $O(N^{-\frac{1}{2}})$ or equivalently that η_{i11} is $O(N^{-\frac{1}{2}})$ for $i=4$, the proof being similar for $i=1$.

---

1) Throughout this section, the sum over the variance component index will run over only 1 and 4 and hence we will not repeat them from now onwards.

From the expression of $\hat{\sigma}_{\varepsilon 11}$ given in (7.34) and from the order results of Appendix 7.A, it is easily verified that :

$$\hat{\sigma}_{\varepsilon 11} = \frac{1}{N(T-1)} u_1' N_4 u_1 + \xi_{\varepsilon 11}$$

where

(7.137) $\xi_{\varepsilon 11} = O(N^{-\frac{1}{2}})$

This means that

(7.138) $\hat{\sigma}_{\varepsilon 11} - \sigma_{\varepsilon 11} = \eta_{\varepsilon 11} = \frac{1}{N(T-1)} u_1' N_1 u_1 - \sigma_{\varepsilon 11} + \xi_{\varepsilon 11}$

Now, we know that

(7.139) $E(\frac{1}{N(T-1)} u_1' N_4 u_1 - \sigma_{\varepsilon 11}) = 0 = O(1)$

Further,

$$V(\frac{1}{N(T-1)} u_1' N_4 u_1 - \sigma_{\varepsilon 11}) = V(\frac{1}{N(T-1)} u_1' N_4 u_1)$$

(7.140) $$= \frac{1}{N^2 (T-1)^2} E(u_1' N_4 u_1 u_1' N_4 u_1) - \sigma_{\varepsilon 11}^2$$

Now, $E(u_1' N_4 u_1 u_1' N_4 u_1)$ can be calculated using the same procedure as that used in Appendix 7.B for calculating expectations of different types of products involving quadratic and linear forms. Doing so, it can be verified that

(7.141) $$E(u_1' N_4 u_1 u_1' N_4 u_1) = \sigma_{\varepsilon 11}^2 \left[(N(T-1))^2 - NT \left(\frac{T-1}{T}\right)^2 \right] + 2 \sigma_{\varepsilon 11}^2 \left[N(T-1) - NT \left(\frac{T-1}{T}\right)^2 \right] + m_{\varepsilon 111}^4 NT \left(\frac{T-1}{T}\right)^2$$

Replacing (7.141) in (7.140) and simplifying, it can be seen that

(7.142) $V\left(\frac{1}{N(T-1)} u_1' N_4 u_1 - \sigma_{\varepsilon 11}\right) = O(\frac{1}{N})$

Combining (7.139) and (7.142) we have :

(7.143) $\frac{1}{N(T-1)} u_1' N_4 u_1 - \sigma_{\varepsilon 11} = O_p(\frac{1}{\sqrt{N}})$

and finally, combining (7.137), (7.138) and (7.144), we conclude that

$$(7.144)\quad \eta_{\varepsilon 11} = O_p(N^{-\frac{1}{2}})$$

Now that we have shown that our expansion is valid, we will substitute it in each of the sums appearing in the expression (7.133) of the feasible G2SLS estimator.

(i) First,

$$(7.145)\quad \sum_i \frac{1}{\hat{\sigma}_{i11}} Z_1'N_iX = \sum_i \frac{1}{\sigma_{i11}} \left(1 - \frac{\eta_{i11}}{\sigma_{i11}} + \frac{\eta_{i11}^2}{\sigma_{i11}^2} - \ldots\right) Z_1'N_iX$$

Using the following order results :

$$(7.146)\quad \frac{\eta_{i11}}{\sigma_{i11}} = O_p(N^{-\frac{1}{2}}) \qquad \text{(cf. (7.144))}$$

$$(7.147)\quad \begin{cases} Z_1'N_iX = O\,(N) \;, \; i=1,4 & \text{(cf. (7.A.31), (7.A.32))} \\ X'N_iX = O\,(N) \;, \; i=1,4 & \text{(cf. (7.A.10), (7.A.11))} \\ X'N_iu_1 = O_p(\sqrt{N}) \;, \; i=1,4 & \text{(cf. (7.A.16), (7.A.17))} \end{cases}$$

we can write

$$(7.148)\quad \sum_i \frac{1}{\hat{\sigma}_{i11}} Z_1'N_iX = (A_1 + A_{\frac{1}{2}} + A_o + \ldots)$$

where[1]

$$(7.149)\quad \begin{cases} A_1 = \dfrac{1}{\sigma_{i11}} Z_1'N_iX \\ A_{\frac{1}{2}} = - \dfrac{1}{\sigma_{i11}} \dfrac{\eta_{i11}}{\sigma_{i11}} Z_1'N_iX \\ A_o = \dfrac{1}{\sigma_{i11}} \dfrac{\eta_{i11}^2}{\sigma_{i11}^2} Z_1'N_iX \end{cases}$$

and so on.

---

1) These notations A_1, $A_{\frac{1}{2}}$, A_o and similar ones to be defined later, replace all the earlier ones of the same kind.

(ii) The next sum in (7.133) is

$$(7.150)\quad \sum_i \frac{1}{\hat{\sigma}_{ill}} X'N_iX = \sum_i \frac{1}{\sigma_{ill}} \left(1 - \frac{\eta_{ill}}{\sigma_{ill}} + \frac{\eta_{ill}^2}{\sigma_{ill}^2} - \ldots\right) X'N_iX$$

Hence

$$\left[\sum_i \frac{1}{\hat{\sigma}_{ill}} X'N_iX\right]^{-1}$$

$$= \left[\sum_i \frac{1}{\sigma_{ill}} \left(X'N_iX - \frac{\eta_{ill}}{\sigma_{ill}} X'N_iX + \frac{\eta_{ill}^2}{\sigma_{ill}^2} X'N_iX - \ldots\right)\right]^{-1}$$

$$= \left\{\left(\sum_i \frac{1}{\sigma_{ill}} X'N_iX\right)\left[I - \left(\sum_i \frac{1}{\sigma_{ill}} X'N_iX\right)^{-1} \left(\sum_i \frac{1}{\sigma_{ill}} \frac{\eta_{ill}}{\sigma_{ill}} X'N_iX - \ldots\right)\right]\right\}^{-1}$$

$$= \left[I - \left(\sum_i \frac{1}{\sigma_{ill}} X'N_iX\right)^{-1}\left(\sum_i \frac{1}{\sigma_{ill}} \frac{\eta_{ill}}{\sigma_{ill}} X'N_iX - \sum_i \frac{1}{\sigma_{ill}} \frac{\eta_{ill}^2}{\sigma_{ill}^2} X'N_iX + \ldots\right)\right]^{-1}$$

$$\left[\sum_i \frac{1}{\sigma_{ill}} X'N_iX\right]^{-1}$$

Expanding the first inverse in series and using the order results given in (7.146), (7.147), we can write :

$$(7.151)\quad \left[\sum_i \frac{1}{\hat{\sigma}_{ill}} X'N_iX\right]^{-1} = \left(B_o + B_{-\frac{1}{2}} + B_{-1} + B_{-\frac{3}{2}} + B_{-2} + \ldots\right)$$

$$\left(\sum_i \frac{1}{\sigma_{ill}} X'N_iX\right)^{-1}$$

defining B_o , $B_{-\frac{1}{2}}$, B_{-1} ... appropriately.

(iii) The sum adjacent to the above one in (7.133) is

$$\sum_i \frac{1}{\hat{\sigma}_{ill}} X'N_iZ_1$$

which is the transpose of (i). Thus, it can be written as

$$A'_1 + A'_{\frac{1}{2}} + A'_o + \ldots$$

Now, let us combine the inverse postmultiplying $(B_o + B_{-\frac{1}{2}} + \ldots)$ in (7.151) with this term and give the following notation to the resulting product :

$$(\sum_i \frac{1}{\sigma_{i11}} X'N_iX)^{-1} (\sum_i \frac{1}{\hat{\sigma}_{i11}} X'N_iZ_1)$$

$$= (\sum_i \frac{1}{\sigma_{i11}} X'N_iX)^{-1} (A_1' + A_{\frac{1}{2}}' + A_o' + \ldots)$$

$$(7.152) \quad = C_o + C_{-\frac{1}{2}} + C_{-1} + \ldots$$

Thus, combining (7.148), (7.151) and (7.152), we can write :

$$Z_1'\hat{\Sigma}_{11}^{-1} X (X'\hat{\Sigma}_{11}^{-1} X)^{-1} X'\hat{\Sigma}_{11}^{-1} Z_1$$

$$= (\sum_i \frac{1}{\hat{\sigma}_{i11}} Z_1'N_iX)(\sum_i \frac{1}{\hat{\sigma}_{i11}} X'N_iX)^{-1}(\sum_i \frac{1}{\hat{\sigma}_{i11}} X'N_iZ_1)$$

$$(7.153) \quad = (A_1+ A_{\frac{1}{2}}+ A_o+ \ldots)(B_o+ B_{-\frac{1}{2}}+ B_{-1}+ \ldots)(C_o+ C_{-\frac{1}{2}}+ C_{-1}+\ldots)$$

By following a similar procedure we will also have :

$$Z_1'\hat{\Sigma}_{11}^{-1} X (X'\hat{\Sigma}_{11}^{-1} X)^{-1} X'\hat{\Sigma}_{11}^{-1} u_1$$

$$= (A_1+ A_{\frac{1}{2}}+ A_o+ \ldots)(B_o+ B_{-\frac{1}{2}}+ B_{-1}+ \ldots)(\sum_i \frac{1}{\sigma_{i11}} X'N_iX)^{-1} .$$

$$(\sum_i \frac{1}{\sigma_{i11}} X'N_iu_1 - \sum_i \frac{1}{\sigma_{i11}} \frac{\eta_{i11}}{\sigma_{i11}} X'N_iu_1 + \sum_i \frac{1}{\sigma_{i11}} \frac{\eta_{i11}^2}{\sigma_{i11}^2} X'N_iu_1+ \ldots)$$

$$(7.154) \quad = (A_1+ A_{\frac{1}{2}}+ A_o+ \ldots)(B_o+ B_{-\frac{1}{2}}+ B_{-1}+ \ldots)(D_{-\frac{1}{2}}+ D_{-1}+ D_{-\frac{3}{2}}+\ldots)$$

Combining (7.153) and (7.154) we finally obtain

$$(7.155) \quad \hat{\alpha}_{1,fG2SLS}- \alpha_1 = \Big[(A_1+ A_{\frac{1}{2}}+ A_o+ \ldots)(B_o+ B_{-\frac{1}{2}}+ B_{-1}+ \ldots)$$

$$(C_o+ C_{-\frac{1}{2}}+ C_{-1}+\ldots)\Big]^{-1}(A_1+ A_{\frac{1}{2}}+ A_o+\ldots)$$

$$(B_o+ B_{-\frac{1}{2}}+ B_{-1}+\ldots)(D_{-\frac{1}{2}}+ D_{-1}+ D_{-\frac{3}{2}}\ldots)$$

Looking at the above expression, we find that we still have an inverse to eliminate by expanding in series. Once this is done, we will then have to perform the following operations :

(i) open up all the brackets one by one and multiply them out,

(ii) truncate the resulting series upto order $O(N^{-1})$,

(iii) calculate expectation, term by term, of the truncated series.

We give in Appendix 7.C , the details of the calculations leading up to the truncated series. In what follows, we start from the truncated series and go on to find expectations.

The series expansion upto order $\frac{1}{N}$ of the bias of the feasible G2SLS estimator (cf. (7.C.37), Appendix 7.C) is given by :

$$\hat{\alpha}_{1,fG2SLS} - \alpha_1 \cong$$

$$(7.156)\quad (A_1^* B_o C_o^*)^{-1} \Big[A_1^* B_o D_{-\frac{1}{2}} + A_{\frac{1}{2}}^* B_o D_{-\frac{1}{2}} + A_1^* B_{-\frac{1}{2}} D_{-\frac{1}{2}} + A_{\frac{1}{2}}^{**} B_o D_{-\frac{1}{2}} + A_1^* B_o D_{-1} \Big] + (A_1^* B_o C_o^*)^{-1} \Big[A_{\frac{1}{2}}^* B_o C_o^* + A_1^* B_o C_{-\frac{1}{2}}^* + A_1^* B_{-\frac{1}{2}} C_o^* + A_{\frac{1}{2}}^{**} B_o C_o^* + A_1^* B_o C_{-\frac{1}{2}}^{**} \Big] (A_1^* B_o C_o^*)^{-1} A_1^* B_o D_{-\frac{1}{2}}$$

where

$$A_1^* B_o C_o^* = \Big(\sum_i \frac{1}{\sigma_{i11}} \bar{Z}_1' N_i X\Big)\Big(\sum_i \frac{1}{\sigma_{i11}} X' N_i X\Big)^{-1}\Big(\sum_i \frac{1}{\sigma_{i11}} X' N_i \bar{Z}_1\Big)$$

$$A_1^* B_o D_{-\frac{1}{2}} = \Big(\sum_i \frac{1}{\sigma_{i11}} \bar{Z}_1' N_i X\Big)\Big(\sum_i \frac{1}{\sigma_{i11}} X' N_i X\Big)^{-1}\Big(\sum_i \frac{1}{\sigma_{i11}} X' N_i u_1\Big)$$

$$A_{\frac{1}{2}}^* B_o D_{-\frac{1}{2}} = \Big(\sum_i \frac{1}{\sigma_{i11}} V_z' N_i X\Big)\Big(\sum_i \frac{1}{\sigma_{i11}} X' N_i X\Big)^{-1}\Big(\sum_i \frac{1}{\sigma_{i11}} X' N_i u_1\Big)$$

$$A_1^* B_{-\frac{1}{2}} D_{-\frac{1}{2}} = -\Big(\sum_i \frac{1}{\sigma_{i11}} \bar{Z}_1' N_i X\Big)\Big(\sum_i \frac{1}{\sigma_{i11}} X' N_i X\Big)^{-1}\Big(\sum_i \frac{1}{\sigma_{i11}} \frac{\eta_{i11}}{\sigma_{i11}} X' N_i X\Big)\Big(\sum_i \frac{1}{\sigma_{i11}} X' N_i X\Big)^{-1}\Big(\sum_i \frac{1}{\sigma_{i11}} X' N_i u_1\Big)$$

$$A^{**}_{\frac{1}{2}} B_o D_{-\frac{1}{2}} = -(\sum_i \frac{1}{\sigma_{i11}} \frac{\eta_{i11}}{\sigma_{i11}} \bar{Z}_1' N_i X)(\sum_i \frac{1}{\sigma_{i11}} X'N_i X)^{-1}(\sum_i \frac{1}{\sigma_{i11}} X'N_i u_1)$$

$$A^*_1 B_o D_{-1} = -(\sum_i \frac{1}{\sigma_{i11}} \bar{Z}_1' N_i X)(\sum_i \frac{1}{\sigma_{i11}} X'N_i X)^{-1}(\sum_i \frac{1}{\sigma_{i11}} \frac{\eta_{i11}}{\sigma_{i11}} X'N_i u_1)$$

$$A^*_{\frac{1}{2}} B_o C^*_o = (\sum_i \frac{1}{\sigma_{i11}} V_z' N_i X)(\sum_i \frac{1}{\sigma_{i11}} X'N_i X)^{-1}(\sum_i \frac{1}{\sigma_{i11}} X'N_i \bar{Z}_1)$$

$$A^*_1 B_o C^*_{-\frac{1}{2}} = (\sum_i \frac{1}{\sigma_{i11}} \bar{Z}_1' N_i X)(\sum_i \frac{1}{\sigma_{i11}} X'N_i X)^{-1}(\sum_i \frac{1}{\sigma_{i11}} X'N_i V_z)$$

$$A^*_1 B_{-\frac{1}{2}} C^*_o = -(\sum_i \frac{1}{\sigma_{i11}} \bar{Z}_1' N_i X)(\sum_i \frac{1}{\sigma_{i11}} X'N_i X)^{-1}(\sum_i \frac{1}{\sigma_{i11}} \frac{\eta_{i11}}{\sigma_{i11}} X'N_i X)$$

$$(\sum_i \frac{1}{\sigma_{i11}} X'N_i X)^{-1}(\sum_i \frac{1}{\sigma_{i11}} X'N_i \bar{Z}_1)$$

$$A^{**}_{\frac{1}{2}} B_o C^*_o = -(\sum_i \frac{1}{\sigma_{i11}} \frac{\eta_{i11}}{\sigma_{i11}} \bar{Z}_1' N_i X)(\sum_i \frac{1}{\sigma_{i11}} X'N_i X)^{-1}(\sum_i \frac{1}{\sigma_{i11}} X'N_i \bar{Z}_1)$$

$$A^*_1 B_o C^{**}_{-\frac{1}{2}} = -(\sum_i \frac{1}{\sigma_{i11}} \bar{Z}_1' N_i X)(\sum_i \frac{1}{\sigma_{i11}} X'N_i X)^{-1}(\sum_i \frac{1}{\sigma_{i11}} \frac{\eta_{i11}}{\sigma_{i11}} X'N_i \bar{Z}_1)$$

Before proceeding to find the expectation of the ten terms of (7.149), let us introduce some more notations which will help us to write them compactly.

$$(7.157) \quad (A^*_1 B_o C^*_o)^{-1} = \left[\bar{Z}_1' \Sigma_{11}^{-1} X (X' \Sigma_{11}^{-1} X)^{-1} X' \Sigma_{11}^{-1} \bar{Z}_1\right]^{-1} \equiv \Phi$$

$$(7.158) \quad (\sum_i \frac{1}{\sigma_{i11}} \bar{Z}_1' N_i X) = \bar{Z}_1' \Sigma_{11}^{-1} X \equiv \Delta_1$$

$$(7.159) \quad (\sum_i \frac{1}{\sigma_{i11}} X' N_i X)^{-1} = (X' \Sigma_{11}^{-1} X)^{-1} \equiv \Delta_2$$

Now, let us consider each term of (7.149) in order, one by one, writing it using notations (7.157) - (7.159) and calculate its expectation :

1)

$$(7.160) \quad E\, \Phi A^*_1 B_o D_{-\frac{1}{2}} = 0$$

2) $$E\,\Phi\,\Big(\sum_i \frac{1}{\sigma_{i11}} V_z' N_i X\Big)\,\Delta_2\,\Big(\sum_i \frac{1}{\sigma_{i11}} X'N_i u_1\Big)$$

$$= E\,\Phi\,\Big(\sum_i \sum_j \frac{1}{\sigma_{i11}}\frac{1}{\sigma_{j11}} \begin{bmatrix} V_1' \\ 0 \end{bmatrix} N_i X\;\Delta_2\; X'N_j u_1$$

$$= E\; -\Phi_1\; L_1'\Gamma^{-1'} \sum_i \sum_j \frac{1}{\sigma_{i11}\sigma_{j11}}\; U'N_i X\;\Delta_2 X'N_j u_1$$

partitioning Φ as $\begin{bmatrix}\Phi_1 & \Phi_2\end{bmatrix}$ appropriately

and writing $V_1 = -\,U\Gamma^{-1}L_1$

$$= \Big[-\Phi_1\; L_1'\Gamma^{-1'} \sum_i \sum_j \frac{1}{\sigma_{i11}\sigma_{j11}}\; E(N_i X\;\Delta_2 X'N_j u_1 u_m')\Big] \quad m=1,\dots,M$$

$$= -\Phi_1\; L_1'\Gamma^{-1'}\Big[\sum_i \sum_j \frac{1}{\sigma_{i11}\sigma_{j11}}\;\mathrm{tr}\; N_i X\;\Delta_2 X'N_j \Sigma_{1m}\Big] \quad m=1,\dots,M$$

$$= -\Phi_1\; L_1'\Gamma^{-1'}\Big[\sum_i \frac{1}{\sigma^2_{i11}}\;\mathrm{tr}\;\Delta_2\; X'N_i X\;\sigma_{i1m}\Big] \quad m=1,\dots,M$$

$$= -\Phi_1\; L_1'\Gamma^{-1'}\Big\{\mathrm{tr}\;\Big[(X'\Sigma_{11}^{-1}X)^{-1}\;\Big(\sum_i \frac{\sigma_{i1m}}{\sigma^2_{i11}}\; X'N_i X\Big)\Big]\Big\} \quad m=1,\dots,M$$

$$(7.161) = -\Phi_1\; L_1'\Gamma^{-1'}\Big\{\mathrm{tr}\;\Big[(X'\Sigma_{11}^{-1}X)^{-1}\;(X'\Sigma_{11}^{-1}\Sigma_{1m}\Sigma_{11}^{-1}X)\Big]\Big\} \quad m=1,\dots,M$$

We notice that the matrix inside the square brackets in (7.161), is in fact

$$E(V_1'\;\Sigma_{11}^{-1}\;X\;(X'\;\Sigma_{11}^{-1}\;X)^{-1}\;X'\;\Sigma_{11}^{-1}\;u_1)$$

(leaving out $L'_1\Gamma^{-1'}$, which appears in front in (7.161)) i.e. it represents the covariance between the explanatory endogenous variables and the errors of the transformed equation. It therefore corresponds to the term containing the factor $(K-1)-(K_1+\tilde{M}_1)$ in the expression of the bias of the covariance 2SLS estimator (7.129).

3) $$-\,E\;\Phi\,\Delta_1\;\Delta_2\;\Big(\sum_i \frac{1}{\sigma_{i11}}\frac{\eta_{i11}}{\sigma_{i11}}\; X'N_i X\Big)\;\Delta_2\;\Big(\sum_i \frac{1}{\sigma_{i11}}\; X'N_i u_1\Big)$$

$$(7.162)\; -E\;\Phi\,\Delta_1\;\Delta_2 \sum_i \sum_j \frac{1}{\sigma^2_{i11}\sigma_{j11}}\; X'N_i X\;\Delta_2\;\eta_{i11}\; X'N_j u_1$$

Now, the expectation of $\eta_{i11} X'N_j u_1$ (i=1,4 ; j=1,4) is calculated separately in Appendix 7.D as it also appears in many of the following terms. Combining the results of I, II, III and IV of this Appendix, we get :

$$(7.163)\quad E(\sum_i \sum_j \eta_{i11} X'N_j u_1) = \frac{1}{NT} X'\iota_{NT}\, m^3_{\varepsilon 111} \qquad \text{(for i=4, j=1)}$$

$$+ \frac{1}{NT} X'\iota_{NT}\, m^3_{1111} \qquad \text{(for i=1, j=1)}$$

Thus, the expectation in (7.162) becomes :

$$(7.164)\quad -\frac{1}{NT}\Phi\Delta_1\Delta_2\left[\frac{m^3_{\varepsilon 111}}{\sigma^2_{\varepsilon 11}\sigma_{111}} X'N_4X\Delta_2X'\iota_{NT} + \frac{m^3_{1111}}{\sigma^2_{111}\sigma_{111}} X'N_1X\Delta_2X'\iota_{NT}\right]$$

4) $$E -\Phi(\sum_i \frac{1}{\sigma_{i11}}\frac{\eta_{i11}}{\sigma_{i11}}\bar{Z}'N_iX)\Delta_2(\sum_i\frac{1}{\sigma_{i11}}X'N_iu_1)$$

$$= E -\Phi\sum_i\sum_j\frac{1}{\sigma^2_{i11}\sigma_{j11}}\bar{Z}'_1N_iX\Delta_2\eta_{i11}X'N_ju_1$$

By using, once again, the results of Appendix 7.D, it can be seen that the above expectation is equal to :

$$(7.165)\quad -\frac{1}{NT}\Phi\left[\frac{m^3_{\varepsilon 111}}{\sigma^2_{\varepsilon 11}\sigma_{111}}\bar{Z}'_1N_4X\Delta_2X'\iota_{NT} + \frac{m^3_{1111}}{\sigma^2_{111}\sigma_{111}}\bar{Z}'N_1X\Delta_2X'\iota_{NT}\right]$$

5) $$-E\,\Phi\Delta_1\Delta_2\sum_i\frac{1}{\sigma^2_{i11}}\eta_{i11}X'N_iu_1$$

$$(7.166)\quad = -\frac{1}{NT}\Phi\Delta_1\Delta_2X'\iota_{NT}\frac{m^3_{1111}}{\sigma^2_{111}}$$

(using the results of Appendix 7.D)

6) $$E - \Phi_1 (\sum_i \frac{1}{\sigma_{i11}} L_1' \Gamma^{-1} U' N_i X) \Delta_2 \Delta_1' \Phi \Delta_1 \Delta_2 (\sum_i \frac{1}{\sigma_{i11}} X' N_i u_1)$$

(writing $-L_1' \Gamma^{-1} U' = V_1'$)

$$= E - \Phi_1 L_1' \Gamma^{-1'} \left[\sum_i \sum_j \frac{1}{\sigma_{i11}\sigma_{j11}} u_m' N_i X \Delta_2 \Delta_1' \Phi \Delta_1 \Delta_2 X' N_j u_1 \right]$$

$m=1,\ldots,M$

$$= - \Phi_1 L_1 \Gamma^{-1'} \left[\mathrm{tr} \sum_i \frac{\sigma_{i1m}}{\sigma^2_{i11}} \Delta_2 \Delta_1' \Phi \Delta_1 \Delta_2 X' N_i X \right] \quad m=1,\ldots,M$$

$$(7.167) = - \Phi_1 L_1 \Gamma^{-1'} \left\{ \mathrm{tr} \left[\Delta_2 \Delta_1' \Phi \Delta_1 \Delta_2 (X' \Sigma_{11}^{-1} \Sigma_{1m} \Sigma_{11}^{-1} X) \right] \right\} \quad m=1,\ldots,M$$

7) $$E \Phi \Delta_1 \Delta_2 \left(\sum_i \frac{1}{\sigma_{i11}} X' N_i V_z \right) \Phi \Delta_1 \Delta_2 \left(\sum_i \frac{1}{\sigma_{i11}} X' N_i u_1 \right)$$

$$= E - \Phi \Delta_1 \Delta_2 \sum_i \sum_j \frac{1}{\sigma_{i11}} \frac{1}{\sigma_{j11}} X' N_i U \Gamma^{-1} L_1 \tilde{\Phi}_1 \Delta_1 \Delta_2 X' N_j u_1$$

partitioning Φ as $\begin{bmatrix} \tilde{\Phi}_1 \\ \tilde{\Phi}_2 \end{bmatrix}$ appropriately

and writing $V_z = [V_1 \; 0]$, $V_1 = -U \Gamma^{-1} L_1$

By following a procedure similar to that of 2) and 6) above, it can be verified that the above expectation is equal to

$$- \Phi \Delta_1 \Delta_2 \sum_i \frac{1}{\sigma^2_{i11}} X' N_i X \Delta_2' \Delta_1' \tilde{\Phi}_1' L_1' \Gamma^{-1'} \begin{bmatrix} \sigma_{i11} \\ \vdots \\ \sigma_{iM1} \end{bmatrix}$$

In the same way, using the results of Appendix 7.D, we can write the expectation of the remaining terms as follows :

8) $$- E \Phi \Delta_1 \Delta_2 (\sum_i \frac{1}{\sigma_{i11}} \frac{\eta_{i11}}{\sigma_{i11}} X' N_i X) \Delta_2 \Delta_1' \Phi \Delta_1 \Delta_2 (\sum_i \frac{1}{\sigma_{i11}} X' N_i u_1)$$

$$(7.169) = \frac{-1}{NT} \Phi \Delta_1 \Delta_2 \left[\frac{m^3_{\varepsilon 111}}{\sigma^2_{\varepsilon 11} \sigma_{111}} X' N_4 X \Delta_2 \Delta_1' \Phi \Delta_1 \Delta_2 X' \iota + \frac{m^3_{1111}}{\sigma^2_{111} \sigma_{111}} X' N_1 X \Delta_2 \Delta_1' \Phi \Delta_1 \Delta_2 X' \iota \right]$$

9) $$-E\,\Phi\left(\sum_i \frac{1}{\sigma_{i11}}\frac{\eta_{i11}}{\sigma_{i11}}\bar{Z}_1'N_iX\right)\Delta_2\Delta_1'\Phi\,\Delta_1\Delta_2\left(\sum_i \frac{1}{\sigma_{i11}}X'N_iu_1\right)$$

$$(7.170) = \frac{-1}{NT}\Phi\left[\frac{m^3_{\varepsilon 111}}{\sigma^2_{\varepsilon 11}\sigma_{111}}\bar{Z}_1'N_4X\,\Delta_2\Delta_1'\,\Phi\,\Delta_1\,\Delta_2\,X'\iota + \frac{m^3_{1111}}{\sigma^2_{111}\sigma_{111}}\bar{Z}_1'N_1X\,\Delta_2\,\Delta_1'\,\Phi\,\Delta_1\,\Delta_2\,X'\iota\right]$$

10) $$-E\,\Phi\,\Delta_1\Delta_2\left(\sum_i \frac{1}{\sigma_{i11}}\frac{\eta_{i11}}{\sigma_{i11}}X'N_i\bar{Z}_1\right)\Phi\,\Delta_1\Delta_2\left(\sum_i \frac{1}{\sigma_{i11}}X'N_iu_1\right)$$

$$(7.171) = \frac{-1}{NT}\Phi\,\Delta_1\,\Delta_2\left[\frac{m^3_{\varepsilon 111}}{\sigma^2_{\varepsilon 11}\sigma_{111}}X'N_4\bar{Z}_1\,\Phi\,\Delta_1\,\Delta_2\,X'\iota + \frac{m^3_{1111}}{\sigma^2_{111}\sigma_{111}}X'N_1\bar{Z}_1\,\Phi\,\Delta_1\,\Delta_2\,X'\iota\right]$$

Thus, the bias of the feasible G2SLS estimator is given by the sum of all these ten results - (7.160), (7.161), (7.164) to (7.171).

Note that we have not assumed normality for deriving any of the results so far. However, if we do so, i.e. if we assume that the error components are normally distributed, then their third moments are zero. In this case, the terms 3), 4), 5), 8), 9) and 10) vanish and the bias of the feasible G2SLS estimator simplifies to :

$$E(\hat{\alpha}_{1,fG2SLS} - \alpha_1) = -\Phi\,L_1\Gamma^{-1'}\left\{tr\left[(X'\Sigma_{11}^{-1}X)^{-1}(X'\Sigma_{11}^{-1}\Sigma_{1m}\Sigma_{11}^{-1}X)\right]\right\}_{m=1,\dots,M}$$

(7.172)

$$-\Phi_1L_1\Gamma^{-1'}\left\{tr\left[\Delta_2\Delta_1'\,\Phi\,\Delta_1\Delta_2\,(X'\Sigma_{11}^{-1}\Sigma_{1m}\Sigma_{11}^{-1}X)\right]\right\}_{m=1,\dots,M}$$

$$-\Phi\,\Delta_1\Delta_2\sum_i\frac{1}{\sigma^2_{i11}}X'N_iX\,\Delta_2'\,\Delta_1'\,\tilde{\Phi}_1'\,L_1'\,\Gamma^{-1'}\begin{bmatrix}\sigma_{i11}\\ \vdots\\ \sigma_{iM1}\end{bmatrix}$$

APPENDIX 7.A : Preliminary Computations of Orders

7.A.1 Some Basic Order Calculations

Order of $\underline{X}'\underline{X}$

From the assumption that

(7.A.1) $\lim_{N\to\infty} \frac{1}{N} \underline{X}'\underline{X} = \underline{R}^{(1)}$, a finite positive definite matrix

we say that

(7.A.2) $\underline{X}'\underline{X} = O(N)$

Order of $\underline{X}'u_m$

By splitting u_m into its two different components, we can write :

$$\underline{X}'u_m = \underline{X}'(I_N \otimes \iota_T)\,\mu^m + \underline{X}'\varepsilon_m$$

Let us consider these two terms one by one.

(i) $\underline{X}'(I_N \otimes \iota_T)\,\mu^m$

Since the elements of μ^m are i.i.d. with mean zero and variance $\sigma_{\mu mm}$ (cf. (3.16), (3.17), pp.49-50), it follows, by applying Central-limit theorem, that

$$\frac{1}{\sqrt{N}}\,\underline{X}'(I_N \otimes \iota_T)\,\mu^m$$

has a limiting normal distribution with zero mean and the following covariance matrix :

$$\lim_{N\to\infty} \frac{1}{N}\,\sigma_{\mu mm}\,\underline{X}'(I_N \otimes \iota_T)\,(I_N \otimes \iota_T')\,\underline{X}$$

provided the above limit exists.

Now, we shall show that the covariance matrix is in fact the one given above and that the limit exists.

$$E(\frac{1}{N}\,\underline{X}'(I_N \otimes \iota_T)\,\mu^m \mu^{m'}\,(I_N \otimes \iota_T')\,\underline{X})$$

$$= \frac{1}{N} \underline{X}'(I_N \otimes \iota_T)\, E(\mu^m \mu^{m'})(I_N \otimes \iota_T')\, \underline{X}$$

$$= \frac{1}{N} \sigma_{\mu\, mm}\, \underline{X}'(I_N \otimes \iota_T)(I_N \otimes \iota_T')\, \underline{X}$$

Hence, the covariance matrix of the limiting distribution is the limit as $N \to \infty$ of the above matrix i.e. the limit of[1)]

$$\frac{1}{N}\begin{bmatrix} x_{211} \cdots x_{21T} \cdots x_{2NT} \\ \vdots \quad \vdots \quad \vdots \\ x_{K11} \cdots x_{K1T} \cdots x_{KNT} \end{bmatrix} \begin{bmatrix} \iota_T & & 0 \\ & \ddots & \\ 0 & & \iota_T \end{bmatrix} \begin{bmatrix} \iota_T' & & 0 \\ & \ddots & \\ 0 & & \iota_T' \end{bmatrix} \begin{bmatrix} x_{211} \cdots x_{K11} \\ \vdots \quad \vdots \\ x_{21T} \cdots x_{K1T} \\ \vdots \quad \vdots \\ x_{2NT} \cdots x_{KNT} \end{bmatrix}$$

i.e.
$$\frac{1}{N}\begin{bmatrix} (\sum_t x_{21t}) \cdots (\sum_t x_{2Nt}) \\ \vdots \qquad \vdots \\ (\sum_t x_{K1t}) \cdots (\sum_t x_{KNt}) \end{bmatrix} \begin{bmatrix} (\sum_t x_{21t}) \cdots (\sum_t x_{K1t}) \\ \vdots \qquad \vdots \\ (\sum_t x_{2Nt}) \cdots (\sum_t x_{KNt}) \end{bmatrix}$$

or
$$\frac{1}{N}\begin{bmatrix} \sum_i \sum_t \sum_{t'} x_{2it}\, x_{2it'} & \cdots & \sum_i \sum_t \sum_{t'} x_{2it}\, x_{Kit'} \\ \vdots & & \vdots \\ \sum_i \sum_t \sum_{t'} x_{Kit}\, x_{2it'} & \cdots & \sum_i \sum_t \sum_{t'} x_{Kit}\, x_{Kit'} \end{bmatrix}$$

Using (7.A.1) we can say that the above matrix tends to a finite matrix as $N \to \infty$. Thus

$$\frac{1}{\sqrt{N}} \underline{X}'(I_N \otimes \iota_T)\, \mu^m$$

has a non-degenerate limiting distribution and hence we say that

(7.A.3) $\underline{X}'(I_N \otimes \iota_T)\, \mu^m = O_p(\sqrt{N})$

(ii) $\underline{X}'\varepsilon_m$

Here, it is straightforward that $\frac{1}{\sqrt{N}} \underline{X}'\varepsilon_m$ has a normal limiting distribution with zero mean and covariance matrix $\sigma_{\varepsilon\, mm} \cdot \underline{R}^{(1)}$ and hence we say that

(7.A.4) $\underline{X}'\varepsilon_m = O_p(\sqrt{N})$

---

1) We leave out $\sigma_{\mu mm}$ as it is a constant.

Finally, combining (7.A.3) and (7.A.4) and using the assumption that the two components μ^m and ε_m are independent, we can say that

(7.A.5) $\underline{X}'u_m = O_p(\sqrt{N})$, $m=1,...,M$

or

(7.A.6) $\underline{X}'U = O_p(\sqrt{N})$

Order of $\underline{X}'N_1\underline{X}$

Using the definition of N_1 given in (7.24), it can be easily verified that a typical element of the above matrix is[1])

$$T \sum_i x_{ji.} \, x_{ki.}$$

$$= T \sum_i \left(\frac{1}{T} \sum_t x_{jit}\right) \left(\frac{1}{T} \sum_t x_{kit}\right)$$

$$= \frac{1}{T} \sum_i \sum_t \sum_{t'} x_{jit} \, x_{kit'}$$

$$= O(N) \text{ using } (7.A.2)$$

Hence,

(7.A.7) $\underline{X}'N_1\underline{X} = O(N)$

Order of $\underline{X}'N_4\underline{X}$

As

$$N_1 + N_4 = I$$

we have

(7.A.8) $\underline{X}'N_4\underline{X} = \underline{X}'\underline{X} - \underline{X}'N_1\underline{X}$

and since both the terms on the RHS of (7.A.8) are O(N), we conclude that

(7.A.9) $\underline{X}'N_4\underline{X} = O(N)$

---

1) A dot in the place of a subscript indicates the average taken over it.

Order of $X'N_1X$

$$X_1' \; N_1 \; X = \begin{bmatrix} \iota_{NT}' \\ \underline{X}' \end{bmatrix} N_1 \begin{bmatrix} \iota_{NT} & \underline{X} \end{bmatrix}$$

$$= \begin{bmatrix} \iota_{NT}' \; N_1 \; \iota_{NT} & \iota_{NT}' \; N_1 \; \underline{X} \\ \underline{X}' \; N_1 \; \iota_{NT} & \underline{X}' \; N_1 \; \underline{X} \end{bmatrix}$$

Now,

$$\iota_{NT}' \; N_1 \; \iota_{NT} = NT = O(N)$$

Next,

$$\iota_{NT}' \; N_1 \; \underline{X} = \iota_{NT}' \; \underline{X} = O(N)$$

as

$$\underset{N\to\infty}{\text{plim}} \frac{1}{NT} \iota_{NT}' \; \underline{X} = r^{(1)'} \qquad \text{(cf. (3.A.5))}$$

And from (7.A.7) we have

$$\underline{X}' \; N_1 \; \underline{X} = O(N)$$

Thus,

$$\text{(7.A.10)} \quad X' \; N_1 \; X = O(N)$$

Order of $X'N_4X$

$$X'N_4X = \begin{bmatrix} \iota_{NT}' \; N_4 \; \iota_{NT} & \iota_{NT}' \; N_4 \; \underline{X} \\ \underline{X}' \; N_4 \; \iota_{NT} & \underline{X}' \; N_4 \; \underline{X} \end{bmatrix}$$

$$= \begin{bmatrix} 0 & 0 \\ 0 & \underline{X}'N_4\underline{X} \end{bmatrix} \qquad \text{as } N_4 \; \iota_{NT} = 0$$

Thus,

$$\text{(7.A.11)} \quad X' \; N_4 \; X = O(N)$$

Order of $\underline{X}'N_1 u_m$:

We have :

$$\underline{X}'N_1 u_m = \underline{X}'N_1 \left[(I_N \otimes \iota_T) \mu^m + \varepsilon_m \right]$$

and a typical element of this vector can be easily seen to be

$$T \sum_i x_{ji.} \mu_{mi} + T \sum_i x_{ji.} \varepsilon_{mi.}$$

Let us consider the above two terms one by one.

First,

$$T \sum_i x_{ji.} \mu_{mi}$$

$$= T \frac{1}{T} \sum_i (\sum_t x_{jit}) \mu_{mi}$$

$$= \frac{1}{T} \sum_i \sum_t x_{jit} \mu_{mi}$$

(7.A.12) $= O_p(\sqrt{N})$ as its expectation = 0

and its variance $= \sum_i \sum_t x_{jit}^2 \sigma_{\mu mm} = O(N)$

Next,

$$T \sum_i x_{ji.} \varepsilon_{mi.}$$

$$= T \frac{1}{T^2} \sum_i (\sum_t x_{jit}) (\sum_{t'} \varepsilon_{mit'})$$

$$= \frac{1}{T} \sum_i \sum_t \sum_{t'} x_{jit} \varepsilon_{mit'}$$

(7.A.13) $= O_p(\sqrt{N})$ as its expectation = 0

and its variance $= \frac{1}{T^2} \sum_i \sum_t x_{jit}^2 \sigma_{\varepsilon mm} = O(N)$

Thus, combining (7.A.12) and (7.A.13), we get :

(7.A.14) $\underline{X}' N_1 u_m = O_p(\sqrt{N})$

Order of $\underline{X}'N_4u_m$

As before, by noting that

$$\underline{X}' N_4 u_m = \underline{X}' u_m - \underline{X}' N_1 u_m$$

it follows that

(7.A.15) $\underline{X}'N_4u_m = O_p(\sqrt{N})$ using (7.A.5) and (7.A.14)

Order of $X'N_1u_m$:

$$X'N_1u_m = \begin{bmatrix} \iota'_{NT} N_1 u_m \\ \underline{X}' N_1 u_m \end{bmatrix} = \begin{bmatrix} \iota'_{NT} u_m \\ \underline{X}' N_1 u_m \end{bmatrix}$$

Now,

$$\iota'_{NT} u_m = \iota'_{NT}\left[(I_N \otimes \iota_T) \mu^m + \varepsilon_m \right]$$

$$= T \sum_i \mu_{mi} + \sum_i \sum_t \varepsilon_{mit}$$

As,

$$E(T \sum_i \mu_{mi}) = 0$$

and

$$V(T \sum_i \mu_{mi}) = T^2 N \sigma_{\mu\, mm} = O(N)$$

we conclude that

$$T \sum_i \mu_{mi} = O_p(\sqrt{N})$$

Similarly, as

$$E(\sum_i \sum_t \varepsilon_{mit}) = 0$$

and

$$V(\sum_i \sum_t \varepsilon_{mit}) = NT \sigma_{\varepsilon\, mm} = O(N)$$

we have

$$\sum_i \sum_t \varepsilon_{mit} = O_p(\sqrt{N})$$

Hence,

$$\iota_{NT}' \, u_m = O_p(\sqrt{N})$$

and therefore

(7.A.16) $X' \, N_1 \, u_m = O_p(\sqrt{N})$ using the above result and (7.A.14)

<u>Order of $X'N_4u_m$</u>

$$X' \, N_4 \, u_m = \begin{bmatrix} \iota_{NT}' \, N_4 \, u_m \\ \underline{X}' \, N_4 \, u_m \end{bmatrix}$$

$$= \begin{bmatrix} 0 \\ \underline{X}' \, N_4 \, u_m \end{bmatrix}$$

with

$$\underline{X}' \, N_4 \, u_m = O_p(\sqrt{N})$$

Hence,

(7.A.17) $X' \, N_4 \, u_m = O_p(\sqrt{N})$

7.A.2 <u>Further Results on Orders</u>

<u>Order of $\bar{Z}_1' \, C \, \bar{Z}_1$</u>

$$\bar{Z}_1' C \, \bar{Z}_1 = \begin{bmatrix} (X \, \Pi_1)' \\ X_1^{*\prime} \end{bmatrix} C \begin{bmatrix} X \, \Pi_1 & X_1^* \end{bmatrix}$$

$$= \begin{bmatrix} \Pi_{*1}' \underline{X}' N_4 \underline{X} (\underline{X}' N_4 \underline{X})^{-1} \underline{X}' N_4 \underline{X} \, \Pi_{*1} & \Pi_{*1}' \underline{X}' N_4 \underline{X} (\underline{X}' N_4 \underline{X})^{-1} \underline{X}' N_4 \underline{X} H_1^* \\ H_1^{*\prime} \underline{X}' N_4 \underline{X} (\underline{X}' N_4 \underline{X})^{-1} \underline{X}' N_4 \underline{X} \, \Pi_{*1} & H_1^{*\prime} \underline{X}' N_4 \underline{X} (\underline{X}' N_4 \underline{X})^{-1} \underline{X}' N_4 \underline{X} H_1^* \end{bmatrix}$$

using the following :

(i) the definition of C in (7.33)

(ii) $X = [\iota \ \underline{X}]$

(iii) $X'N_4 = \begin{bmatrix} 0 \\ \underline{X}'N_4 \end{bmatrix}$ as $\iota' N_4 = 0$

(iv) $\Pi_1' X' N_4 = \begin{bmatrix} \Pi_{o1} & \Pi_{*1}' \end{bmatrix} \begin{bmatrix} 0 \\ \underline{X}'N_4 \end{bmatrix} = \Pi_{*1}' \underline{X}' N_4$

(v) $X_1^* = \underline{X} \ H_1^*$

Thus,

$$\bar{Z}_1' C \ \bar{Z}_1 = \begin{bmatrix} \Pi_{*1}' \ \underline{X}'N_4\underline{X} \ \Pi_{*1} & \Pi_{*1}' \ \underline{X}'N_4\underline{X} \ H_1^* \\ H_1^{*\prime} \ \underline{X}'N_4\underline{X} \ \Pi_{*1} & H_1^{*\prime} \ \underline{X}'N_4\underline{X} \ H_1^* \end{bmatrix}$$

$= O(N)$

using (7.A.9) and the fact that Π_{*1}, H_1^* are O(1) as they are finite constant matrices

Therefore

(7.A.18) $(\bar{Z}_1' \ C \ \bar{Z}_1)^{-1} = O(\frac{1}{N})$

<u>Order of $u_1' \ C \ \bar{Z}_1$</u>

$$u_1' C \ \bar{Z}_1 = \begin{bmatrix} u_1' N_4 \underline{X} (\underline{X}'N_4\underline{X})^{-1} \underline{X}'N_4\underline{X} \ \Pi_{*1} & u_1' N_4 \underline{X} (\underline{X}'N_4\underline{X})^{-1} \underline{X}'N_4\underline{X} \ H_1^* \end{bmatrix}$$

$$= \begin{bmatrix} u_1' N_4 \underline{X} \ \Pi_{*1} & u_1' N_4 \underline{X} \ H_1^* \end{bmatrix}$$

(7.A.19) $= O_p(\sqrt{N})$ using (7.A.15)

<u>Order of $u_1' C \ V_z$</u>

$$u_1' C \ V_z = \begin{bmatrix} u_1' N_4 \underline{X} (\underline{X}'N_4\underline{X})^{-1} \underline{X}'N_4 \ V_1 & u_1' N_4 \underline{X} (\underline{X}'N_4\underline{X})^{-1} \underline{X}'N_4 \ 0 \end{bmatrix}$$

$= \begin{bmatrix} u_1' N_4 \underline{X} (\underline{X}'N_4\underline{X})^{-1} \ \underline{X}'N_4 V_1 & 0 \end{bmatrix}$ using the definitions of C (7.33) and V_z (7.37)

$= -\begin{bmatrix} u_1' N_4 \underline{X} (\underline{X}'N_4\underline{X})^{-1} \ \underline{X}'N_4 \ U \ \Gamma^{-1} \ L_1 & 0 \end{bmatrix}$ using (7.60)

$= \begin{bmatrix} O_p(\sqrt{N}) \ O(\frac{1}{N}) \ O_p(\sqrt{N}) & O(1) \end{bmatrix}$ using (7.A.15), (7.A.9)

(7.A.20) $= O_p(1)$

Order of $V_z'C\ \bar{Z}_1$

As in the previous cases, we write :

$$V_z'C\ \bar{Z}_1 = \begin{bmatrix} V_1'N_4\underline{X}(\underline{X}'N_4\underline{X})^{-1}\underline{X}'N_4\underline{X}\ \Pi_{*1} & V_1'N_4\underline{X}(\underline{X}'N_4\underline{X})^{-1}\underline{X}'N_4\underline{X}\ H_1^* \\ 0 & 0 \end{bmatrix}$$

$$= \begin{bmatrix} -L_1'\Gamma^{-1'}U'N_4\underline{X}\ \Pi_{*1} & -L_1'\Gamma^{-1'}U'N_4\underline{X}\ H_1^* \\ 0 & 0 \end{bmatrix}$$

$$(7.A.21) \quad = O_p(\sqrt{N})$$

Order of $V_z'C\ V_z$

$$V_z'C\ V_z = \begin{bmatrix} V_1'N_4\underline{X}(\underline{X}'N_4\underline{X})^{-1}\underline{X}'N_4V_1 & 0 \\ 0 & 0 \end{bmatrix}$$

$$= \begin{bmatrix} L_1'\Gamma^{-1'}U'N_4\underline{X}(\underline{X}'N_4\underline{X})^{-1}\underline{X}'N_4U\ \Gamma^{-1}L_1 & 0 \\ 0 & 0 \end{bmatrix}$$

$$(7.A.22) \quad = O_p(1)$$

Orders of $\bar{Z}_1'N_1u_1$, $\bar{Z}_1'N_4u_1$

$$\bar{Z}_1'N_1u_1 = \begin{bmatrix} \Pi_{*1}'\ \underline{X}'N_1u_1 & H_1^{*'}\ \underline{X}'N_1u_1 \end{bmatrix}$$

$$(7.A.23) \quad = O_p(\sqrt{N})$$

Similarly, it is easily seen that

$$(7.A.24) \quad \bar{Z}_1'N_4u_1 = O_p(\sqrt{N})$$

Orders of $V_z'N_1u_1$, $V_z'N_4u_1$

First,

$$V_z'N_1u_1 = \begin{bmatrix} V_1'N_1u_1 & 0 \end{bmatrix}$$

$$= \begin{bmatrix} -L_1'\Gamma^{-1'}U'N_1u_1 & 0 \end{bmatrix}$$

Now,

$$E\,U'N_1u_1 = E\begin{bmatrix} u_1'N_1u_1 \\ \vdots \\ u_M'N_1u_1 \end{bmatrix} = E\begin{bmatrix} \mathrm{tr}\ N_1u_1u_1' \\ \vdots \\ \mathrm{tr}\ N_1u_1u_M' \end{bmatrix}$$

$$= \begin{bmatrix} \mathrm{tr}\ N_1\Sigma_{11} \\ \vdots \\ \mathrm{tr}\ N_1\Sigma_{1M} \end{bmatrix} = \begin{bmatrix} \sigma_{111}\ \mathrm{tr}\ N_1 \\ \vdots \\ \sigma_{11M}\ \mathrm{tr}\ N_1 \end{bmatrix} = \begin{bmatrix} \sigma_{111}\ N \\ \vdots \\ \sigma_{11M}\ N \end{bmatrix}$$

$$= O(N) \qquad \text{since } \sigma_{1mm'} = \sigma_{\varepsilon mm'} + T\,\sigma_{\mu mm'} = O(1)$$

Thus

$$(7.A.25)\quad V_z'N_1u_1 = O_p(N)$$

Next,

$$V_z'N_4u_1 = \begin{bmatrix} V_1'N_4u_1 & 0 \end{bmatrix} = \begin{bmatrix} -\,L_1'\Gamma^{-1'}U'N_4u_1 & 0 \end{bmatrix}$$

and

$$E\,U'N_4u_1 = \begin{bmatrix} E\ \mathrm{tr}\ N_4u_1u_1' \\ \vdots \\ E\ \mathrm{tr}\ N_4u_1u_M' \end{bmatrix}$$

$$= \begin{bmatrix} \sigma_{\varepsilon 11}\ \mathrm{tr}\ N_4 \\ \vdots \\ \sigma_{\varepsilon 1M}\ \mathrm{tr}\ N_4 \end{bmatrix} = \begin{bmatrix} \sigma_{\varepsilon 11}\ N(T-1) \\ \vdots \\ \sigma_{\varepsilon 1M}\ N(T-1) \end{bmatrix}$$

$$= O(N)$$

Thus

$$(7.A.26)\quad V_z'N_4u_1 = O_p(N)$$

<u>Orders of $\bar{Z}_1'N_1X$, $\bar{Z}_1'N_4X$</u>

$$\bar{Z}_1'N_1X = \begin{bmatrix} \Pi_1'X' \\ X_1^{*'} \end{bmatrix} N_1 \begin{bmatrix} \iota_{NT} & \underline{X} \end{bmatrix}$$

$$= \begin{bmatrix} \Pi_1' \begin{bmatrix} \iota_{NT}' \\ \underline{X}' \end{bmatrix} N_1\,\iota_{NT} & \Pi_1' \begin{bmatrix} \iota_{NT}' \\ \underline{X}' \end{bmatrix} N_1\,\underline{X} \\ H_1^{*'}\,\underline{X}'\,N_1\,\iota_{NT} & H_1^{*'}\,\underline{X}'\,N_1\,\underline{X} \end{bmatrix}$$

$$= \begin{bmatrix} \Pi_1' \begin{bmatrix} NT \\ \underline{X}' \iota_{NT} \end{bmatrix} & \Pi_1' \begin{bmatrix} \iota_{NT}' \underline{X} \\ \underline{X}' N_1 \underline{X} \end{bmatrix} \\ H_1^{*'} \underline{X}' \iota_{NT} & H_1^{*'} \underline{X}' N_1 \underline{X} \end{bmatrix}$$

(7.A.27) $= O(N)$ using (7.A.7) and that $\underline{X}' \iota_{NT} = O(N)$

$$\bar{Z}_1' N_4 X = \begin{bmatrix} \Pi_1' X' \\ X_1^{*'} \end{bmatrix} N_4 \begin{bmatrix} \iota_{NT} & \underline{X} \end{bmatrix}$$

$$= \begin{bmatrix} \Pi_1' X' N_4 \iota_{NT} & \Pi_1' X' N_4 \underline{X} \\ H_1^{*'} \underline{X}' N_4 \iota_{NT} & H_1^{*'} \underline{X}' N_4 \underline{X} \end{bmatrix}$$

$$= \begin{bmatrix} 0 & \Pi_1' \begin{bmatrix} \iota_{NT}' \\ \underline{X}' \end{bmatrix} N_4 \underline{X} \\ 0 & H_1^{*'} \underline{X}' N_4 \underline{X} \end{bmatrix} \qquad \text{as } N_4 \iota_{NT} = 0$$

$$= \begin{bmatrix} 0 & \Pi_1' \begin{bmatrix} 0 \\ \underline{X}' N_4 \underline{X} \end{bmatrix} \\ 0 & H_1^{*'} \underline{X}' N_4 \underline{X} \end{bmatrix} \qquad \text{again as } N_4 \iota_{NT} = 0$$

(7.A.28) $= O(N)$ using (7.A.9)

<u>Orders of $V_z' N_1 X$, $V_z' N_4 X$</u>

$$V_z' N_1 X = \begin{bmatrix} V_1' \\ 0' \end{bmatrix} N_1 \begin{bmatrix} \iota_{NT} & \underline{X} \end{bmatrix}$$

$$= \begin{bmatrix} V_1' N_1 \iota_{NT} & V_1' N_1 \underline{X} \\ 0 & 0 \end{bmatrix}$$

$$= \begin{bmatrix} V_1' \iota_{NT} & V_1' N_1 \underline{X} \\ 0 & 0 \end{bmatrix}$$

$$= \begin{bmatrix} - L_1' \Gamma^{-1'} U' \iota_{NT} & - L'_1 \Gamma^{-1'} U' N_1 \underline{X} \\ 0 & 0 \end{bmatrix}$$

(7.A.29) $= O_p(\sqrt{N})$ using (7.A.14) and that $\iota_{NT}' U = O_p(\sqrt{N})$

$$V_z' N_4 X = \begin{bmatrix} V_1' N_4 \iota_{NT} & V_1' N_4 \underline{X} \\ 0 & 0 \end{bmatrix}$$

$$= \begin{bmatrix} 0 & -L_1'\Gamma^{-1'}U'N_4\underline{X} \\ 0 & 0 \end{bmatrix}$$

(7.A.30) $= O_p(\sqrt{N})$ using (7.A.15)

<u>Orders of $Z_1'N_1X$, $Z_1'N_4X$</u>

$Z_1'\ N_1\ X \;= (\bar{Z}_1' + V_z')\ N_1\ X$

$= \bar{Z}_1'N_1X + V_z'N_1X$

$= O(N) + O_p(\sqrt{N})$ using (7.A.27) and (7.A.29)

(7.A.31) $= O_p(N)$

$Z_1'\ N_4\ X \;= (\bar{Z}_1' + V_z')\ N_4\ X$

$= \bar{Z}_1'N_4X + V_z'N_4X$

$= O(N) + O_p(\sqrt{N})$ using (7.A.28) and (7.A.30)

(7.A.32) $= O_p(N)$

APPENDIX 7.B : Derivations of Expectations

7.B.1 Expectation of $u_1'\bar{F}\ U'N_4u_1$

$E(u_1'\bar{F}\ U'N_4u_1)$ can be written as

$$E\ u_1'\begin{bmatrix}\bar{F}_1 & \ldots & \bar{F}_M\end{bmatrix}\begin{bmatrix}u_1' \\ \vdots \\ u_M'\end{bmatrix} N_4\ u_1 \tag{7.B.1}$$

where $\bar{F}_j$ of dimension (NTx1) is the j-th column of $\bar{F}, j=1,\ldots,M$ or $\bar{F} = \begin{bmatrix}\bar{F}_1\ldots\bar{F}_M\end{bmatrix}$

$$= E\ u_1' \sum_k \bar{F}_k\ u_k'\ N_4\ u_1$$

$$= E \sum_k u_1'\ \bar{F}_k\ u_k'\ N_4\ u_1 \tag{7.B.2}$$

From our assumption on the error component structure of u_m, we have

$$u_m = (I_N \otimes \iota_T)\ \mu^m + \varepsilon_m \quad , \quad m=1,\ldots,M \tag{7.B.3}$$

Hence we can write (7.B.2) as

$$E \sum_k \left[\mu^{1'}\ (I_N \otimes \iota_T') + \varepsilon_1'\right] \bar{F}_k \left[(I_N \otimes \iota_T)\ \mu^k + \varepsilon_k\right]'$$
$$N_4 \left[(I_N \otimes \iota_T)\ \mu^1 + \varepsilon_1\right]$$

$$= E \sum_k \left[\mu^{1'}\ (I_N \otimes \iota_T')\ \bar{F}_k\ \mu^{k'}\ (I_N \otimes \iota_T') + \varepsilon_1'\ \bar{F}_k\ \varepsilon_k'\right] N_4$$
$$\left[(I_N \otimes \iota_T)\ \mu^1 + \varepsilon_1\right]^{1)}$$

$$= E \sum_k \left[\mu^{1'}\ (I_N \otimes \iota_T')\ \bar{F}_k\ \mu^{k'}(I_N \otimes \iota_T')\ N_4\ (I_N \otimes \iota_T)\ \mu^1 + \varepsilon_1'\ \bar{F}_k\ \varepsilon_k'\ N_4\ \varepsilon_1\right] \tag{7.B.4}$$

1) While expanding products of quadratic forms (and/or linear forms), we leave out cross-products between different components as their expectation is zero. The same rule will be followed in future.

$$= E \sum_k \varepsilon_1' \bar{F}_k \varepsilon_k' N_4 \varepsilon_1 \qquad \text{as } N_4(I_N \otimes \iota_T) = 0$$

$$= E \sum_k (\sum_i \sum_t \varepsilon_{1it} \bar{F}_{kit})(\sum_i \sum_t \sum_{i'} \sum_{t'} N^4_{it,i't'} \varepsilon_{kit} \varepsilon_{1i't'})$$

where $\bar{F}_{kit}$ and $N^4_{it,i't'}$
denote typical elements of $\bar{F}_k$
and N_4 respectively

(7.B.5) $$= E \sum_k \left(\sum_i \sum_{i'} \sum_{i''} \sum_t \sum_{t'} \sum_{t''} \bar{F}_{kit} N^4_{i't',i''t''} \varepsilon_{1it} \varepsilon_{ki't'} \varepsilon_{1i''t''} \right)$$

Case 1 : All i indices different, all t indices different.
$E(\varepsilon_{1it} \varepsilon_{ki't'} \varepsilon_{1i''t''}) = 0$

Case 2 : Two i indices same, the third one different and two t indices same, the third one different. Eg.
$E(\varepsilon_{1it} \varepsilon_{kit} \varepsilon_{1i''t''}) = E(\varepsilon_{1it} \varepsilon_{kit}) E(\varepsilon_{1i''t''}) = 0$

Case 3 : All the three i indices equal and all the three t indices equal.
$E(\varepsilon_{1it} \varepsilon_{kit} \varepsilon_{1it}) = m^3_{\varepsilon 11k}$

where $m^3_{\varepsilon 11k}$ denotes $E(\varepsilon_{1it}\varepsilon_{kit}\varepsilon_{1it})$ (m^3 indicates that it is a moment of the third order, the index ε stands for the random variable of which it is the third moment and the indices 1,1,k for the structural equations to which the three errors belong; the same notation will be used whenever a third moment is encountered).

All other cases (like "two i indices same, the third one different where all the t indices are equal" etc.) lead to zero expectation.

Hence, the expectation in (7.B.5) is equal to :

(7.B.6) $$\sum_k \sum_i \sum_t \bar{F}_{kit} N^4_{it,it} m^3_{\varepsilon 11k}$$

$$= \sum_k m^3_{\varepsilon 11k} \frac{T-1}{T} \sum_i \sum_t \bar{F}_{kit} \qquad \text{as } N^4_{it,it} = \frac{T-1}{T}$$

Now,

$$\sum_i \sum_t \bar{F}_{kit} = \bar{F}_k' \iota_{NT}$$

$$= -(\bar{P}_1 L_1' \gamma^k)' \iota_{NT}$$ using $\bar{F} = P_1 L_1' \Gamma^{-1}{}'$ and denoting the k-th column of $\Gamma^{-1}{}'$ as γ^k

$$= -\gamma^{k'} L_1 \bar{P}_1' \iota_{NT}$$

$$= -\gamma^{k'} L_1 (C \bar{Y}_1 \bar{W}_{11} + C X_1^* \bar{W}_{21})' \iota_{NT}$$

(7.B.7) by writing
$$\bar{P} = C \bar{Z}_1 (\bar{Z}_1' C \bar{Z}_1)^{-1} = C \begin{bmatrix} \bar{Y}_1 X_1^* \end{bmatrix} \begin{bmatrix} \bar{W}_{11} \bar{W}_{12} \\ \bar{W}_{21} \bar{W}_{22} \end{bmatrix} = \begin{bmatrix} \bar{P}_1 & \bar{P}_2 \end{bmatrix}$$

(7.B.8) $= 0$ as $C' \iota_{NT} = 0$

Therefore, combining (7.B.5), (7.B.6) and (7.B.8), we get

$$E \sum_k \varepsilon_1' \bar{F}_k \varepsilon_k' N_4 \varepsilon_1 = 0$$

and consequently,

(7.B.9) $E\, u_1' \bar{F} U' N_4 u_1 = 0$, thus proving result (7.62).

7.B.2 Expectation of $u_1' \bar{F}\, U'\, \bar{F}\, U'\, N_4 u_1$

$$E\, u_1' (\bar{F}_1 \ldots \bar{F}_M) \begin{bmatrix} u_1' \\ \vdots \\ u_M' \end{bmatrix} (\bar{F}_1 \ldots \bar{F}_M) \begin{bmatrix} u_1' \\ \vdots \\ u_M' \end{bmatrix} N_4 u_1$$

$$= E\, (u_1' \bar{F}_1 \ldots u_1' \bar{F}_M) \begin{bmatrix} u_1' \bar{F}_1 & \ldots & u_1' \bar{F}_M \\ \vdots & & \vdots \\ u_M' \bar{F}_1 & \ldots & u_M' \bar{F}_M \end{bmatrix} \begin{bmatrix} u_1' N_4 u_1 \\ \vdots \\ u_M' N_4 u_1 \end{bmatrix}$$

$$= E(\sum_k u_1' \bar{F}_k u_k' \bar{F}_1 \ldots \sum_k u_1' \bar{F}_k u_k' \bar{F}_M) \begin{bmatrix} u_1' N_4 u_1 \\ \vdots \\ u_M' N_4 u_1 \end{bmatrix}$$

$$(7.B.10) = E \sum_k \sum_\ell u_1' \bar{F}_k \, u_k' \bar{F}_\ell \, u_\ell' N_4 \, u_1$$

Now, by substituting the error component structure for u_1, u_k, u_ℓ we obtain

$$E \sum_k \sum_\ell \Big[\mu^{1'} (I_N \otimes \iota_T') \bar{F}_k \mu^{k'} (I_N \otimes \iota_T') \bar{F}_\ell \mu^{\ell '} (I_N \otimes \iota_T') N_4 (I_N \otimes \iota_T) \mu^1$$

$$+ \varepsilon_1' \bar{F}_k \, \varepsilon_k' \bar{F}_\ell \, \varepsilon_\ell' N_4 \, \varepsilon_1 \Big]$$

(cross-products between different error components are omitted in the multiplication, see footnote on p.280 for reason)

$$= E \sum_k \sum_\ell \varepsilon_1' \bar{F}_k \, \varepsilon_k' \bar{F}_\ell \, \varepsilon_\ell' N_4 \, \varepsilon_1$$

as $N_4 (I_N \otimes \iota_T) = 0$

$$= E \sum_k \sum_\ell \Big\{ \Big(\sum_i \sum_t \varepsilon_{1it} \bar{F}_{kit}\Big)\Big(\sum_i \sum_t \varepsilon_{kit} \bar{F}_{\ell it}\Big)\Big(\sum_i \sum_t \sum_{i'} \sum_{t'} \varepsilon_{\ell it} N^4_{it,i't'} \varepsilon_{1i't'}\Big) \Big\}$$

$$= E \sum_k \sum_\ell \Big\{ \sum_i \sum_{i'} \sum_j \sum_{j'} \sum_t \sum_{t'} \sum_s \sum_{s'} \bar{F}_{kit} \bar{F}_{\ell i't'} N^4_{js,j's'} \varepsilon_{1it} \varepsilon_{ki't'} \varepsilon_{\ell js} \varepsilon_{1j's'} \Big\}$$

$$(7.B.11) = E \sum_k \sum_\ell \Big\{ \sum_i \sum_{i'} \sum_j \sum_{j'} \sum_t \sum_{t'} \sum_s \sum_{s'} \bar{F}^{k\ell}_{it,i't'} N^4_{js,j's'} \varepsilon_{1it} \varepsilon_{ki't'} \varepsilon_{\ell js} \varepsilon_{1j's'} \Big\}$$

where $\bar{F}^{k\ell}_{it,i't'} = \bar{F}_{kit} \bar{F}_{\ell i't'}$

Here, the only cases which will give non-zero expectations are the following :

Case 1 : The i indices are equal by pairs, each pair being different from the other and the same for t indices.

Sub-cases :

i) $i=i'$, $j=j'$, $t=t'$, $s=s'$ but $i \neq j$, $t \neq s$

$E(\varepsilon_{1it} \varepsilon_{kit} \varepsilon_{\ell js} \varepsilon_{1js}) = \sigma_{\varepsilon 1k} \sigma_{\varepsilon \ell 1}$

and the resulting sum is

$$\sum_i \sum_{j \neq i} \sum_t \sum_{s \neq t} \bar{F}^{k\ell}_{it,it} N^4_{js,js} \sigma_{\varepsilon 1k} \sigma_{\varepsilon \ell 1}$$

$$= \sigma_{\varepsilon 1k} \sigma_{\varepsilon \ell 1} \Big(\mathrm{tr}\, \bar{F}^{k\ell} \, \mathrm{tr}\, N_4 - \sum_i \sum_t \bar{F}^{k\ell}_{it,it} N^4_{it,it}\Big)$$

ii) i=j, i'=j', t=s, t'=s', i≠i', t≠t'

$E(\varepsilon_{1it}\ \varepsilon_{ki't'}\ \varepsilon_{\ell it}\ \varepsilon_{1i't'}) = \sigma_{\varepsilon 1\ell}\ \sigma_{\varepsilon k1}$

and the resulting sum is

$$\sum_i \sum_{j\neq i} \sum_t \sum_{s\neq t} \bar{F}^{k\ell}_{it,js}\ N^4_{it,js}\ \sigma_{\varepsilon 1\ell}\ \sigma_{\varepsilon k1}$$

$$= \sigma_{\varepsilon 1\ell}\ \sigma_{\varepsilon k1}\ (\mathrm{tr}\ \bar{F}^{k\ell}\ N_4 \quad - \sum_i \sum_t \bar{F}^{k\ell}_{it,it}\ N^4_{it,it})$$

iii) i=j', i'=j, t=s', t'=s, i≠i', t≠t'

$E(\varepsilon_{1it}\ \varepsilon_{ki't'}\ \varepsilon_{\ell i't'}\ \varepsilon_{1it}) = \sigma_{\varepsilon 11}\ \sigma_{\varepsilon k\ell}$

and the resulting sum is

$$\sum_i \sum_{j\neq i} \sum_t \sum_{s\neq t} \bar{F}^{k\ell}_{it,js}\ N^4_{js,it}\ \sigma_{\varepsilon 11}\ \sigma_{\varepsilon k\ell}$$

$$= \sigma_{\varepsilon 11}\ \sigma_{\varepsilon k\ell}\ (\mathrm{tr}\ \bar{F}^{k\ell}\ N'_4 \quad - \sum_i \sum_t \bar{F}^{k\ell}_{it,it}\ N^4_{it,it})$$

Case 2 : All i indices equal and all t indices equal.

$E(\varepsilon_{1it}\ \varepsilon_{kit}\ \varepsilon_{\ell it}\ \varepsilon_{1it}) = m^4_{\varepsilon 11k\ell}$

and the resulting sum is

$$\sum_i \sum_t \bar{F}^{k\ell}_{it,it}\ N^4_{it,it}\ m^4_{\varepsilon 11k\ell}$$

$$= m^4_{\varepsilon 11k\ell} \sum_i \sum_t \bar{F}^{k\ell}_{it,it}\ N^4_{it,it}$$

All other cases lead to zero expectations.

Thus (7.B.11) becomes :

$$\sum_k \sum_\ell \sigma_{\varepsilon 1k}\ \sigma_{\varepsilon \ell 1}\ (\mathrm{tr}\ \bar{F}^{k\ell}\ \mathrm{tr}\ N_4 - \sum_i \sum_t \bar{F}^{k\ell}_{it,it}\ N^4_{it,it}) + \sigma_{\varepsilon 1\ell}\ \sigma_{\varepsilon k1}\ (\mathrm{tr}\ \bar{F}^{k\ell}\ N_4 - \sum_i \sum_t \bar{F}^{k\ell}_{it,it}\ N^4_{it,it}) + \sigma_{\varepsilon 11}\ \sigma_{\varepsilon k\ell}\ (\mathrm{tr}\ \bar{F}^{k\ell}\ N'_4 - \sum_i \sum_t \bar{F}^{k\ell}_{it,it} N^4_{it,it}) + m^4_{\varepsilon 11k\ell} \sum_i \sum_t \bar{F}^{k\ell}_{it,it}\ N^4_{it,it} \qquad (7.B.12)$$

Now,

$$\mathrm{tr}\ \bar{F}^{k\ell} = \mathrm{tr}\ \bar{F}_k\ \bar{F}'_\ell$$

$$= \mathrm{tr}\ \bar{P}_1 L_1 \gamma^k\ \gamma^{\ell\,'}\ L'_1 \bar{P}'_1$$

$$= \text{tr}\ \gamma^{\ell'} L_1' \bar{P}_1' \bar{P}_1 L_1 \gamma^k$$

Let us expand $\bar{P}_1'\bar{P}_1$:

$$\bar{P}_1'\bar{P}_1 = (C\ \bar{Y}_1\bar{W}_{11} + C\ X_1^*\bar{W}_{21})'(C\ \bar{Y}_1\bar{W}_{11} + C\ X_1^*\ \bar{W}_{21})$$

using (7.B.7)

$$= (W_{11}'\bar{Y}_1' + \bar{W}_{21}'\ X_1^{*\prime})\ C(\bar{Y}_1\bar{W}_{11} + X_1^*\bar{W}_{21}) \quad \text{as } C'C = C$$

$$= \begin{bmatrix} \bar{W}_{11}' & \bar{W}_{21}' \end{bmatrix} (\bar{Z}_1'C\ \bar{Z}_1) \begin{bmatrix} \bar{W}_{11} \\ \bar{W}_{21} \end{bmatrix}$$

(7.B.13) $$= \bar{W}_{11} \qquad \text{recalling that } \bar{W} = (\bar{Z}_1'C\ \bar{Z}_1)^{-1}$$

Hence,

(7.B.14) $$\text{tr}\ \bar{F}^{k\ell} = \text{tr}\ \gamma^{\ell'} L_1' \bar{W}_{11} L_1 \gamma^k$$

Let us denote it as $\bar{R}_{k\ell}$; the reason will become clear in a short while when we will see that it is also a typical element of a matrix denoted as $\bar{R}$.

Thus

(7.B.15) $$\text{tr}\ \bar{F}^{k\ell}\ \text{tr}\ N_4 = \bar{R}_{k\ell}\, N(T-1)$$

Next,

$$\text{tr}\ \bar{F}^{k\ell}\ N_4 = \text{tr}\ \bar{P}_1 L_1 \gamma^k \gamma^{\ell'} L_1' \bar{P}_1' N_4$$

$$= \text{tr}\ \gamma^{\ell'} L_1' \bar{P}_1' N_4 \bar{P}_1 L_1 \gamma^k$$

But

$$\bar{P}_1' N_4 \bar{P}_1 = (\bar{W}_{11}'\bar{Y}_1' + \bar{W}_{21}' X_1^{*\prime})\ C' N_4 C(\bar{Y}_1\bar{W}_{11} + X_1^*\bar{W}_{21})$$

$$= (\bar{W}_{11}'\bar{Y}_1' + \bar{W}_{21}' X_1^{*\prime}) C(\bar{Y}_1\bar{W}_{11} + X_1^*\bar{W}_{21}) \quad \text{as } C'N_4C = C$$

$$= \bar{W}_{11} \qquad \text{as seen above in (7.B.13)}$$

Thus, we see that

(7.B.16) $\operatorname{tr} \bar{F}^{k\ell} N_4 = \operatorname{tr} \bar{F}^{k\ell} = \bar{R}_{k\ell}$

Next

(7.B.17) $\operatorname{tr} \bar{F}^{k\ell} N_4' = \operatorname{tr} \bar{F}^{k\ell} = \bar{R}_{k\ell}$ as $N_4 = N_4'$ and using (7.B.16)

Finally, we need to simplify the following expression appearing in (7.B.12) :

$$\sum_i \sum_t \bar{F}^{k\ell}_{it,it} N^4_{it,it} = \frac{(T-1)}{T} \sum_i \sum_t \bar{F}^{k\ell}_{it,it}$$

$$= \frac{T-1}{T} \operatorname{tr} \bar{F}^{k\ell}$$

(7.B.17) $$= \frac{T-1}{T} \bar{R}_{k\ell}$$

Inserting (7.B.14), (7.B.15), (7.B.16), (7.B.17), and (7.B.18) in (7.B.12) yields :

$$E(u_1'\bar{F}\ U'N_4U'\bar{F}\ u_1)$$

$$= \sum_k \sum_\ell \Big\{ \sigma_{\varepsilon 1k}\, \sigma_{\varepsilon \ell 1}\, \bar{R}_{k\ell} \left(N(T-1) - \frac{T-1}{T}\right)$$

$$+ \sigma_{\varepsilon 1\ell}\, \sigma_{\varepsilon k1}\, \bar{R}_{k\ell} \frac{1}{T} + \sigma_{\varepsilon 11}\, \sigma_{\varepsilon k\ell}\, \bar{R}_{k\ell} \frac{1}{T}$$

$$+ m^4_{\varepsilon 11k\ell}\, \bar{R}_{k\ell} \frac{(T-1)}{T} \Big\}$$

thus proving result (7.65).

7.B.3 Expectation of $u_1'\bar{H}\ U\ \bar{R}\ U'N_4u_1$

$$E\ u_1'\bar{H}\ U\ \bar{R}\ U'N_4u_1$$

$$= E\ u_1'\bar{H} \begin{bmatrix} u_1 & \cdots & u_M \end{bmatrix} \begin{bmatrix} \bar{R}_{11} & \cdots & \bar{R}_{1M} \\ \vdots & & \vdots \\ \bar{R}_{M1} & \cdots & \bar{R}_{MM} \end{bmatrix} \begin{bmatrix} u_1' \\ \vdots \\ u_M' \end{bmatrix} N_4u_1$$

$$= E \begin{bmatrix} u_1'\bar{H}u_1 & \cdots & u_1'\bar{H}u_M \end{bmatrix} \begin{bmatrix} \bar{R}_{11} & \cdots & \bar{R}_{1M} \\ \vdots & & \vdots \\ \bar{R}_{M1} & \cdots & \bar{R}_{MM} \end{bmatrix} \begin{bmatrix} u_1'N_4u_1 \\ \vdots \\ u_M'N_4u_1 \end{bmatrix}$$

$$= E \sum_k \sum_\ell u_1'\bar{H}\ u_k\bar{R}_{k\ell}\ u_\ell' N_4 u_1$$ [1]

$$= E \sum_k \sum_\ell \bar{R}_{k\ell}\ u_1'\bar{H}\ u_k\ u_\ell' N_4 u_1$$

$$= E \sum_k \sum_\ell \bar{R}_{k\ell}\Big[\mu^{1}{}'(I_N \otimes \iota_T')\bar{H}(I_N \otimes \iota_T)\mu^k \mu^\ell{}'(I_N \otimes \iota_T')N_4$$

$$(I_N \otimes \iota_T)\mu^1 + \varepsilon_1'\ \bar{H}\ \varepsilon_k \varepsilon_\ell'\ N_4\ \varepsilon_1\Big]$$

(by introducing the error component structure and leaving our cross-products)

$$(7.B.19) \quad = E \sum_k \sum_\ell \bar{R}_{k\ell}\ \varepsilon_1'\ \bar{H}\ \varepsilon_k\ \varepsilon_\ell'\ N_4\ \varepsilon_1 \qquad \text{as } N_4(I_N \otimes \iota_T) = 0$$

$$= E \sum_k \sum_\ell \bar{R}_{k\ell}\ \Big(\sum_i \sum_t \sum_{i'} \sum_{t'} \varepsilon_{1it}\ \bar{H}_{it,i't'}\ \varepsilon_{ki't'}\Big)$$

$$\Big(\sum_j \sum_s \sum_{j'} \sum_{s'} \varepsilon_{\ell js}\ N^4_{js,j's'}\ \varepsilon_{1j's'}\Big)$$

$$(7.B.20) \quad = E \sum_k \sum_\ell \bar{R}_{k\ell} \Big\{ \sum_i \sum_{i'} \sum_j \sum_{j'} \sum_t \sum_{t'} \sum_s \sum_{s'} \bar{H}_{it,i't'}\ N^4_{js,j's'}\ \varepsilon_{1it}\ \varepsilon_{ki't'}\ \varepsilon_{\ell js}\ \varepsilon_{1j's'} \Big\}$$

Here again we have an expression similar to (7.B.11) of Section 7.B.2 and following the same procedure as for this term, we will get the above expectation to be :

$$\sum_k \sum_\ell \bar{R}_{k\ell} \Big\{ \sigma_{\varepsilon 1k}\sigma_{\varepsilon \ell 1}\ (\text{tr } \bar{H}\ \text{tr } N_4 - \sum_i \sum_t \bar{H}_{it,it}\ N^4_{it,it}) +$$

$$(7.B.21) \qquad \sigma_{\varepsilon 1\ell}\sigma_{\varepsilon k1}\ (\text{tr } \bar{H}\ N_4 - \sum_i \sum_t \bar{H}_{it,it}\ N^4_{it,it}) +$$

$$\sigma_{\varepsilon 11}\ \sigma_{\varepsilon k\ell}\ (\text{tr } \bar{H}\ N_4' - \sum_i \sum_t \bar{H}_{it,it}\ N^4_{it,it}) +$$

$$m^4_{\varepsilon 11k\ell} \sum_i \sum_t \bar{H}_{it,it}\ N^4_{it,it} \Big\}$$

---

1) Note that $\bar{R}_{k\ell}$, which is a typical element of $\bar{R} = \Gamma^{-1}{}'L_1\bar{W}_{11}L_1'\Gamma^{-1}$ is given by $\gamma^{k}{}'L_1\bar{W}_{11}L_1'\gamma^\ell = \text{tr } \gamma^{k}{}'L_1\bar{W}_{11}L_1'\gamma^\ell = \text{tr } \bar{F}^{k\ell}$, thus justifying the notation $\bar{R}_{k\ell}$ used for representing tr $\bar{F}^{k\ell}$ in 7.B.2.

Now,

$$\text{tr } \bar{H} = \text{tr } C\ \bar{Z}_1(\bar{Z}_1'C\ \bar{Z}_1)^{-1}\ \bar{Z}_1'C$$

$$= \text{tr } I_{(K_1+\tilde{M}_1-1)}$$

(7.B.22) $$= (K_1+\tilde{M}_1-1)$$

$$\text{tr}\bar{H}N_4 = \text{tr } C\ \bar{Z}_1(\bar{Z}_1'C\ \bar{Z}_1)^{-1}\ \bar{Z}_1'C\ N_4$$

$$= \text{tr } C\ \bar{Z}_1(\bar{Z}_1'C\ \bar{Z}_1)^{-1}\ \bar{Z}_1'C \qquad \text{as } C\ N_4=C$$

(7.B.23) $$= \text{tr } \bar{H} = K_1 + \tilde{M}_1-1$$

(7.B.24) $$\text{tr}\bar{H}N_4' = \text{tr } \bar{H}\ N_4 = \text{tr } \bar{H} = K_1 + \tilde{M}_1-1$$

and finally,

(7.B.25) $$\sum_i \sum_t \bar{H}_{it,it} N^4_{it,it} = \frac{T-1}{T} \sum_i \sum_t \bar{H}_{it,it} = \frac{T-1}{T}\ \text{tr } \bar{H}$$

$$= \frac{T-1}{T}\ (K_1+\tilde{M}_1-1)$$

Replacing the above results in (7.B.21), we get :

$$E\ u_1'\bar{H}\ U\ \bar{R}\ U'N_4u_1 = \sum_k \sum_\ell \bar{R}_{k\ell}\Big\{\sigma_{\varepsilon 1k}\ \sigma_{\varepsilon\ell 1}(K_1+\tilde{M}_1-1)(N(T-1) - \frac{T-1}{T}) +$$

$$\sigma_{\varepsilon 1\ell}\sigma_{\varepsilon k1}(K_1+\tilde{M}_1-1)(1 - \frac{T-1}{T}) + \sigma_{\varepsilon 11}\sigma_{\varepsilon k\ell}(K_1+\tilde{M}_1-1)(1 - \frac{T-1}{T})$$

$$+ m^4_{\varepsilon 11k\ell}\ (K_1+\tilde{M}_1-1)(\frac{T-1}{T})\Big\}$$

which proves result (7.71).

7.B.4 Expectation of $u_1'C\ U\ \bar{R}\ U'N_4u_1$

$u_1'C\ U\ \bar{R}\ U'N_4u_1$ has the same form as $u_1'\bar{H}\ U\ \bar{R}\ U'N_4u_1$ whose expectation was calculated in 7.B.3 above. Hence, the expectation of $u_1'C\ U\ \bar{R}\ U'N_4u_1$ will be the same as expression (7.B.21) except that $\bar{H}$ will be replaced by C everywhere. Thus

$$Eu_1'CU\bar{R}U'N_4u_1 =$$

$$\sum_k \sum_\ell \bar{R}_{k\ell} \Big\{ \sigma_{\varepsilon 1k}\sigma_{\varepsilon\ell 1} (trC\ trN_4 - \sum_i \sum_t C_{it,it}N^4_{it,it})$$

$$+ \sigma_{\varepsilon 1\ell}\ \sigma_{\varepsilon k1}(tr\ C\ N_4 - \sum_i \sum_t C_{it,it}\ N^4_{it,it}) + \tag{7.B.26}$$

$$+ \sigma_{\varepsilon 11}\ \sigma_{\varepsilon k\ell}(tr\ C\ N_4' - \sum_i \sum_t C_{it,it}\ N^4_{it,it}) +$$

$$+ m^4_{\varepsilon 11k\ell} \sum_i \sum_t C_{it,it}\ N^4_{it,it} \Big\}$$

Now,

(7.B.27) $tr\ C = tr\ N_4'\underline{X}(\underline{X}'N_4\underline{X})^{-1}\underline{X}'N_4 = tr\ I_{K-1} = (K-1)$

(7.B.28) $tr\ C\ N_4' = tr\ C\ N_4 = tr\ C = (K-1)$

and

(7.B.29) $\sum_i \sum_t C_{it,it}\ N^4_{it,it} = \frac{T-1}{T}\ tr\ C = \frac{T-1}{T}\ (K-1)$

Therefore,

$$E\ u_1'CU\bar{R}U'N_4u_1 = \sum_k \sum_\ell \bar{R}_{k\ell}\Big\{\sigma_{\varepsilon 1k}\ \sigma_{\varepsilon\ell 1}(K-1)(N(T-1)- \frac{T-1}{T}) +$$

$$\sigma_{\varepsilon 1\ell}\sigma_{\varepsilon k1}\ (K-1)\ \frac{1}{T} + \sigma_{\varepsilon 11}\sigma_{\varepsilon k\ell}\ (K-1)\ \frac{1}{T}$$

$$+ m^4_{\varepsilon 11k\ell}\ (K-1)\ \frac{T-1}{T}\Big\}$$

which is the same as result (7.74) of 7.3.2.

7.B.5 Expectation of $u_1'\bar{F}\ U'N_4U\ \bar{F}'u_1$

$$E\ u_1'\bar{F}\ U'N_4U\ \bar{F}'u_1$$

$$= E\ u_1'[\ \bar{F}_1\ \cdots\ \bar{F}_M] \begin{bmatrix} u_1' \\ \vdots \\ u_M' \end{bmatrix} N_4\ [u_1\ \cdots\ u_M\ ] \begin{bmatrix} \bar{F}_1' \\ \vdots \\ \bar{F}_M' \end{bmatrix} u_1$$

$$= E\ [u_1'\bar{F}_1\ \cdots\ u_1'\bar{F}_M] \begin{bmatrix} u_1'N_4u_1 & \cdots & u_1'N_4u_M \\ \vdots & & \vdots \\ u_M'N_4u_1 & \cdots & u_M'N_4u_M \end{bmatrix} \begin{bmatrix} \bar{F}_1u_1 \\ \vdots \\ \bar{F}_M'u_1 \end{bmatrix}$$

$$= E \sum_k \sum_\ell u_1' \bar{F}_k u_k' N_4 u_\ell \bar{F}_\ell' u_1$$

$$= E \sum_{k\ell}\sum \Big\{ \mu^{1\prime}(I_N \otimes \iota_T')\bar{F}_k \mu^{k\prime}(I_N \otimes \iota_T') N_4 (I_N \otimes \iota_T)\mu^\ell \bar{F}_\ell' (I_N \otimes \iota_T)\mu^1$$

$$+ \varepsilon_1' \bar{F}_k \varepsilon_k' N_4 \varepsilon_\ell \bar{F}_\ell' \varepsilon_1 \Big\}$$

$$= E \sum_k \sum \varepsilon_1' \bar{F}_k \varepsilon_k' N_4 \varepsilon_\ell \bar{F}_\ell' \varepsilon_1 \qquad \text{as } N_4 (I_N \otimes \iota_T) = 0$$

$$= E \sum_k \sum_\ell \Big(\sum_i \sum_t \varepsilon_{1it} \bar{F}_{kit}\Big)\Big(\sum_i \sum_t \sum_{i'} \sum_{t'} N^4_{it,i't'} \varepsilon_{kit}\, \varepsilon_{\ell i't'}\Big)$$

$$\Big(\sum_i \sum_t \bar{F}_{\ell it} \varepsilon_{1it}\Big)$$

$$(7.B.30) \quad = E \sum_k \sum_\ell \Big(\sum_i \sum_{i'} \sum_j \sum_{j'} \sum_t \sum_{t'} \sum_s \sum_{s'} \bar{F}^{k\ell}_{it,i't'} N^4_{js,j's'} \varepsilon_{1it} \varepsilon_{1i't'} \varepsilon_{kjs} \varepsilon_{\ell j's'}\Big)$$

$$\text{where } \bar{F}^{k\ell}_{it,i't'} = \bar{F}_{kit} \bar{F}_{\ell i't'}$$

This expression is the same as (7.B.11) of Section 7.B.2 and its value is therefore given by the final result of 7.B.2 which is exactly the result given in (7.89).

APPENDIX 7.C : Order Calculations Involved in the Determination of the Bias of the Feasible G2SLS Estimator

In this appendix, we show how we get to the truncated series given in (7.156), starting from expression (7.155) of the bias of the feasible G2SLS estimator. The latter expression is as follows :

$$(7.C.1)\quad \hat{\alpha}_{1,fG2SLS} - \alpha_1 = \left[(A_1 + A_{\frac{1}{2}} + A_o + \ldots)(B_o + B_{-\frac{1}{2}} + B_{-1} + \ldots)(C_o + C_{-\frac{1}{2}} + C_{-1} + \ldots)\right]^{-1} (A_1 + A_{\frac{1}{2}} + A_o + \ldots)(B_o + B_{-\frac{1}{2}} + B_{-1} + \ldots)(D_{-\frac{1}{2}} + D_{-1} + D_{-\frac{3}{2}} + \ldots)$$

Let us first consider the "inverse term" i.e. the expression with the inverse sign in (7.C.1) :

$$\left[(A_1 + A_{\frac{1}{2}} + A_o + \ldots)(B_o + B_{-\frac{1}{2}} + B_{-1} + \ldots)(C_o + C_{-\frac{1}{2}} + C_{-1} + \ldots)\right]^{-1}$$

$$= \Big[A_1 B_o C_o + A_{\frac{1}{2}} B_o C_o + A_o B_o C_o + A_1 B_{-\frac{1}{2}} C_o + A_{\frac{1}{2}} B_{-\frac{1}{2}} C_o$$
$$+ A_o B_{-\frac{1}{2}} C_o + A_1 B_{-1} C_o + A_{\frac{1}{2}} B_{-1} C_o + A_o B_{-1} C_o + A_1 B_o C_{-\frac{1}{2}} +$$
$$+ A_{\frac{1}{2}} B_o C_{-\frac{1}{2}} + A_o B_o C_{-\frac{1}{2}} + A_1 B_{-\frac{1}{2}} C_{-\frac{1}{2}} + A_{\frac{1}{2}} B_{-\frac{1}{2}} C_{-\frac{1}{2}} + A_o B_{-\frac{1}{2}} C_{-\frac{1}{2}} +$$
$$+ A_1 B_{-1} C_{-\frac{1}{2}} + A_{\frac{1}{2}} B_{-1} C_{-\frac{1}{2}} + A_o B_{-1} C_{-\frac{1}{2}} + A_1 B_o C_{-1} + A_{\frac{1}{2}} B_o C_{-1} +$$
$$+ A_o B_o C_{-1} + A_1 B_{-\frac{1}{2}} C_{-1} + A_{\frac{1}{2}} B_{-\frac{1}{2}} C_{-1} + A_o B_{-\frac{1}{2}} C_{-1} + A_1 B_{-1} C_{-1} +$$
$$+ A_{\frac{1}{2}} B_{-1} C_{-1} + A_o B_{-1} C_{-1} + \ldots\Big]^{-1}$$

$$(7.C.2)\quad = \left[A_1 B_o C_o + D_{\frac{1}{2}}\right]^{-1}$$

where $D_{\frac{1}{2}}$ denotes all the terms of the inverse above except for $A_1B_oC_o$, with its leading term of order $O(N^{\frac{1}{2}})$.

Now,

$$A_1 = (\sum_i \frac{1}{\sigma_{i11}} Z_1'N_iX)$$

$$= (\sum_i \frac{1}{\sigma_{i11}} (\bar{Z}_1' + V_z') N_iX)$$

$$(7.C.3)\quad = A_1^* + A_{\frac{1}{2}}^*$$

where

$$A_1^* = \sum_i \frac{1}{\sigma_{i11}} \bar{Z}_1' N_i X \quad (7.C.4)$$

and

$$A_{\frac{1}{2}}^* = \sum_i \frac{1}{\sigma_{i11}} V_z' N_i X \quad (7.C.5)$$

Similarly,

$$C_o = (\sum_i \frac{1}{\sigma_{i11}} X'N_iX)^{-1} (\sum_i \frac{1}{\sigma_{i11}} X'N_iZ_1)$$

$$= (\sum_i \frac{1}{\sigma_{i11}} X'N_iX)^{-1} (\sum_i \frac{1}{\sigma_{i11}} X'N_i(\bar{Z}_1 + V_z))$$

$$= C_o^* + C_{-\frac{1}{2}}^* \quad (7.C.6)$$

where

$$C_o^* = (\sum_i \frac{1}{\sigma_{i11}} X'N_iX)^{-1} (\sum_i \frac{1}{\sigma_{i11}} X'N_i\bar{Z}_1) \quad (7.C.7)$$

$$C_{-\frac{1}{2}}^* = (\sum_i \frac{1}{\sigma_{i11}} X'N_iX)^{-1} (\sum_i \frac{1}{\sigma_{i11}} X'N_iV_z) \quad (7.C.8)$$

Thus,

$$A_1B_oC_o = A_1^*B_oC_o^* + A_{\frac{1}{2}}^*B_oC_o^* + A_1^*B_oC_{-\frac{1}{2}}^* + A_{\frac{1}{2}}^*B_oC_{-\frac{1}{2}}^*$$

$$= E_1 + E_{\frac{1}{2}} + E_o \quad \text{(say)} \quad (7.C.9)$$

Hence, the "inverse term" becomes :

$$\left[A_1B_oC_o + D_{\frac{1}{2}}\right]^{-1} = \left[E_1 + E_{\frac{1}{2}} + E_o + D_{\frac{1}{2}}\right]^{-1}$$

combining (7.C.2) and (7.C.9)

$$= (E_1 + E_{\frac{1}{2}}^*)^{-1}$$

(7.C.10) writing $E_{\frac{1}{2}} + E_o + D_{\frac{1}{2}} \equiv E_{\frac{1}{2}}^*$

$$= (I + E_1^{-1} E_{\frac{1}{2}}^*)^{-1} E_1^{-1}$$

$$= (I + F_{-\frac{1}{2}})^{-1} E_{-1}$$

(7.C.11) $\left\{ \begin{array}{ll} \text{writing } E_1^{-1} E^*_{\frac{1}{2}} \equiv F_{-\frac{1}{2}} \\ \text{and } \quad E_1^{-1} \equiv E_{-1} \end{array} \right.$

(7.C.12) $$= (I - F_{-\frac{1}{2}} + F_{-\frac{1}{2}} F_{-\frac{1}{2}} - F_{-\frac{1}{2}} F_{-\frac{1}{2}} F_{-\frac{1}{2}} + \ldots) E_{-1}$$

Now, let us examine the term post multiplying the "inverse term" in (7.C.1) :

(7.C.13) $(A_1 + A_{\frac{1}{2}} + A_o + \ldots)\ (B_o + B_{-\frac{1}{2}} + B_{-1} + \ldots)\ (D_{-\frac{1}{2}} + D_{-1} + D_{-\frac{3}{2}} + \ldots)$

Multiplying out the three brackets above, we will have a series with the leading term of order $O(\sqrt{N})$ and successive terms with orders decreasing by half a unit each time. Let us denote this series as :

(7.C.14) $(G_{\frac{1}{2}} + G_o + G_{-\frac{1}{2}} + G_{-1} + \ldots)$

Combining (7.C.12) and (7.C.14), we can write :

(7.C.15) $$\hat{\alpha}_{1,fG2SLS} - \alpha_1 = (I - F_{-\frac{1}{2}} + F_{-\frac{1}{2}} F_{-\frac{1}{2}} + \ldots) E_{-1} (G_{\frac{1}{2}} + G_o + G_{-\frac{1}{2}} + \ldots)$$

In order to calculate its expectation upto order $\frac{1}{N}$, we need to retain only terms upto order $\frac{1}{N}$ in the above expression. Thus, we finally have

$$\hat{\alpha}_{1,fG2SLS} - \alpha_1 \cong (I - F_{-\frac{1}{2}}) E_{-1} G_{\frac{1}{2}} + I E_{-1} G_o$$

$$= E_{-1} G_{\frac{1}{2}} - F_{-\frac{1}{2}} E_{-1} G_{\frac{1}{2}} + E_{-1} G_o$$

(7.C.16) $$= E_{-1} (G_{\frac{1}{2}} + G_o) - F_{-\frac{1}{2}} E_{-1} G_{\frac{1}{2}}$$

Now, we have to see what E_{-1} , $F_{-\frac{1}{2}}$, $G_{\frac{1}{2}}$, G_o stand for and trace back to their expressions in terms of A's, B's and C's.

I. From (7.C.11) and (7.C.9) it is immediate that

(7.C.17) $E_{-1} \equiv E_1^{-1} = (A_1^* B_o C_o^*)^{-1}$

II. Next, from (7.C.11), we have

$$F_{-\frac{1}{2}} = E_1^{-1} E^*_{\frac{1}{2}}$$

$$= E_1^{-1} (E_{\frac{1}{2}} + E_o + D_{\frac{1}{2}}) \qquad \text{using (7.C.10)}$$

Since we need only terms of order $N^{-\frac{1}{2}}$, we can leave out E_o; we are then left with

(7.C.18) $E_1^{-1} (E_{\frac{1}{2}} + D_{\frac{1}{2}})$

Now,

(7.C.19) $E_1^{-1} = (A_1^* B_o C_o^*)^{-1}$

(7.C.20) $E_{\frac{1}{2}} = A^*_{\frac{1}{2}} B_o C_o^* + A_1^* B_o C^*_{-\frac{1}{2}}$

and $D_{\frac{1}{2}}$ represents all the terms of expression (7.C.2) excluding $A_1 B_o C_o$. Among these terms, we have to choose only those which are of order exactly equal to $N^{\frac{1}{2}}$. These can be verified to be the following :

(7.C.21) $D_{\frac{1}{2}} \cong A_{\frac{1}{2}} B_o C_o + A_1 B_{-\frac{1}{2}} C_o + A_1 B_o C_{-\frac{1}{2}}$

But we have not yet reached the end point of the "trace-back" operation, as A_1, $A_{\frac{1}{2}}$, C_o and $C_{-\frac{1}{2}}$ also contain terms of order less than that of their leading terms, indicated by their indices. These lesser order terms have, therefore, yet to be eliminated.

(i) First, let us consider $A_{\frac{1}{2}} B_o C_o$.

From the definition of $A_{\frac{1}{2}}$ in (7.149) :

$$A_{\frac{1}{2}} = - \sum_i \frac{1}{\sigma_{ill}} \frac{\eta_{ill}}{\sigma_{ill}} Z_1' N_i X$$

$$= - \sum_i \frac{1}{\sigma_{i11}} \frac{\eta_{i11}}{\sigma_{i11}} (\bar{Z}_1' + V_z') N_i X$$

(7.C.22) $$= A_{\frac{1}{2}}^{**} + A_o^{**}$$

where

(7.C.23) $$A_{\frac{1}{2}}^{**} = - \sum_i \frac{1}{\sigma_{i11}} \frac{\eta_{i11}}{\sigma_{i11}} \bar{Z}_1' N_i X$$

and

(7.C.24) $$A_o^{**} = A_{\frac{1}{2}} - A_{\frac{1}{2}}^{**}$$

Thus, in the term

$$A_{\frac{1}{2}} B_o C_o = (A_{\frac{1}{2}}^{**} + A_o^{**}) B_o C_o$$

we retain only

$$= A_{\frac{1}{2}}^{**} B_o C_o$$

But from (7.C.6) we also have

$$C_o = C_o^* + C_{-\frac{1}{2}}^*$$

Hence, even in the expression

$$A_{\frac{1}{2}}^{**} B_o C_o$$

$$= A_{\frac{1}{2}}^{**} B_o (C_o^* + C_{-\frac{1}{2}}^*)$$

we need to take only

(7.C.25) $$A_{\frac{1}{2}}^{**} B_o C_o^*$$

(ii) Now, let us look at $A_1 B_{-\frac{1}{2}} C_o$.

Knowing that (see (7.C.3))

$$A_1 = A_1^* + A_{\frac{1}{2}}^*$$

and that (see (7.C.6))

$$C_o = C_o^* + C_{-\frac{1}{2}}^*$$

we conclude that the only term of order $N^{\frac{1}{2}}$ in $A_1 \, B_{-\frac{1}{2}} \, C_o$ is

(7.C.26) $A_1^* \, B_{-\frac{1}{2}} \, C_o^*$

(iii) Finally, we turn to $A_1 \, B_o \, C_{-\frac{1}{2}}$.

We have already noted that (cf. (7.C.3)) :

$$A_1 = A_1^* + A_{\frac{1}{2}}^*$$

Thus, in the term

$$A_1 \, B_o \, C_{-\frac{1}{2}} = A_1^* \, B_o \, C_{-\frac{1}{2}} + A_{\frac{1}{2}}^* \, B_o \, C_{-\frac{1}{2}}$$

we straight away eliminate the second part.

Now,

$$C_{-\frac{1}{2}} = (\sum_i \frac{1}{\sigma_{i11}} X'N_i X)^{-1} \quad A'_{\frac{1}{2}} \qquad \text{from (7.152)}$$

$$= -(\sum_i \frac{1}{\sigma_{i11}} X'N_i X)^{-1} (\sum_i \frac{1}{\sigma_{i11}} \frac{\eta_{i11}}{\sigma_{i11}} X'N_i Z_1)$$

using (7.149)

$$= -(\sum_i \frac{1}{\sigma_{i11}} X'N_i X)^{-1} \left\{ \sum_i \frac{1}{\sigma_{i11}} \frac{\eta_{i11}}{\sigma_{i11}} X'N_i (\bar{Z}_1 + V_z) \right\}$$

$$(7.C.27) = C^{**}_{-\frac{1}{2}} + C^{**}_{-1}$$

where

$$(7.C.28)\quad C^{**}_{-\frac{1}{2}} = \left(\sum_i \frac{1}{\sigma_{i11}} X'N_iX\right)^{-1} \left\{\sum_i \frac{1}{\sigma_{i11}} \frac{\eta_{i11}}{\sigma_{i11}} X'N_i\bar{Z}_1\right\}$$

and

$$(7.C.29)\quad C^{**}_{-1} = C_{-\frac{1}{2}} - C^{**}_{-\frac{1}{2}}$$

Thus

$$A^*_1 B_o C_{-\frac{1}{2}} = A^*_1 B_o (C^{**}_{-\frac{1}{2}} + C^{**}_{-1})$$

from which we retain only

$$(7.C.30)\quad A^*_1 B_o C^{**}_{-\frac{1}{2}}$$

Therefore, putting together the retained terms of (i), (ii) and (iii) above, namely (7.C.25), (7.C.26) and (7.C.30), we have :

$$(7.C.31)\quad D_{\frac{1}{2}} \cong A^{**}_{\frac{1}{2}} B_o C^*_o + A^*_1 B_{-\frac{1}{2}} C^*_o + A^*_1 B_o C^{**}_{-\frac{1}{2}}$$

Hence,

$$F_{-\frac{1}{2}} = E_1^{-1} (E_{\frac{1}{2}} + D_{\frac{1}{2}})$$

$$(7.C.32)\quad \cong (A^*_1 B_o C^*_o)^{-1}\Big[A^*_{\frac{1}{2}} B_o C^*_o + A^*_1 B_o C^*_{-\frac{1}{2}} + A^*_1 B_o C^{**}_{-\frac{1}{2}} + A^{**}_{\frac{1}{2}} B_o C^*_o + A^*_1 B_{-\frac{1}{2}} C^*_o\Big]$$

III. The third element to be looked at is $G_{\frac{1}{2}}$. $G_{\frac{1}{2}}$ represents the terms of order $N^{\frac{1}{2}}$ in the product given in (7.C.13). By looking carefully at the three series forming the product, we see that there is in fact only one term of order $N^{\frac{1}{2}}$, namely

(7.C.33) $G_{\frac{1}{2}} = A_1 \; B_o \; D_{-\frac{1}{2}}$

But, using the split-up (7.C.3) of A_1 :

$$A_1 = A_1^* + A_{\frac{1}{2}}^*$$

we can write

(7.C.34) $A_1 \; B_o \; D_{-\frac{1}{2}} = A_1^* \; B_o \; D_{-\frac{1}{2}} + A_{\frac{1}{2}}^* \; B_o \; D_{-\frac{1}{2}}$

where the first term is of order $N^{\frac{1}{2}}$ and the second one of order 1 . Thus, we retain only the first one for $G_{\frac{1}{2}}$ i.e.

(7.C.35) $G_{\frac{1}{2}} = A_1^* \; B_o \; D_{-\frac{1}{2}}$

However, the second term, being of order 1, will form part of G_o which is also required for our calculations (cf. expression (7.C.16)).

IV. The final element to be examined is G_o. G_o stands for the terms of order 1 in the same product as the one considered in III i.e. the one given in (7.C.13). As mentioned in III above, we already have

$$A_{\frac{1}{2}}^* \; B_o \; D_{-\frac{1}{2}}$$

in G_o. In addition, we have the following terms of order 1 in the product of (7.C.13) :

$$A_1 \; B_{-\frac{1}{2}} \; D_{-\frac{1}{2}} + A_{\frac{1}{2}} \; B_o \; D_{-\frac{1}{2}} + A_1 \; B_o \; D_{-1}$$

Using the two split-ups

$$A_1 = A_1^* + A_{\frac{1}{2}}^* \qquad \text{(cf. (7.C.3))}$$

and

$$A_{\frac{1}{2}} = A_{\frac{1}{2}}^{**} + A_o^{**} \qquad \text{(cf. (7.C.22))}$$

it can be verified that the terms of order exactly equal to 1 will only be :

$$A_1^* B_{-\frac{1}{2}} D_{-\frac{1}{2}} + A_{\frac{1}{2}}^{**} B_o D_{-\frac{1}{2}} + A_1^* B_o D_{-1}$$

Therefore,

$$G_o = A_{\frac{1}{2}}^* B_o D_{-\frac{1}{2}} + A_1^* B_{-\frac{1}{2}} D_{-\frac{1}{2}} + A_{\frac{1}{2}}^{**} B_o D_{-\frac{1}{2}} + A_1^* B_o D_{-1} \tag{7.C.36}$$

Finally, replacing the relevant expressions of E_{-1}, $F_{-\frac{1}{2}}$, $G_{\frac{1}{2}}$ and G_o derived in I, II, III and IV above, in (7.C.16), we obtain, upto order $O(\frac{1}{N})$:

$$\begin{aligned}
\hat{\alpha}_{1,fG2SLS} - \alpha_1 \cong \\
& (A_1^* B_o C_o^*)^{-1} \Big[A_1^* B_o D_{-\frac{1}{2}} + A_{\frac{1}{2}}^* B_o D_{-\frac{1}{2}} + A_1^* B_{-\frac{1}{2}} D_{-\frac{1}{2}} \\
& + A_{\frac{1}{2}}^{**} B_o D_{-\frac{1}{2}} + A_1^* B_o D_{-1} \Big] + (A_1^* B_o C_o^*)^{-1} \Big[A_{\frac{1}{2}}^* B_o C_o^* \\
& + A_1^* B_o C_{-\frac{1}{2}}^* + A_1^* B_{-\frac{1}{2}} C_o^* + A_{\frac{1}{2}}^{**} B_o C_o^* + A_1^* B_o C_{-\frac{1}{2}}^{**} \Big] \\
& (A_1^* B_o C_o^*)^{-1} A_1^* B_o D_{-\frac{1}{2}}
\end{aligned} \tag{7.C.37}$$

APPENDIX 7.D : Expectation of $\eta_{i11}\ X'N_ju_1$ for i=1,4 and j=1,4

Let us recall, from our order calculations in Section 7.4.3, that we need to retain an expansion only upto $O(N^{-\frac{1}{2}})$ for η_{i11}. This is given by :

$$(7.D.1)\quad \eta_{i11} = \hat{\sigma}_{i11} - \sigma_{i11} \cong \frac{1}{n_i}\Big[u_1'N_iu_1 + u_1'C\,\bar{Z}_1(\bar{Z}_1'C\,\bar{Z}_1)^{-1}V_z'N_iu_1\Big] - \sigma_{i11}$$

with

$$n_i = \begin{cases} N & \text{for } i=1 \\ N(T-1) & \text{for } i=4 \end{cases}$$

Let us write (7.D.1) compactly as :

$$(7.D.2)\quad \eta_{i11} = \rho_{i11} - \sigma_{i11}$$

with ρ_{i11} denoting the bracketed expression of (7.D.1).

Noting that,

$$(7.D.3)\quad E(\sigma_{i11}\ X'N_ju_1) = 0 \qquad \forall\ i,j$$

we can write :

$$(7.D.4)\quad E(\eta_{i11}\ X'N_ju_1) = E(\rho_{i11}\ X'N_ju_1)$$

Now, let us derive the expectation of $\rho_{i11}X'N_ju_1$ for all possible combinations of i and j .

I. <u>i=4, j=1</u> :

$E(\rho_{411}\ X'N_1u_1)$:

From the definition of ρ_{i11}, we have :

$E(\rho_{411}\ X'N_1u_1) =$

$$(7.D.5)\quad E\Big[\frac{1}{N(T-1)}(u_1'N_4u_1X'N_1u_1 + u_1'C\bar{Z}_1(\bar{Z}_1'C\bar{Z}_1)^{-1}V_z'N_4u_1X'N_1u_1)\Big]$$

Let us consider the above two terms one by one :

(i) $E(u_1'N_4u_1X'N_1u_1)$

$$= E\Big[u_1'N_4u_1\bar{S}_{k1}'u_1\Big]\ k=1,\dots,K$$

$$(7.D.6)\qquad \text{where } \bar{S}_1 = N_1X = \Big[\bar{S}_{11}\ \dots\ \bar{S}_{K1}\Big]$$

$$= E\left[\varepsilon_1' N_4 \varepsilon_1 \bar{S}_{k1}' \varepsilon_1\right] \quad k=1,\dots,K$$

as $N_4 u_1 = N_4 \varepsilon_1$ and covariance between ε_1 and μ^1 is zero.

$$= E\left[\sum_i \sum_{i'} \sum_{i''} \sum_t \sum_{t'} \sum_{t''} N^4_{it,i't'} \bar{S}^{k1}_{i''t''} \varepsilon_{1it} \varepsilon_{1i't'} \varepsilon_{1i''t''}\right]$$

$$k=1,\dots,K$$

$$= \left[\sum_i \sum_t N^4_{it,it} \bar{S}^{k1}_{it} m^3_{\varepsilon 111}\right] \quad k=1,\dots,K$$

using result (7.B.6) of Appendix 7.B (and where $m^3_{\varepsilon 111}$ denotes the third moment of ε_{1it})

$$\text{(7.D.7)} \quad = \left[\frac{T-1}{T} \sum_i \sum_t \bar{S}^{k1}_{it} m^3_{\varepsilon 111}\right] \quad k=1,\dots,K$$

Now,

$$\sum_i \sum_t \bar{S}^{k1}_{it} = \bar{S}_{k1}' \iota_{NT} = x_k' N_1 \iota_{NT}$$

where x_k is the k-th column of X

$$\text{(7.D.8)} \quad = x_k' \iota_{NT}$$

Thus,

$$E \frac{1}{N(T-1)} u_1' N_4 u_1 X' N_1 u_1 = \frac{1}{N(T-1)} \frac{T-1}{T} X' \iota_{NT} m^3_{\varepsilon 111}$$

$$\text{(7.D.9)} \quad = \frac{1}{NT} X' \iota_{NT} m^3_{\varepsilon 111}$$

(ii) The second term of (7.D.5) is

$$E(u_1' C \bar{Z}_1 (\bar{Z}_1' C \bar{Z}_1)^{-1} V_z' N_4 u_1 X' N_1 u_1)$$

$$\text{(7.D.10)} \quad = E\left[u_1' \bar{F} U' N_4 u_1 x_k' N_1 u_1\right] \quad k=1,\dots,K$$

using the definition of $\bar{F}$ given in (7.61)

(7.D.11) $= E\left[\sum_{\ell} \varepsilon_1' \bar{F}_{\ell} \varepsilon_{\ell}' N_4 \varepsilon_1 \bar{S}_{k1}' \varepsilon_1\right]$ $k=1,\ldots,K$

where $\bar{F}_{\ell}$ is the ℓ-th column of $\bar{F}$ and where we have used the results that $N_4 u_1 = N_4 \varepsilon_1$ and that there is no covariance between the two components μ and ε.

The above expression is of the same form as that of Section 7.B.5 (a product involving two linear forms and one quadratic form), which is in turn equal to (7.B.12). Hence the expectation in (7.D.11) will be equal to the expression (7.B.12) with $\bar{F}^{k\ell}$ replaced by $\bar{F}_{\ell}\ \bar{S}_{k1}'$. Thus we need to calculate the following :

- $\operatorname{tr} \bar{F}_{\ell}\ \bar{S}_{k1}'$
- $\operatorname{tr} \bar{F}_{\ell}\ \bar{S}_{k1}'\ N_4$ and
- $\sum_i \sum_t (\bar{F}_{S1}^{k\ell})_{it,it}\ N_{it,it}^4$

denoting

(7.D.12) $\bar{F}_{\ell}\ \bar{S}_{k1}' \equiv \bar{F}_{S1}^{k\ell}$

Now,

$$\operatorname{tr} \bar{F}_{\ell}\ \bar{S}_{k1}' = \operatorname{tr} \bar{P}_1\ L_1 \gamma^{\ell}\ x_k'\ N_1$$

$$= \operatorname{tr} x_k'\ N_1\ \bar{P}_1\ L_1 \gamma^{\ell}$$

(7.D.13) $$= 0 \qquad \text{as } N_1 \bar{P}_1 = 0$$

Next,

(7.D.14) $\operatorname{tr} \bar{F}_{\ell}\ \bar{S}_{k1}' N_4 = \operatorname{tr} \bar{P}_1 L_1 \gamma^{\ell} x_k' N_1 N_4 = 0$ as $N_1 N_4 = 0$

Finally,

$$\sum_i \sum_t (\bar{F}_{\ell}\ \bar{S}_{k1}')_{it,it}\ N_{it,it}^4$$

$$= \frac{1}{N(T-1)} \sum_i \sum_t (\bar{F}^{k\ell}_{S1})_{it,it}$$

$$(7.D.15) \quad = \frac{1}{N(T-1)} \operatorname{tr} \bar{F}^{k\ell}_{S1} = 0$$

Hence,

$$(7.D.16) \quad E\left[\sum_\ell \varepsilon_1' \bar{F}_\ell \varepsilon_\ell' N_4 \varepsilon_1 \bar{S}_{k1}' \varepsilon_1\right] = 0$$

Therefore, combining (7.D.15), (7.D.9) and (7.D.16), we get :

$$(7.D.17) \quad E(\rho_{411} X'N_1u_1) = \frac{1}{NT} X'\iota_{NT} m^3_{\varepsilon 111}$$

II. <u>i=4, j=4</u> :

$$E(\rho_{411} X'N_4u_1) =$$

$$(7.D.18) \; E\left[\frac{1}{N(T-1)}(u_1'N_4u_1X'N_4u_1 + u_1'C\bar{Z}_1(\bar{Z}_1'C\bar{Z}_1)^{-1}V_z'N_4u_1X'N_4u_1)\right]$$

(a) First term of (7.D.18) :

By following the same procedure as for (i) of I above, we will have :

$$(7.D.19) \; E(u_1'N_4u_1X'N_4u_1) = \left[\frac{T-1}{T} \sum_i \sum_t \bar{s}^{k4}_{it} m^3_{\varepsilon 111}\right] k=1,\ldots,K$$

But

$$(7.D.20) \; \sum_i \sum_t \bar{s}^{k4}_{it} = \bar{S}_{k4}' \iota_{NT} = x_k'N_4\iota_{NT} = 0 \qquad \text{as } N_4\iota_{NT} = 0$$

Thus,

$$(7.D.21) \; E \frac{1}{N(T-1)} u_1'N_4u_1X'N_4u_1 = 0$$

(b) Second term of (7.D.18) :

$$E(u_1'C\,\bar{Z}_1(\bar{Z}_1'C\,\bar{Z}_1)^{-1} V_z'N_4u_1X'N_4u_1)$$

$$(7.D.22) = \left[E\, u_1'\bar{F}U'N_4u_1x_k'N_4u_1\right] k=1,\ldots,K$$ using the definition of $\bar{F}$ in (7.61)

This expression is the same as (7.D.10) except that X'N4 replaces X'N1 of (7.D.10). Thus we have to repeat the same procedure as in (ii) of I with the above change. It means that we need to calculate the following :

- $\text{tr } \bar{F}_\ell \; \bar{S}'_{k4}$
- $\text{tr } \bar{F}_\ell \; \bar{S}'_{k4} \; N_4$ and
- $\sum_i \sum_t (\bar{F}^{k\ell}_{S4})_{it,it} \; N^4_{it,it}$

denoting

$$\bar{F}_\ell \; \bar{S}'_{k4} \equiv \bar{F}^{k\ell}_{S4}$$

First,

$$\text{tr } \bar{F}_\ell \; \bar{S}'_{k4} = \text{tr } \bar{P}_1 \; L_1 \; \gamma^\ell \; x'_k \; N_4$$

$$= \text{tr } x'_k \; N_4 \; \bar{P}_1 \; L_1 \; \gamma^\ell$$

(7.D.23) $$= \text{tr } x'_k \; \bar{P}_1 \; L_1 \; \gamma^\ell \qquad \text{as } N_4\bar{P}_1 = \bar{P}_1$$

Now,

$$x'_k\bar{P}_1 = x'_k \; (C \; \bar{Y}_1 \; \bar{W}_{11} + C \; X^*_1 \; \bar{W}_{21}) \qquad \text{using (7.B.7)}$$

(7.D.24) $$= 0$$

which is verified by expanding the following identity :

(7.D.25) $$(\bar{Z}'_1 C \bar{Z}_1)(\bar{Z}'_1 C \bar{Z}_1)^{-1} = \begin{bmatrix} \bar{Y}'_1 \; C \; \bar{Y}_1 & \bar{Y}'_1 \; C \; X^*_1 \\ X^{*\prime}_1 C \; \bar{Y}_1 & X^{*\prime}_1 C \; X^*_1 \end{bmatrix} \begin{bmatrix} \bar{W}_{11} & \bar{W}_{12} \\ \bar{W}_{21} & \bar{W}_{22} \end{bmatrix} = I$$

Thus,

(7.D.26) $\text{tr } \bar{F}_\ell \; \bar{S}'_{k4} = 0$

Next,

$$\text{tr } \bar{F}_\ell \; \bar{S}'_{k4} \; N_4 = \text{tr } \bar{F}_\ell \; x'_k \; N_4 \; N_4$$

$$= \text{tr } \bar{F}_\ell \; x'_k \; N_4$$

$$= \text{tr } \bar{F}_\ell \; \bar{S}'_{k4}$$

(7.D.27) $$= 0 \qquad \text{using (7.D.6)}$$

$$\sum_i \sum_t (\bar{F}^{k\ell}_{S4})_{it,it} \; N^4_{it,it}$$

(7.D.28) $= \frac{1}{N(T-1)} \, tr \, \bar{F}^{k\ell}_{S4} = 0$ using (7.D.26)

Therefore

(7.D.29) $E(\frac{1}{N(T-1)} u_1'C \, \bar{Z}_1 \, (\bar{Z}_1'C \, \bar{Z}_1)^{-1} V_z'N_4u_1X'N_4u_1) = 0$

Combining (7.D.18), (7.D.21) and (7.D.29), we get :

(7.D.30) $E(\rho_{411} \; X'N_4u_1) = 0$

III. <u>i=1, j=1</u> :

$E(\rho_{111} \; X'N_1u_1)$

(7.D.31) $= E \, \frac{1}{N}(u_1'N_1u_1X'N_1u_1 + u_1'C \, \bar{Z}_1(\bar{Z}_1'C \, \bar{Z}_1)^{-1} V_z'N_1u_1X'N_1u_1)$

(i) $E \, \frac{1}{N} \, u_1'N_1u_1X'N_1u_1$

$= E \, \frac{1}{N} \, (\varepsilon_1'N_1\varepsilon_1X'N_1\varepsilon_1 + \mu^{1\prime}(I \otimes \iota')N_1(I \otimes \iota)\mu^1X'N_1(I \otimes \iota)\mu^1$

omitting cross-products whose expectation is zero

(7.D.32) $= E \, \frac{1}{N} \, (\varepsilon_1'N_1\varepsilon_1X'N_1\varepsilon_1 + \mu^{1\prime}\bar{N}_1\mu^1X'\tilde{N}_1\mu^1)$

where

(7.D.33) $\bar{N}_1 = (I \otimes \iota') \, N_1 \, (I \otimes \iota)$

(7.D.34) $\tilde{N}_1 = N_1 \, (I \otimes \iota)$

Following the same procedure as in (i) of I (page 300) it can be easily verified that the above expectation is equal to

$$\frac{1}{N}\left[\sum\sum_{it} N^1_{it,it}\bar{S}^{kl}_{it} \, m^3_{\varepsilon 111} + \sum_i \bar{N}^1_{i,i} \; \tilde{S}^{kl}_i \; m^3_{\mu 111}\right] \; k=1,\ldots,K$$

with $\tilde{S}'_{kl} = x_k' \, \tilde{N}_1$

(7.D.35) $= \frac{1}{N}\frac{1}{T} \left[X'\iota_{NT} \; m^3_{\varepsilon 111} + X'\iota_{NT} \; T \; m^3_{\mu 111}\right]$

using (7.D.8) and the results:

$\bar{N}^1_{i,i} = T$ and $\sum_i \tilde{S}^{kl}_i = x_k' \, \iota_{NT}$

$$= \frac{1}{NT} X'\iota_{NT} (m^3_{\varepsilon 111} + T\, m^3_{\mu 111})$$

(7.D.36) $= \frac{1}{NT} X'\iota_{NT}\, m^3_{1111}$

denoting

(7.D.37) $m^3_{\varepsilon 111} + T\, m^3_{\mu 111} \equiv m^3_{1111}$

(ii) $E\left[\frac{1}{N} u_1'C\, \bar{Z}_1(\bar{Z}_1'C\, \bar{Z}_1)^{-1} V_z'N_1u_1X'N_1u_1\right]$

Here again, by adopting the same procedure as in (ii) of I (page 301), it can be seen that this expectation vanishes.

Combining (i) and (ii) above yields

(7.D.38) $E(\rho_{111}\, X'N_1u_1) = \frac{1}{NT} X'\iota_{NT}\, m^3_{1111}$

IV. <u>i=1, j=4</u> :

$E(\rho_{111}\, X'N_4u_1)$

(7.D.39) $= E\,\frac{1}{N} (u_1'N_1u_1X'N_4u_1 + u_1'C\, \bar{Z}_1(\bar{Z}_1'C\, \bar{Z}_1)^{-1}V_z'N_1u_1X'N_4u_1)$

(a) $E\,\frac{1}{N}\, u_1'N_1u_1X'N_4u_1$

$= E\left[\frac{1}{N} \sum_i \sum_t N^1_{it,it}\, \bar{S}^{k4}_{it}\, m^3_{\varepsilon 111}\right]$ k=1,...,K using the procedure of (i) of I (page 300)

$= E\left[\frac{1}{N}\frac{1}{T} \sum_i \sum_t \bar{S}^{k4}_{it}\, m^3_{\varepsilon 111}\right]$ k=1,...,K

(7.D.40) $= 0$ using (7.D.20)

(b) $E\,\frac{1}{N}\, u_1'C\, \bar{Z}_1(\bar{Z}_1'C\, \bar{Z}_1)^{-1} V_z'N_1u_1X'N_4u_1$

This expectation can be seen to be equal to zero, following a similar procedure as in (ii) of I of page 301.

Combining (a) and (b) above, we conclude that

(7.D.41) $E(\rho_{111}\, X'N_4u_1) = 0$

CHAPTER 8

APPLICATION TO A MODEL OF RESIDENTIAL ELECTRICITY DEMAND

8.1 The Model

An empirical work involving the estimation of a simultaneous equation model with error components is presented in this chapter. This application concerns the behaviour of households regarding electricity consumption and uses data collected from a national household survey conducted in the U.S.A.

Our model of household demand for electricity is largely based on the one proposed by Garbacz in a study using national data [15]. It is a system of three structural equations with three endogenous variables, namely :

KWH = Number of kilowatt-hours of electricity used by the household per year ;

PEL = Average price paid by the household for electricity (dollars per kilowatt-hour) ;

APP = An index of the size of appliance stock of the household ;

and six basic exogenous variables, namely :

HDD = Heating Degree Days (base 65°F) ;

CDD = Cooling Degree Days (base 65°F) ;

ALLEL = Dummy variable for all-electric households (1 if all-electric, 0 otherwise) ;

INC = Annual Family Income in dollars ;

POF = Average Price paid by the households for all fuels except electricity (dollars per BTU) ;

SIZE = Number of household members.

The Price Equation

As electricity is sold at decreasing block rates, the average price paid by each customer (in our case, the household) is inversely related to the quantity consumed. Therefore, the price equation is written as :

$$\text{(8.1)} \quad \log(PEL) = a_1 + a_2 \log(KWH) + a_3 (ALLEL) + u_1$$

As in [15], we have also added a dummy variable for all-electric residences, in order to account for any reduction in the rate schedule that would lead to a lower average price beyond a certain level of consumption.

The Demand Equation

The demand equation is formulated as follows :

$$\text{(8.2)} \quad \log(KWH) = b_1 + b_2 \log(PRICE) + b_3 \log(APP) + b_4 \log(CDD) + b_5 \log(HDD) + u_2$$

The residential consumption of electricity is mainly determined by the volume of electrical appliances and devices and the intensity of their usage. Hence, an index of the total appliance size appears as an explanatory variable of the demand for electricity. The measurement of this variable is described while discussing the appliance stock equation.

The intensity of usage of the various electrical appliances may change in response to variations in electricity priçe. Therefore, price is another important factor explaining demand.

Two types of price can be envisaged namely, the marginal price (denoted as MP) and the average price (PEL). The marginal price is the incremental cost to the consumer (household) of using an additional kilowatt-hour. The data used provides two marginal rates for each household - the winter rate and the summer rate. Since there was no major difference in the results using one or the other, we only present the results obtained using the winter rate. The average price paid by the household for electricity is calculated by dividing the annual amount paid in dollars by the number of kilowatt-hours consumed in that year. In our paper, we present results of estimation of both the specifications of the demand equation (i.e. one using average price and the other using marginal price). Note that in the second specification marginal price is an exogenous variable and replaces average price only in the demand equation. Thus the simultaneous nature of the system is conserved.

The third major element in the determination of electricity consumption is the weather. Weather affects electricity consumption both through electric heating and electric air-conditioning requirements. Thus, the relevant climate variables are the heating degree days and cooling degree days. The heating degree days are the number of degrees the average temperature is below the base temperature, and the cooling degree days are the number of degrees the average daily temperature is above the base temperature.

Garbacz also introduces income as an explanatory variable in the demand equation. However, we removed it after preliminary estimations of the equation with income which gave unsatisfactory results with non-significant coefficients and poor R^2 values. This may be due to the strong collinearity between income and appliance size. Hence, we decided to include income only in the appliance stock equation, where it seems more appropriate.

The Appliance Stock Equation

This equation explains the stock of appliances held by the household, in terms of the average price of electricity and certain exogenous variables, thus completing the three-equation system determining demand, price and equipment.

The appliance index is calculated using Table 8.1 taken from [15], which gives the index of the major electrical applicances based on their typical usage. The estimates for the first eleven items are those prepared by the Response Analysis Corporation, Princeton while those of the last three items are computed by Garbacz.

The exogenous variables that appear in the appliance stock equation are the total family income, the number of household members (size) and the average price of alternate fuels which are utility gas, fuel oil/kerosene and liquified petroleum gas. The combined average price of these fuels is computed by dividing the total amount of dollars paid for all the three fuels by the sum of the annual use of the three fuels expressed in BTU.

Thus the proposed appliance stock equation is

$$(8.3) \quad \log(APP) = c_1 + c_2 \log(PEL) + c_3 \log(INC) + c_4 \log(POF) + c_5 \log(SIZE) + u_3$$

8.2 The Data

The source of the data used in our work is the public use tape of the U.S. Residential Energy Consumption Survey for the year April 1982 to March 1983, which is a national survey covering a representative sample of 4660 households from all over the United States. However, households from Hawai and Alaska have been removed from the public use tape for confidentiality reasons. Data regarding the household characteristics (such as income, size) were collected by means of personal interviews in 95% of the cases and by mail questionnaires for the remaining 5%. On the other hand, data regarding the energy consumption and expenditure (such as quantity used, cost etc.) were obtained from the records provided by the households' fuel suppliers.

In our study, it was not possible to use the entire data for two reasons. Firstly, only 2806 of the 4660 household records contained data on marginal rates. Further, a selection had to be made based on whether each household paid for all its uses of electricity. In other words, we excluded all those households for which, the electricity payment was included in the rent or was made by a third party or was part of a billing pool scheme, for one or more electricity uses. In this way, we ensured that the household is perfectly aware of the exact amount paid and hence can control its consumption of electricity. Further, as also pointed out in the companion report on the survey [42] , households whose energy costs are included in the rent, do not feel the immediate effect of energy prices of reduced consumption in their monthly bill, since their rent does not usually vary from month to month. Therefore, the operation of free market forces is not possible in the case of these households. However, it is to be noted that, as 76% of households who pay directly to the supplier, are in the "$10'000 or more" income category (see [42] , page 13), this selection criterion may cause a slight bias towards higher-income groups. This may also be the reason for income not being significant in the demand equation.

The above selection procedure combined with the necessity of having an equal number of households in each region for our model, left us with a potential sample of 1080 households.

The second limiting factor of the total number of households used in the estimation, was the memory space available in the computer programs used. This point is dealt with in more detail in the following section, while discussing the programming aspect.

8.3 Estimation Methods

Since our data concerns households of different regions within a country, specific "regional" effects (or location effects) can be introduced in our model, to reflect differences in prices and behavior over regions. There are two variables relating to location in the survey, namely, the Census region (a broad segmentation) and the Census division (a finer classification of areas). We have taken the Census division to be representing a "region" for our model. There are nine Census divisions - New England, Middle Atlantic, East North Central, West North Central, South Atlantic, East South Central, West South Central, Mountain and Pacific. It is to be noted that ideally the number of regions should be large to ensure consistency of estimators. In our case, only nine regions are involved and this should be kept in mind while interpreting our results. A greater number of regions could not be considered in our study due to lack of data relating to finer regional classification of households.

Now, to come back to the introduction of error component structure in our model, it consists in splitting the error term of each structural equation into two components - a regional effect (assumed to be the same for all households in a particular region but different from one region to the other) and an overall random disturbance term. Thus, our system becomes :

$$(8.4) \quad \log(PEL)_{ih} = a_1 + a_2 \log(KWH)_{ih} + a_3 \log(ALLEL)_{ih} + \mu_{1i} + \varepsilon_{1ih}$$

$$(8.5) \quad \log(KWH)_{ih} = b_1 + b_2 \log(PRICE)_{ih} + b_3 \log(APP)_{ih} + b_4 \log(CDD)_{ih} + b_5 \log(HDD)_{ih} + \mu_{2i} + \varepsilon_{2ih}$$

$$(8.6) \quad \log(APP)_{ih} = c_1 + c_2 \log(PEL)_{ih} + c_3 \log(INC)_{ih} + c_4 \log(POF)_{ih} + c_5 \log(SIZE)_{ih} + \mu_{3i} + \varepsilon_{3ih}$$

where i represents the region and h the household.

The above three equations form a simultaneous equation model with error component structure (with only one specific effect). Therefore, any of the structural estimation techniques developed in the earlier chapters, can be used for estimating this model. In our application, we have employed the covariance 2SLS (with Method 2), the feasible generalised 2SLS and the feasible generalised 3SLS. In addition, for comparison purposes, we have also estimated the model without the regional effects i.e. by classical 2SLS and 3SLS.

Now, we turn to the problem of programming our estimation methods. As the estimation methods are all developed by the author herself, no ready-made computer package was available for implementation. In other words, all the estimation procedures were also programmed by the author. The programs selecting the households who paid for all their uses, was written in Fortran language and executed on the computer UNIVAC 1108, whereas the estimation methods were programmed in Matlab language and executed on the IBM/AT personal computer. (Matlab is a matrix-computation program designed by the Department of Computer Science of the University of New Mexico).

At this point, it should be mentioned that the total number of households had to be limited to 900 (100 in each region) for reasons of memory availability in Matlab. A higher number of households per region led to a memory occupation exceeding the maximum possible amount. As the main objective of our study is to illustrate our methodology and as the number of regions is fixed anyway, 100 households per region is sufficient for our purpose. However, an in-depth work on the optimisation of memory utilisation of our estimation programs, will definitely to useful for providing greater flexibility on the size of the sample. In fact, we can even go a step further to modify our programs in such a way that they can be executed easily in an interactive manner by any potential user.

8.4 Results

Since we had to select 900 households from a total of 1080, for estimation purposes, these households were chosen at random, imposing the only condition of an equal number per region. This enabled us to estimate the model for different sub-samples obtained with different sequences of random numbers. We decided to present results relating to only one sub-sample as no major differences occur among results of different sub-samples.

Let us add that the two different specifications tested in the case of the demand equation, lead to two different systems of equations, which are named Model I and Model II in the results. In Model I, PRICE is taken to be the average price and in Model II, it is the marginal price.

Price Equation

The estimated price equation (see Table 8.2 and Table 8.3) confirms the inverse relationship between price and quantity and hence the need for a simultaneous equation system for explaining the residential demand for electricity. The coefficient of the quantity of electricity consumed is well below one, in the range 0.1 to 0.2 (in absolute value) for the estimations with regional effects (covariance 2SLS,fG2SLS,fG3SLS). The value is nearer to -0.1 in Model I and nearer to -0.2 in Model II. The ordinary 2SLS and 3SLS estimates of this coefficient turn out to be non-significant in Model I (Table 8.2).

There is also a marked difference in the R^2 values between the generalised and the ordinary/classical results; the generalised estimations consistently result in higher R^2 values. This clearly shows that differences over regions are important in the determination of average prices and these differences are well captured by the regional effects. Let us also note that the coefficient of ALLEL is negative as postulated and is in the range -0.1 to -0.2.

Demand Equation

Now, let us turn to the estimated demand equation. As all the equations are in log-linear form, the elasticities are directly given by the corresponding coefficients. From our estimation results (see Table 8.4, Table 8.5), we observe that the price-elasticity of demand is negative but non-significant in all the estimations, whether it relates to marginal price or average price. Since this coefficient represents the sensitivity of intensity of usage of electrical appliances with respect to price or in other words, the sensitivity of demand to price in the short-run, the above result implies that demand is insensitive to price changes, in the short run. However, as we will see while examining derived elasticities of demand, average price does affect demand in the long run through its influence on the stock of appliances which in turn affects demand.

The appliance stock elasticity of demand turns out to be highly significant and ranges from 1.3 to 1.7, the system methods generally yielding slightly higher values than the single-equation methods. However, there is no marked difference between the generalised estimations (with regional effects) and the classical estimations.

The coefficients of the weather variables are positive as expected but turn out to be non-significant in the limited-information methods (cov2SLS,fG2SLS,O2SLS). They become significant when all the information is used in the estimation i.e. when estimated by fG3SLS or O3SLS. It may be argued that the two variables, HDD and CDD, are strongly negatively correlated and hence including both of them in the same equation may lead to multi-collinearity problems. Therefore, we re-estimated the equation by the same methods with only one of the two, namely CDD, as the climate variable. The reason for choosing CDD rather than HDD is that only 16% of the total households heat by electricity (see [42]) whereas almost all air-conditioners are run by electricity. This re-estimation does not significantly change the values obtained for the remaining coefficients neither does it improve the R^2 value. Hence, we decided to retain both of them in the equation.

Appliance Stock Equation

The estimations of the appliance stock equation gave relatively low R^2 values. This may be due to the nature of the variable itself, being difficult to measure and liable to great fluctuations. Another reason may be that an important variable, namely a price index for appliances, is absent from the equation because of lack of data. However, our results show that price of electricity, income and price of alternate fuels are all significant in the determination of appliance size (see Tables 8.6 and 8.7).

The elasticity of appliance stock with respect to price of electricity is above unity in Model I (except in the fG3SLS estimation) and around 0.8 in Model II (except in the O3SLS estimation). The elasticity with respect to income is well below unity (0.16 to 0.18, in most cases). The impact of the price of alternate fuels is also inelastic. The size of the household is not generally significant and even exhibits the wrong sign in certain cases. Further, it can be noted that both the generalised and the ordinary/classical estimations yield similar results in this case.

Derived Elasticities of Demand

Before describing the different elasticities of demand that can be derived using our model, let us make the following observation. Since price (average or marginal) is uniformly non-significant in all the estimations of the demand equation, it is only logical to respecify the demand equation without any price variable on its right hand side. The resulting model, which is different from models I and II, is named Model III and its estimations using different methods are given in Tables 8.8, 8.9 and 8.10. Note that the single-equation estimations of the price and the appliance equations of Model III are the same as those of Model I. All the elasticities discussed below are based on this Model III.

"Long-Run" Price Elasticity of Demand

As mentioned earlier, the appliance stock equation enables us to calculate a price elasticity of demand in the long-run. This is given by multiplying the coefficient of (average) price in this equation by the coefficient of appliance in the demand equation. As can be seen from Table 8.11, estimates of this value range from -1.9 to -2.4 except in the O3SLS case where it is equal to -1.4.

"Total" Price Elasticity of Demand

The so-called total price elasticity of demand measures the effect on demand of a parallel shift in the price schedule i.e. of an exogenous change in average price at all levels of consumption. This exogenous change represents a change in the constant term of the price equation i.e. in a_1. Its effect can be directly obtained from the reduced form of the model.

The reduced form is, by definition, the solution of our system for log (PEL KWH APP) in terms of all the exogenous variables. This can be derived as follows. The three equations of Model III can be written in matrix form as :

$$(8.7)\quad (\text{PEL KWH APP})\ \Gamma + (\text{CONST CDD HDD ALLEL INC POF SIZE})\ B + (u_1\ u_2\ u_3) = 0$$

where 'log' is omitted in front of each variable to simplify notations and where

$$(8.8) \qquad \Gamma = \begin{bmatrix} -1 & 0 & c_2 \\ a_2 & -1 & 0 \\ 0 & b_3 & -1 \end{bmatrix} \quad ; \quad B = \begin{bmatrix} a_1 & b_1 & c_1 \\ 0 & b_4 & 0 \\ 0 & b_5 & 0 \\ a_3 & 0 & 0 \\ 0 & 0 & c_3 \\ 0 & 0 & c_4 \\ 0 & 0 & c_5 \end{bmatrix}$$

Therefore,

$$(8.9) \quad (\text{PEL KWH APP}) = -\,(\text{CONST CDD HDD ALLEL INC POF SIZE})\, B\Gamma^{-1} - (u_1\ u_2\ u_3)\,\Gamma^{-1}$$

The results of calculations of $-\Gamma^{-1}$ and $-B\Gamma^{-1}$ are as follows :

$$(8.10) \qquad -\Gamma^{-1} = \frac{1}{1-a_2b_3c_2} \begin{bmatrix} 1 & b_3c_2 & c_2 \\ a_2 & 1 & a_2c_2 \\ a_2b_3 & b_3 & 1 \end{bmatrix}$$

$$-B\Gamma^{-1} = \frac{1}{1-a_2b_3c_2} \begin{bmatrix} a_1+a_2b_1+c_1a_2b_3 & a_1b_3c_2+b_1+c_1b_3 & a_1c_2+b_1a_2c_2+c_1 \\ a_2b_4 & b_4 & b_4a_2c_2 \\ a_2b_5 & b_5 & b_5a_2c_2 \\ a_3 & a_3b_3c_2 & a_3c_2 \\ c_3a_2b_3 & c_3b_3 & c_3 \\ c_4a_2b_3 & c_4b_3 & c_4 \\ c_5a_2b_3 & c_5b_3 & c_5 \end{bmatrix}$$

Now, it can be verified that the total price elasticity is given by the factor multiplying a_1 in the element in the first row and second column of $- B\Gamma^{-1}$. This is given by

$$\frac{b_3 c_2}{1-a_2 b_3 c_2}$$

Its estimates vary between -1.6 and -2.4 depending on the method of estimation but are all well above unity in absolute value (see Table 8.11).

Income Elasticity of Demand

The income elasticity of demand is given by the coefficient of income in the reduced form equation for log(KWH) and is thus equal to

$$\frac{c_3 b_3}{1-a_2 b_3 c_2}$$

The estimates of income elasticity corresponding to the different estimation methods can also be found in Table 8.11. Their values are close to 0.3 in the case of generalised estimations and around 0.2 in the case of classical estimations. Thus we see that demand is relatively inelastic with respect to income.

To conclude, our present application has brought forth some valuable information on the implications of the use of error components in simultaneous equations. Firstly, it has provided favourable evidence on the advantages of the introduction of random specific effects in presence of double-indexed data, though a finer regional classification of our households, for instance into the different States comprising the U.S.A., would have been more suitable for our purpose. Secondly, we note that the introduction of regional effects has proved to be highly beneficial for the price equation whereas it has had a somewhat neutral effect on the other two equations. This leads us to suggest that while specifying a model, it may not be necessary to have an error component structure in all the structural equations but only in those where it seems to be relevant. For instance, in our demand equation, the regional differences were probably taken adequate account of by the climate variables which made the regional effects redundant. It may be noted that in case a structural equation does not include error components, the generalised 2SLS estimation of that equation is simply the "classical" 2SLS estimation.

Table 8.1

Appliance	Index Value (million Btu per year on average)
Clothes washer	1
Electric dishwasher	4
Electric clothesdryer	11
Freezer	16
Microwave oven	2
Electric oven/range	8
Refrigerator	
Manual or automatic defrost	16
Frost-free	25
Small Electric Appliances	1
Room conditioning (heat)	44.5
Room conditioning (cool)	9.5
Hot water heater	17

Source : Garbacz [15], page 126.

Table 8.2 Estimation Results - Model I - Price Equation*

Method of Estimation \ Estimated Coefficient of	Constant Term	log KWH	log ALLEL	R^2	Estimations of Variance Components
Cov2SLS	- 1.682	- 0.109 (0.022)	- 0.151 (0.029)	0.252	$\hat{\sigma}_\varepsilon = 0.055$ $\hat{\sigma}_1 = 2.522$ $\hat{\sigma}_\mu = 0.025$
fG2SLS	- 1.745 (0.202)	-0.101 (0.022)	- 0.157 (0.029)	0.246	
fG3SLS	- 1.671 (0.187)	- 0.109 (0.022)	- 0.214 (0.028)	0.254	
O2SLS	- 2.842 (0.253)	0.025 (0.029)	- 0.301 (0.037)	0.053	$\hat{\sigma}^2 = 0.089$
O3SLS	- 2.881 (0.246)	0.031 (0.028)	- 0.366 (0.036)	0.036	

* Figures inside parentheses are estimations of asymptotic standard deviations.

Table 8.3 Estimation Results - Model II - Price Equation*

Method of Estimation \ Estimated Coefficient of	Constant Term	log KWH	log ALLEL	R^2	Estimations of Variance Components
Cov2SLS	- 0.963	- 0.191 (0.021)	- 0.084 (0.028)	0.290	$\hat{\sigma}_{\varepsilon} = 0.052$ $\hat{\sigma}_{1} = 2.395$ $\hat{\sigma}_{\mu} = 0.023$
fG2SLS	- 0.962 (0.191)	- 0.191 (0.021)	- 0.085 (0.028)	0.290	
fG3SLS	- 0.927 (0.190)	- 0.195 (0.021)	- 0.126 (0.027)	0.292	
O2SLS	- 0.787 (0.195)	- 0.211 (0.022)	- 0.106 (0.031)	0.291	$\hat{\sigma}^2 = 0.067$
O3SLS	- 0.761 (0.193)	- 0.213 (0.022)	- 0.160 (0.029)	0.287	

* Figures inside parentheses are estimations of asymptotic standard deviations.

Table 8.4 Estimation Results - Model I - Demand Equation*

Method of Estimation \ Estimated Coefficient of	Constant Term	log PEL	log APP	log CDD	log HDD	R^2	Estimations of Variance Components
Cov2SLS	1.468	- 0.379	1.279	0.116	0.028	0.427	$\hat{\sigma}_{\varepsilon} = 0.342$
		(0.726)	(0.223)	(0.109)	(0.058)		$\hat{\sigma}_1 = 1.252$
							$\hat{\sigma}_{\mu} = 0.009$
fG2SLS	2.909	0.194	1.418	0.056	0.014	0.312	
	(1.439)	(0.496)	(0.151)	(0.068)	(0.056)		
fG3SLS	- 1.999	- 0.615	1.417	0.203	0.231	0.320	
	(1.335)	(0.487)	(0.150)	(0.065)	(0.050)		
O2SLS	2.295	- 0.121	1.292	0.107	0.010	0.411	$\hat{\sigma}^2 = 0.321$
	(0.846)	(0.201)	(0.082)	(0.033)	(0.047)		
O3SLS	0.744	- 0.048	1.449	0.195	0.073	0.313	
	(0.753)	(0.198)	(0.080)	(0.030)	(0.041)		

* Figures inside parentheses are estimations of asymptotic standard deviations.

Table 8.5 **Estimation Results - Model II - Demand Equation***

Method of Estimation \ Estimated Coefficient of	Constant Term	log MP	log APP	log CDD	log HDD	R^2	Estimations of Variance Components
Cov2SLS	2.285	- 0.0002	1.401	0.064	0.029	0.343	$\hat{\sigma}_\varepsilon = 0.391$
		(0.079)	(0.075)	(0.045)	(0.062)		$\hat{\sigma}_1 = 1.535$
							$\hat{\sigma}_\mu = 0.011$
fG2SLS	2.239	- 0.034	1.378	0.082	0.021	0.358	
	(0.770)	(0.074)	(0.076)	(0.038)	(0.058)		
fG3SLS	- 1.121	- 0.021	1.672	0.174	0.208	0.147	
	(0.652)	(0.070)	(0.068)	(0.034)	(0.051)		
O2SLS	2.380	- 0.069	1.334	0.097	0.002	0.381	$\hat{\sigma}^2 = 0.337$
	(0.612)	(0.062)	(0.070)	(0.027)	(0.045)		
O3SLS	1.069	- 0.009	1.583	0.131	0.029	0.227	
	(0.485)	(0.058)	(0.062)	(0.022)	(0.036)		

* Figures inside parentheses are estimations of asymptotic standard deviations.

Table 8.6 Estimation Results - Model I - Appliance Stock Equation*

Estimated Coefficient of / Method of Estimation	Constant Term	log PEL	log INC	log POF	log SIZE	R^2	Estimations of Variance Components
Cov2SLS	0.304	- 1.348	0.168	0.262	0.002	0.176	$\hat{\sigma}_{\varepsilon} = 0.258$
		(0.132)	(0.023)	(0.083)	(0.033)		$\hat{\sigma}_1 = 2.419$
							$\hat{\sigma}_{\mu} = 0.022$
fG2SLS	0.318	- 1.429	0.167	0.304	- 0.002	0.151	
	(0.561)	(0.138)	(0.023)	(0.079)	(0.033)		
fG3SLS	- 1.273	- 0.802	0.158	0.457	0.146	-52.51	
	(0.210)	(0.081)	(0.019)	(0.041)	(0.026)		
O2SLS	1.351	- 1.550	0.167	0.571	- 0.006	0.116	$\hat{\sigma}^2 = 0.266$
	(0.411)	(0.140)	(0.023)	(0.071)	(0.034)		
O3SLS	- 0.002	- 1.601	0.166	0.357	0.156	0.053	
	(0.347)	(0.132)	(0.018)	(0.061)	(0.026)		

* Figures inside parentheses are estimations of asymptotic standard deviations.

Table 8.7 Estimation Results - Model II - Appliance Stock Equation*

Method of Estimation / Estimated Coefficient of	Constant Term	log PEL	log INC	log POF	log SIZE	R^2	Estimations of Variance Components
Cov2SLS	1.931	- 0.780	0.183	0.316	0.020	0.280	$\hat{\sigma}_\varepsilon = 0.237$
		(0.085)	(0.022)	(0.079)	(0.032)		$\hat{\sigma}_1 = 0.932$
							$\hat{\sigma}_\mu = 0.007$
fG2SLS	2.083	- 0.778	0.181	0.341	0.019	0.281	
	(0.438)	(0.080)	(0.022)	(0.072)	(0.032)		
fG3SLS	- 0.818	- 0.729	0.126	0.395	0.142	-45.81	
	(0.207)	(0.051)	(0.017)	(0.036)	(0.025)		
O2SLS	2.347	- 0.785	0.175	0.385	0.018	0.282	$\hat{\sigma}^2 = 0.216$
	(0.343)	(0.066)	(0.020)	(0.059)	(0.030)		
O3SLS	1.098	- 1.122	0.122	0.233	0.130	0.226	
	(0.269)	(0.064)	(0.017)	(0.045)	(0.023)		

* Figures inside parentheses are estimations of asymptotic standard deviations.

Table 8.8 Estimation Results - Model III - Price Equation*

Estimated Coefficient of / Method of Estimation	Constant Term	log KWH	log ALLEL	R^2	Estimations of Variance Components
Cov2SLS	- 1.682	- 0.109 (0.022)	- 0.151 (0.029)	0.252	$\hat{\sigma}_{\varepsilon} = 0.055$ $\hat{\sigma}_{1} = 2.522$ $\hat{\sigma}_{\mu} = 0.025$
f2GSLS	- 1.745 (0.202)	- 0.101 (0.022)	- 0.157 (0.029)	0.246	
f3GSLS	- 1.768 (0.201)	- 0.097 (0.022)	- 0.228 (0.028)	0.243	
O2SLS	- 2.842 (0.253)	0.025 (0.029)	- 0.301 (0.037)	0.053	$\hat{\sigma}^2 = 0.089$
O3SLS	- 2.931 (0.244)	- 0.037 (0.028)	- 0.373 (0.036)	0.023	

* Figures inside parentheses are estimations of asymptotic standard deviations.

Table 8.9 Estimation Results - Model III - Demand Equation*

Method of Estimation \ Estimated Coefficient of	Constant Term	log APP	log CDD	log HDD	R^2	Estimations of Variance Components
Cov2SLS	2.344	1.390	0.063	0.028	0.349	$\hat{\sigma}_\varepsilon = 0.388$
		(0.071)	(0.043)	(0.062)		$\hat{\sigma}_1 = 1.528$
						$\hat{\sigma}_\mu = 0.011$
fG2SLS	2.415	1.367	0.078	0.020	0.363	
	(0.766)	(0.071)	(0.038)	(0.057)		
fG3SLS	- 1.311	1.635	0.184	0.249	0.173	
	(0.670)	(0.064)	(0.035)	(0.052)		
O2SLS	2.649	1.323	0.095	0.0002	0.385	$\hat{\sigma}^2 = 0.335$
	(0.619)	(0.066)	(0.027)	(0.045)		
O3SLS	0.701	1.472	0.201	0.077	0.293	
	(0.508)	(0.061)	(0.024)	(0.038)		

* Figures inside parentheses are estimations of asymptotic standard deviations.

Table 8.10 Estimation Results - Model III - Appliance Stock Equation*

Estimated Coefficient of / Method of Estimation	Constant Term	log PEL	log INC	log POF	log SIZE	R^2	Estimations of Variance Components
Cov2SLS	0.304	- 1.348	0.168	0.262	0.002	0.176	$\hat{\sigma}_\varepsilon = 0.258$
		(0.132)	(0.023)	(0.083)	(0.033)		$\hat{\sigma}_1 = 2.419$
							$\hat{\sigma}_\mu = 0.022$
fG2SLS	0.318	- 1.429	0.167	0.304	- 0.002	0.151	
	(0.561)	(0.138)	(0.023)	(0.079)	(0.033)		
fG3SLS	- 1.265	- 0.829	0.148	0.455	0.153	-52.51	
	(0.205)	(0.079)	(0.017)	(0.040)	(0.025)		
O2SLS	1.351	- 1.550	0.167	0.571	- 0.006	0.116	$\hat{\sigma}^2 = 0.266$
	(0.411)	(0.140)	(0.023)	(0.071)	(0.034)		
O3SLS	0.030	- 1.604	0.164	0.362	0.158	0.051	
	(0.320)	(0.129)	(0.018)	(0.057)	(0.025)		

* Figures inside parentheses are estimations of asymptotic standard deviations.

Table 8.11 Estimates of "Derived" Elasticities of Demand - Model III

Estimation Method \ Type of Elasticity	"Long-run" Price Elasticity	Total Price Elasticity	Income Elasticity
Cov2SLS	- 1.874	- 2.355	0.293
fG2SLS	- 1.953	- 2.434	0.284
fG3SLS	- 1.355	- 1.561	0.279
O2SLS	- 2.051	- 1.951	0.210
O3SLS	- 2.361	- 2.171	0.222

APPENDIX 8.A : Computer Programs of Estimation Methods

```
%    COVARIANCE  2SLS  ESTIMATION  AND
%
%    AOV  ESTIMATION  OF  VARIANCE  COMPONENTS
%
casesen
diary  output.cov
load  var
format  long
%
[  nt,  mk1  ]  =  size(z)
n  =  9
t  =  nt/n
%
%    COVARIANCE  TRANSFORMATION  MATRIX
%
for  i=1:n
imx  =  ones(t,1)*(ones(1,t)*x((i-1)*t+1:(i-1)*t+t,:)/t);
axot((i-1)*t+1  :  (i-1)*t+t,  :)  =  imx;
end
qx  =  x  -  axot;
save  tqx  qx  axot
clear  imx;
%
%    COVARIANCE  2SLS  ESTIMATION  OF  THE  COEFFICIENTS
%
var1  =  z'*qx*inv(qx'*x);
var2  =  qx'*z;
varal  =  inv(var1*var2);
clear  var2;
qxy  =  qx'*y;
alst  =  varal*var1*qxy;
clear  qxy;  clear  var1;
const    =  ones(1,nt)*(y  -  z*alst)/nt;
alpha  =  [  const
                    alst  ]
clear  qx;  clear  axot;
%
%    RESIDUALS  OF  COVARIANCE  ESTIMATION
%
u  =  y  -  z*alst  -  ones(nt,1)*const;
%
%    AOV  ESTIMATION  OF  VARIANCE  COMPONENTS
%
for  i=1:n
imu  =  ones(t,1)*(ones(1,t)*u((i-1)*t+1:(i-1)*t+t,:)/t);
auot((i-1)*t+1  :  (i-1)*t+t,  :)  =  imu;
end
qu  =  u  -  auot;
clear  imu;
sigeps  =  qu'*u/((n-1)*(t-1))
sig1  =  u'*auot/(n-1)
save  aov  sigeps  sig1
save  ucov  qu  auot
%
%
%
%
```

```
%    ESTIMATION  OF  ASYMPTOTIC  VARIANCE  OF  COEFFICIENTS
%
varal  =  sigeps  *  varal;
sdal  =  diag(varal);
sdal  =  sqrt(sdal)
save  covres  alpha    sdal
%
%    CORRELATION  MATRIX  OF  VARIABLES  OF  THE  EQUATION
%
yz  =  [  y  z  ];
yzmean  =  (ones(1,nt)*yz)/nt;
yzbar  =  ones(nt,1)*yzmean;
yzyz  =  (yz-yzbar)'*(yz-yzbar);
vary  =  yzyz(1,1);
save  vary  vary
dyzyz  =  diag(diag(yzyz));
dyzyz  =  sqrt(dyzyz);
corr  =  inv(dyzyz)*yzyz*inv(dyzyz)
%
%    ESTIMATION  OF  INDIVIDUAL  EFFECTS
%
for  i  =  1:n
imy(i,1)  =  ones(1,t)*y((i-1)*t+1:(i-1)*t+t)/t;
end
for  j  =  1:mk1
for  i  =  1:n
imz(i,j)  =  ones(1,t)*z((i-1)*t+1:(i-1)*t+t,j)/t;
end
end
sigmu  =  (sig1  -sigeps)/t
for  i  =  1:n
mu(i,1)  =  ((t*sigmu)/sig1)*(imy(i,1)-const-imz(i,:)*alst);
vmu((i-1)*t+1:(i-1)*t+t,1)  =  mu(i,1)*ones(t,1);
end
%
%    COEFFICIENT  OF  DETERMINATION  (  R  squared  )
%
r2  =  1  -(u'*u)/vary
```

```
%    GENERALISED  2SLS  ESTIMATION
%
casesen
diary  output.g2s
%
%    ESTIMATION  OF  VARIANCE  COMPONENTS  BY  AOV
%
%    USING  COV2SLS  ESTIMATION
%
load  aov
load  var
[nt,km1]  =  size(x)
n  =  9
t  =  nt/n
x1  =  [  ones(nt,1)  x(:,:)  ];
clear  x;
x  =  x1;
clear  x1;
z1  =  [  ones(nt,1)  z(:,:)  ];
clear  z;
z  =  z1;
clear  z1;
%
%    G2SLS  ESTIMATION  OF  THE  COEFFICIENTS
%
for  i=1:n
imx  =  ones(t,1)*(ones(1,t)*x((i-1)*t+1:(i-1)*t+t,:)/t);
axot((i-1)*t+1  :  (i-1)*t+t,  :)  =  imx;
end
clear  imx;
save  tqxc  x  axot
pack
v1  =  x'*z/sigeps  -  axot'*z/sigeps  +  axot'*z/sig1  ;
v2  =  inv(x'*x/sigeps  -  x'*axot/sigeps  +  x'*axot/sig1);
varal  =  inv(v1'*v2*v1);
v3  =  x'*y/sigeps  -  axot'*y/sigeps  +  axot'*y/sig1;
alpha  =  varal*v1'*v2*v3
%
%    ESTIMATION  OF  ASYMPTOTIC  STANDARD  DEVIATION  OF  COEFFICIENTS
%
sdal  =  sqrt(diag(varal))
save  g2sres  alpha  sdal
%
%    ESTIMATION  OF  INDIVIDUAL  EFFECTS
%
for  i  =  1:n
imy(i,1)  =  ones(1,t)*y((i-1)*t+1:(i-1)*t+t)/t;
end
[nt,mk]  =  size(z)
for  j  =  1:mk
for  i  =  1:n
imz(i,j)  =  ones(1,t)*z((i-1)*t+1:(i-1)*t+t,j)/t;
end
end
sigmu  =  (sig1-sigeps)/t
for  i  =  1:n
mu(i,1)  =  ((t*sigmu)/sig1)*(imy(i,1)-imz(i,:)*alpha);
vmu((i-1)*t+1:(i-1)*t+t,1)  =  mu(i,1)*ones(t,1);
end
```

```
%    COEFFICIENT  OF  DETERMINATION  (  R  squared  )
%
load  vary
u  =  y  -  z*alpha;
r2  =  1  -  (u'*u)/vary
```

```
%    COV2SLS  ESTIMATION  OF  ALL  THE  THREE  EQUATIONS  AND  STORAGE  OF
%    THE  VARIABLES  RELEVANT  FOR  THE  ESTIMATION  OF  SIGMA  MATRIX
%
casesen
%
%    COV2SLS  OF  EQUATION  1
%
eq1var
save  var1  z
cov2sls
save  aov1  sigeps  sig1
qu1  =  qu;
auot1  =  auot;
save  ucov1  qu1  auot1
clear
%
%    COV2SLS  OF  EQUATION  2
%
eq2var
save  var2  z
cov2sls
save  aov2  sigeps  sig1
qu2  =  qu;
auot2  =  auot;
save  ucov2  qu2  auot2
clear
%
%    COV2SLS  OF  EQUATION  3
%
eq3var
save  var3  z
cov2sls
save  aov3  sigeps  sig1
qu3  =  qu;
auot3  =  auot;
save  ucov3  qu3  auot3
clear
```

```
%    DEFINITION  OF  THE  MATRIX  OF  ALL  THE  ENDOGENOUS  VARIABLES
%
%    OF  THE  SYSTEM
%
casesen
load  data1
load  data2
bigy  =  [  log(dt1a(:,5))  log(dt1a(:,7))  log(dt2a(:,4))
                  log(dt1b(:,5))  log(dt1b(:,7))  log(dt2b(:,4))  ];
save  bigy  bigy
```

```
%    AOV  ESTIMATION  OF  SIGMA  MATRIX  IN  THE  SEM  -  EC  CASE
%
diary  output.sig
casesen
%
%    ESTIMATION  OF  SIGEPS  AND  SIG1  MATRICES,
%
%    ELEMENT  BY  ELEMENT
%
load  bigy
[nt,m]  =  size(bigy)
n  =  9
t  =  nt/n
clear  bigy
load  aov1
seps11  =  sigeps
sig111  =  sig1
load  aov2
seps22  =  sigeps
sig122  =  sig1
load  aov3
seps33  =  sigeps
sig133  =  sig1
load  ucov1
load  ucov2
load  ucov3
seps12  =  qu1'*qu2/(n*(t-1))
seps13  =  qu1'*qu3/(n*(t-1))
seps23  =  qu2'*qu3/(n*(t-1))
seps  =  [  seps11  seps12  seps13
               seps12  seps22  seps23
               seps13  seps23  seps33  ]
sig112  =  auot1'*auot2/n
sig113  =  auot1'*auot3/n
sig123  =  auot2'*auot3/n
sig1  =  [sig111  sig112  sig113
             sig112  sig122  sig123
             sig113  sig123  sig133]
save  sigres  seps  sig1
```

```
%    GENERALISED  3SLS  ESTIMATION
%
diary  output.g3s
casesen
%
%    DEFINITION  OF  THE  VARIOUS  MATRICES  NEEDED
%
load  bigy
[nt,m]  =  size(bigy)
n  =  9
t  =  nt/n
load  var1
z1  =  [ones(nt,1)    z];
load  var2
z2  =  [ones(nt,1)    z];
load  var3
z3  =  [ones(nt,1)    z];
clear  z;
load  sigres
load  tqxc
[nt,k]  =  size(x)
[nt,mk1]  =  size(z1)
[nt,mk2]  =  size(z2)
[nt,mk3]  =  size(z3)
save  zzz  z1  z2  z3
pack
for  i=1:3*k
xsy(i,1)  =  0;
end
for  i=1:3*k
for  j  =1:mk1+mk2+mk3
xsz(i,j)  =  0;
end
for  j  =  1:3*k
xsx(i,j)  =  0;
end
end
pack
xsz(1:k,  1:mk1)  =    x'*z1/seps(1,1)  -  axot'*z1/seps(1,1)  ..
                                     +  axot'*z1/sig1(1,1);
xsz(k+1:2*k,  mk1+1:mk1+mk2)  =  x'*z2/seps(2,2)  -  axot'*z2/seps(2,2)..
                                                  +  axot'*z2/sig1(2,2);
xsz(2*k+1:3*k,  mk1+mk2+1:mk1+mk2+mk3)  =  x'*z3/seps(3,3)  ..
                                                  +  axot'*z3/seps(3,3)..
                                                  +  axot'*z3/sig1(3,3);
for  i=1:3
for  j=1:3
xsx((i-1)*k+1:(i-1)*k+k,  (j-1)*k+1:(j-1)*k+k)  =  ..
                          seps(i,j)*(x'*x-x'*axot-axot'*x+axot'*axot)..
                          /(seps(i,i)*seps(j,j))..
                          +sig1(i,j)*axot'*axot/(sig1(i,i)*sig1(j,j));
end
end
for  i  =1:3
xsy((i-1)*k+1:(i-1)*k+k,1)  =  x'*bigy(:,i)/seps(i,i)    ..
                                            -  axot'*bigy(:,i)/seps(i,i)..
                                            +  axot'*bigy(:,i)/sig1(i,i);
end
%
%
```

```
%    G3SLS  ESTIMATION  OF  ALL  THE  COEFFICIENTS
%
alpha  =  inv(xsz'*inv(xsx)*xsz)*xsz'*inv(xsx)*xsy
%
%    ESTIMATION  OF  THE  ASYMPTOTIC  VARIANCE MATRIX  OF  COEFFICIENTS
%
varal  =  inv(xsz'*inv(xsx)*xsz);
%
%    ASYMPTOTIC  STANDARD  DEVIATIONS  OF  COEFFICIENTS
%
sdal  =  sqrt(diag(varal))
save  g3sres  alpha  sdal
%
%    INDIVIDUAL  EFFECTS  FOR  EACH  EQUATION
%
clear
load  zzz
[nt,mk1]  =  size(z1)
[nt,mk2]  =  size(z2)
[nt,mk3]  =  size(z3)
n  =  9
t  =  nt/n
load  bigy
[nt,m]  =  size(bigy)
load  sigres
load  g3sres
for  j  =  1:m
for  i  =  1:n
imy(i,j)  =  ones(1,t)*bigy((i-1)*t+1:(i-1)*t+t,j)/t;
end
end
for  j  =  1:mk1
for  i  =  1:n
imz1(i,j)  =  ones(1,t)*z1((i-1)*t+1:(i-1)*t+t,j)/t;
end
end
for  j  =  1:mk2
for  i  =  1:n
imz2(i,j)  =  ones(1,t)*z2((i-1)*t+1:(i-1)*t+t,j)/t;
end
end
for  j  =  1:mk3
for  i  =  1:n
imz3(i,j)  =  ones(1,t)*z3((i-1)*t+1:(i-1)*t+t,j)/t;
end
end
for  i  =  1:m
for  j  =  1:m
sigmu(i,j)  =  (sig1(i,j)  -  seps(i,j))/t;
end
end
for  i  =  1:n
mu1(i,1)  =  ((t*sigmu(1,1))/sig1(1,1))*(imy(i,1)-imz1(i,:)*alpha(1:mk1));
mu2(i,1)  =  ((t*sigmu(2,2))/sig1(2,2))*(imy(i,2)-imz2(i,:)..
                                         *alpha(mk1+1:mk1+mk2));
mu3(i,1)  =  ((t*sigmu(3,3))/sig1(3,3))*(imy(i,3)-imz3(i,:)..
                                         *alpha(mk1+mk2+1:mk1+mk2+mk3));
vmu1((i-1)*t+1:(i-1)*t+t,1)  =  mu1(i,1)*ones(t,1);
vmu2((i-1)*t+1:(i-1)*t+t,1)  =  mu2(i,1)*ones(t,1);
vmu3((i-1)*t+1:(i-1)*t+t,1)  =  mu3(i,1)*ones(t,1);
end
```

```
%    COEFFICIENT  OF  DETERMINATION  FOR  EACH  EQUATION
%
%    TOTAL  VARIATIONS  OF  ENDOGENOUS  VARIABLES
%
yb  =  ones(nt,1)*(ones(1,nt)*bigy)/nt;
yy  =  (bigy-yb)'*(bigy-yb);
for  i  =1:3
vary(i,1)  =  yy(i,i);
end
%
%    RESIDUALS  OF  EACH  EQUATION
%
u1  =  bigy(:,1)  -  z1*alpha(1:mk1);
u2  =  bigy(:,2)  -  z2*alpha(mk1+1:mk1+mk2);
u3  =  bigy(:,3)  -  z3*alpha(mk1+mk2+1:mk1+mk2+mk3);
%
%    COEFFICIENTS  OF  DETERMINATION
%
r21  =    1  -  (u1'*u1)/vary(1,1)
r22  =  1  -  (u2'*u2)/vary(2,1)
r23  =  1  -  (u3'*u3)/vary(3,1)
```

```
%    ORDINARY  2SLS  ESTIMATION
%
diary  output.o2s
casesen
load  var
[nt,km1]  =  size(x)
n  =  9
t  =  nt/n
x1  =  [ones(nt,1)  x(:,:)  ];
clear  x;
x  =  x1;
clear  x1;
z1  =  [ones(nt,1)  z(:,:)  ];
clear  z;
z  =  z1;
clear  z1;
%
%    ORDINARY  2SLS  ESTIMATION  OF  COEFFICIENTS
%
alpha  =  inv(z'*x*inv(x'*x)*x'*z)*z'*x*inv(x'*x)*x'*y
%
%    RESIDUAL  VARIANCE
%
u  =  y  -  z*alpha;
sig2  =  u'*u/nt
%
%    ESTIMATION  OF  ASYMPTOTIC  VARIANCE  OF  ALPHA(O2SLS)
%
varal  =  sig2*inv(z'*x*inv(x'*x)*x'*z);
sdal  =  sqrt(diag(varal))
save  o2sres  alpha  sdal
%
%    COEFFICIENT  OF  DETERMINATION
%
yb  =  ones(nt,1)*(ones(1,nt)*y)/nt;
vary  =  (y-yb)'*(y-yb);
r2  =  1  -  (u'*u)/vary
```

```
%    O2SLS  ESTIMATION  OF  ALL  THE  THREE  EQUATIONS  AND  STORAGE  OF
%
%    THE  VARIABLES  RELEVANT  FOR  THE  ESTIMATION  OF  THE  SIGMA  MATRIX
%
%    O2SLS  OF  EQUATION  1
%
eq1var
clear
o2sls
u1  =  u;
save  o2su1  u1
save  osig1  sig2
clear
%
%    O2SLS  OF  EQUATION  2
%
eq2var
clear
o2sls
u2  =  u;
save  o2su2  u2
save  osig2  sig2
clear
%
%    O2SLS  OF  EQUATION  3
%
eq3var
clear
o2sls
u3  =  u;
save  o2su3  u3
save  osig3  sig2
clear
```

```
%    ESTIMATION  OF  SIGMA  MATRIX  IN  THE  CLASSICAL/ORDINARY  CASE
%
diary  output.osig
casesen
load  bigy
[nt,m]  =  size(bigy)
n  =  9
t  =  nt/n
clear  bigy
load  osig1
sig(1,1)  =  sig2;
load  osig2
sig(2,2)  =  sig2;
load  osig3
sig(3,3)  =  sig2;
load  o2su1
load  o2su2
load  o2su3
sig(1,2)  =  u1'*u2/nt;
sig(1,3)  =  u1'*u3/nt;
sig(2,3)  =  u2'*u3/nt;
sig(2,1)  =  sig(1,2);
sig(3,1)  =  sig(1,3);
sig(3,2)  =  sig(2,3);
save  osigres  sig
sig
```

```
%    ORDINARY  3SLS  ESTIMATION
%
diary  output.o3s
casesen
%
%    DEFINITION  OF  ALL  THE  MATRICES  NEEDED
%
load  bigy
[nt,m]  =  size(bigy)
y1  =  bigy(:,1);
y2  =  bigy(:,2);
y3  =  bigy(:,3);
clear  bigy
load  var1
z1  =  [ones(nt,1)  z];
load  var2
z2  =  [ones(nt,1)  z];
load  var3
z3  =  [ones(nt,1)  z  ];
load  tqxc
clear  axot
[nt,k]  =  size(x)
[nt,mk1]  =  size(z1)
[nt,mk2]  =  size(z2)
[nt,mk3]  =  size(z3)
load  osigres
sinv  =  inv(sig);
%
%    ORDINARY  3SLS  ESTIMATION  OF  ALL  THE  COEFFICIENTS
%
xz1  =  x'*z1;
xz2  =  x'*z2;
xz3  =  x'*z3;
xxi  =  inv(x'*x);
var1  =  [  sinv(1,1)*xz1'*xxi*xz1      sinv(1,2)*xz1'*xxi*xz2  ..
                    sinv(1,3)*xz1'*xxi*xz3
                    sinv(2,1)*xz2'*xxi*xz1      sinv(2,2)*xz2'*xxi*xz2  ..
                    sinv(2,3)*xz2'*xxi*xz3
                    sinv(3,1)*xz3'*xxi*xz1      sinv(3,2)*xz3'*xxi*xz2  ..
                    sinv(3,3)*xz3'*xxi*xz3  ];
varal  =  inv(var1);
clear  var1;
xy1  =  x'*y1;
xy2  =  x'*y2;
xy3  =  x'*y3;
zxy  =  [  sinv(1,1)*xz1'*xxi*xy1+sinv(1,2)*xz1'*xxi*xy2+..
                sinv(1,3)*xz1'*xxi*xy3
                sinv(2,1)*xz2'*xxi*xy1+sinv(2,2)*xz2'*xxi*xy2+..
                sinv(2,3)*xz2'*xxi*xy3
                sinv(3,1)*xz3'*xxi*xy1+sinv(3,2)*xz3'*xxi*xy2+..
                sinv(3,3)*xz3'*xxi*xy3  ];
alpha  =  varal*zxy
%
%    ASYMPTOTIC  STANDARD  DEVIATIONS  OF  COEFFICIENTS
%
sdal  =  sqrt(diag(varal))
save  o3sres  alpha  sdal
%
%
%
```

```
%    COEFFICIENT  OF  DETERMINATION  FOR  EACH  EQUATION
%
%    RESIDUALS  OF  EACH  EQUATION
%
u1  =  y1  -  z1*alpha(1:mk1);
u2  =  y2  -  z2*alpha(mk1+1:mk1+mk2);
u3  =  y3  -  z3*alpha(mk1+mk2+1:mk1+mk2+mk3);
%
%    TOTAL  VARIATIONS  OF  ENDOGENOUS  VARIABLES
%
y1b  =  ones(nt,1)*(ones(1,nt)*y1)/nt;
y2b  =  ones(nt,1)*(ones(1,nt)*y2)/nt;
y3b  =  ones(nt,1)*(ones(1,nt)*y3)/nt;
vy1  =  (y1-y1b)'*(y1-y1b);
vy2  =  (y2-y2b)'*(y2-y2b);
vy3  =  (y3-y3b)'*(y3-y3b);
%
%    COEFFICIENTS  OF  DETERMINATION
%
r21  =  1  -  (u1'*u1)/vy1
r22  =  1  -  (u2'*u2)/vy2
r23  =  1  -  (u3'*u3)/vy3
```

CHAPTER 9

CONCLUSIONS

The use of error components as a means of pooling time series and cross section data in the estimation of economic models has known a constant development in recent years, both at the theoretical and empirical levels. The rapid growth of statistical data, especially of data concerning different units over time, and their increasing accessibility to economic researchers has been an important reason for this development. Another reason is the great progress made in data processing techniques on computers.

In this book, we have presented and extensively analysed the combination of error components and simultaneous equations. Basically, the error component structure accounts for the cross-sectional and temporal heterogeneity of panel data by splitting the error term of the regression equation into different components - a specific unit effect, a specific time effect and a residual disturbance term. In our model, we have assumed the above error structure in each structural equation. Further, as in the classical simultaneous equation model, the errors are correlated across equations. In our case, the correlation is as follows. There is non-zero correlation between the unit effects of two different structural equations if both concern the same unit and between the time effects of two different equations if they relate to the same period. Consequently, the combined error term of any structural equation is correlated not only with that of another equation but also with that of the same equation over time for the same unit and with that of the same equation for different units in the same time period.

Upon deriving the stochastic properties of the errors of the reduced form of the system, it is seen that the reduced form errors are also of error component structure. Thus, since there is no correlation between the explanatory variables and the errors of the reduced form, the reduced form is a model of seemingly unrelated regressions with error components, an extension due to Avery.

Three important methods of estimating the reduced form are discussed. The first one is the covariance estimation, which consists in transforming each reduced form equation by the so-called covariance transformation (which eliminates the specific effects) and then estimating the system by OLS. The second method is the GLS estimation method, which is made feasible by a prior estimation of the variance components using analysis of variance (AOV) formulae. The residuals of the covariance estimation are used for this purpose. The third method is obviously the maximum likelihood estimation of the unconstrained reduced form, assuming normality. An iteration procedure is outlined to solve the system of first-order conditions of maximisation, as these are highly non-linear and hence, no analytical expression of the solution can be derived. It is shown that all the above estimators are consistent and all the three share the same limiting distribution.

In general, the estimation of the reduced form does not permit the identification of structural parameters (except in very special cases, which we will see later). Hence, one has to find ways of estimating the structural equations directly. It is well known that GLS (or OLS, as the case may be) cannot be applied to any structural equation as such because of the non-zero correlation between the explanatory endogenous variables and the errors of the equation. One way of overcoming this problem is to follow the instrumental variable (IV) approach. This approach consists in premultiplying the structural equation by a suitable instruments matrix (which eliminates the covariance between the explanatory variables and the errors, at least in limit) and then apply GLS (or OLS) on the transformed equation. The two stage least squares (2SLS) procedure proposed by Theil in the classical case is an IV method and uses the matrix of exogenous variables as instruments. In our case, this method in not efficient because of the non-scalar covariance matrix of error components. Hence, we have proposed a generalisation of Theil's 2SLS which selects the "best" instruments in a class of all linear transformations of the matrix of exogenous variables. The term "best" is used in the sense that premultiplying the structural equation by the chosen instruments and applying GLS on the transformed equation, minimises the trace and determinant of the asymptotic covariance matrix of the resulting estimator and also gives the minimal positive definite asymptotic covariance matrix.

It turns out that the "best" set of instruments is given by transforming the exogenous variables matrix by the inverse of the variance-covariance matrix of the errors of the equation under consideration. Thus, our "generalised" 2SLS (G2SLS) consists in transforming any structural equation by the corresponding matrix of "best" instruments and then performing GLS.

Before applying the above method (G2SLS) to any equation, the variance-covariance matrix of the errors of that equation has to be estimated. This is again done by means of analysis of variance, which in turn requires an estimation of the residuals of the equation in question. For this purpose, the structural equation is first estimated by what we call the covariance 2SLS estimation method. In fact, we propose two such methods. The first one consists in replacing the explanatory endogenous variables of the equation by an appropriate estimation (through a consistent estimation of the reduced form), transforming the equation by the covariance transformation and then applying OLS. The second method is a special case of the generalised 2SLS in which, instead of transforming the exogenous variables matrix by the inverse of the corresponding variance-covariance matrix, we transform it by the covariance transformation matrix and use it as instruments for estimating the equation. Incidentally, both these covariance 2SLS estimators of structural coefficients are identical if we use the reduced form covariance estimation of the explanatory endogenous variables in the first case and both are consistent.

Once the covariance estimation of the structural coefficients is obtained, residuals are computed and using these residuals the variance components of the equation are estimated. The consistency of these variance component estimators and of the resulting <u>feasible</u> G2SLS estimator is proved.

A more important result is that the pure G2SLS estimator, the feasible G2SLS estimator and the two covariance 2SLS estimators are all asymptotically equivalent, in the sense that they all have the same limiting distribution. This result has been established by deriving all the relevant limiting distributions in a rigorous manner.

The G2SLS method described above is a single equation method i.e. is one which estimates only one structural equation at a time. The extension of this to a simultaneous estimation of the whole system is straightforward and can be made in an analogous manner to that from classical 2SLS and 3SLS. Thus, in our "generalised" 3SLS method, we premultiply each structural equation by the corresponding "best" instruments matrix and estimate the whole system by GLS. This method is made feasible by a prior estimation of the variance components by AOV, for which either the covariance 2SLS or the feasible G2SLS residuals can be used. The limiting distribution of the feasible G3SLS estimator is also derived in full detail.

The various instrumental variables methods proposed so far do not assume any specific distribution for the random terms. However, if we make the additional assumption that the error components are normally distributed, then we can also apply the maximum likelihood principle for estimating our model. In fact, we have examined at length the constrained full information maximum likelihood (FIML) estimation of the structural form. As in the case of the reduced form, the first-order conditions of the constrained maximisation problem are highly non-linear and do not allow for an analytical derivation of the solution. However, we have reformulated them in a convenient way so as to obtain a partial solution for the coefficient parameters in terms of the covariance parameters and vice-versa. Based on this, an iteration method is suggested to arrive at a numerical solution. Our procedure is inspired from the one proposed by Pollock in the classical case. The limiting distribution of the FIML estimator is also derived in detail and is seen to be the same as that of the feasible G3SLS estimator. Regarding the limited information maximum likelihood (LIML) method, we have limited ourselves to showing that it is the FIML of a "reduced" model.

Now, let us turn to the special case that we mentioned earlier, in which the structural parameters can be indirectly estimated from the reduced form parameters. This case, which is known as the just-identified case, is one in which the "a priori" restrictions on the structural parameters are such as to enable their identification from the reduced form parameters. It may be added that the conditions for the just-identification of a structural equation in our model are the same as those of the classical model.

Several interesting results have been obtained regarding the indirect estimation of the structural parameters. In the case of a single just-identified equation, the indirect estimator derived using the covariance estimator of the reduced form is exactly equal to the covariance 2SLS estimator (the second method). On the other hand, when the feasible GLS estimator of the reduced form is used, the resulting indirect estimator of the structural coefficients is asymptotically equivalent to the corresponding feasible G2SLS estimator. Further, when the whole system is just-identified, the G3SLS reduces to G2SLS equation by equation and both are asymptotically equivalent to the indirect estimator (using either covariance or feasible GLS of the reduced form).

The asymptotic properties of estimators are valid in practice only when the sample size is sufficiently large. Now, what happens if this is not the case ? How do the different estimators perform in small or finite samples ? Essentially, there are two ways of obtaining answers to the above questions. One is by the so-called Monte Carlo study and the other is by analytical derivation of the finite sample distribution. In our research, we have adopted the second approach and we have confined ourselves to examining only the finite sample bias. For the reduced form, we have shown that the AOV estimators of variance components are unbiased and that the feasible GLS estimator of the coefficients is also unbiased, if its mean exists.

In the case of the structural form, the exact moments are impossible to determine. However, they can be approximated upto any order by an expansion in series. To this effect, we have followed an approach similar to that proposed by Nagar in the classical simultaneous equation case. We have approximated the bias of the structural covariance components estimators as well as that of the covariance 2SLS and the feasible G2SLS coefficient estimators to the order of N^{-1} (N representing the number of cross-sectional units), keeping the number of time periods (T) fixed. The expressions of the bias that we obtain have interesting similarities to those of the bias of the residual variance estimator and of the 2SLS estimator obtained by Nagar in the classical case.

At the end, we have applied our estimation techniques to a simultaneous equation model of residential electricity demand using data concerning households of different regions of the United States. This model consists of three structural equations explaining demand, price and appliance stock. A specific random "regional" effect is introduced in each of the three equations to take account of the differences in household behaviour and electricity prices over regions. Thus, we are in presence of a simultaneous equation model with error components, with only one specific effect. This model was estimated by covariance 2SLS, feasible G2SLS, feasible G3SLS as well as by (ordinary) 2SLS and 3SLS for comparison purposes.

Our results are seen to produce satisfactory evidence of the usefulness of our new methodology and to provide an interesting illustration of the type of situations in which our theoretical model may be appropriate to represent the phenomenon under consideration.

In any field, there is always scope for further research. As far as our topic is concerned, we see the following extensions. Firstly, all our results have been derived assuming absence of lagged endogenous variables in the system. It would therefore be interesting to see to what extent the same results hold if lagged endogenous variables are included in the model. In particular, the assumptions on the limits of the sample moment matrices of the predetermined variables have to be re-examined carefully in this case.

Another aspect which could be studied is the case of a recursive or block-recursive system in order to verify whether the complete system can be partitioned into several smaller equation (interdependent) systems as in the classical case.

In our work, the LIML was just shown to be a special case of the FIML without going deeper into it. However, it may be worthwhile analysing the LIML further, especially from the point of view of developing tests of exogeneity.

Finally, we propose a few extensions which may contribute towards improving the applicability of our model in empirical research, namely, the estimation of a simultaneous equation model with error components in presence of unbalanced (non-overlapping) pooled data, estimation of a simultaneous equation model with the error components only in some equations and lastly, the development of a suitable user-oriented computer package for estimating both single equations and simultaneous equations with error components.

REFERENCES

[1] AMEMIYA, T. (1971). The Estimation of Variances in a Variance-Components Model. International Economic Review, 12, 1-13.

[2] ANDERSON, T.W. and C. HSIAO (1982). Formulation and Estimation of Dynamic Models using Panel Data. Journal of Econometrics, 18, 47-82.

[3] AVERY, R.B. (1977). Error Component Models and Seemingly Unrelated Regressions. Econometrica, 45, 199-209.

[4] BALESTRA, P. (1978). Determinant and Inverse of a Sum of Matrices with Applications in Economics and Statistics. Document de travail, 24, Institut de Mathématiques Economiques de Dijon, France.

[5] BALESTRA, P. (1983). La Dérivation Matricielle. Collection de l'Institut de Mathématiques Economiques de Dijon, 12, Sirey, Paris.

[6] BALESTRA, P. and M. NERLOVE (1966). Pooling Cross-Section and Time-Series Data in the Estimation of a Dynamic Model : The Demand for Natural Gas. Econometrica, 34, 585-612.

[7] BALTAGI, B.H. (1980). On Seemingly Unrelated Regressions with Error Components. Econometrica, 48, 1547-1551.

[8] BALTAGI, B.H. (1981). Pooling : An Experimental Study of Alternative Testing and Estimation Procedures in a Two-way Error Component Model. Journal of Econometrics, 17, 21-49.

[9] BALTAGI, B.H. (1981). Simultaneous Equations with Error Components. Journal of Econometrics, 17, 189-200.

[10] BERZEC, K. (1979). The Error Components Models : Conditions for the Existence of Maximum Likelihood Estimates. Journal of Econometrics, 10, 99-102.

[11] BHATTACHARYA, R.N. and J.K. GHOSH (1978). On the Validity of the Formal Edgeworth Expansion. The Annals of Statistics, 6, 434-451.

[12] BIORN, E. (1981). Estimating Economic Relations from Incomplete Cross-section/Time-Series Data. Journal of Econometrics, 16, 221-236.

[13] DON, F.J.H. (1985). The Use of Generalized Inverses in Restricted Maximum Likelihood. Linear Algebra and its Applications, 70.

[14] FULLER, W.A. and G.E. BATTESE (1974). Estimation of Linear Models with Crossed Error Structure. Journal of Econometrics, 2, 67-78.

[15] GARBACZ, C. (1983). A Model of Residential Demand for Electricity using a National Household Sample. Energy Economics, 5, 124-128.

[16] GRAYBILL, F.A. (1961). An Introduction to Linear Statistical Models. McGraw-Hill Book Company, Inc., New York.

[17] HALVORSEN, R. (1975). Residential Demand for Electric Energy. The Review of Economics and Statistics, LVII, 12-18.

[18] HAUSMANN, J.A. and W.E. TAYLOR (1981). Panel Data and Unobservable Individual Effects. Econometrica, 49, 1377-1398.

[19] HENDERSON, H.V. and S.R. SEARLE (1979). Vec and Vech Operators for Matrices, with Some Uses in Jacobians and Multivariate Statistics. The Canadian Journal of Statistics, 7, 65-81.

[20] HOCH, I. (1962). Estimation of Production Function Parameters Combining Time Series and Cross-Section Data. Econometrica, 30, 34-53.

[21] HSIAO, C. (1975). Some Estimation Methods for a Random Coefficient Model. Econometrica, 43, 305-325.

[22] JUDGE, G.G., W.E. GRIFFITHS, R.C. HILL and T.C. LEE (1980). The Theory and Practice of Econometrics. John Wiley and Sons, Inc., New York.

[23] KAKWANI, N.C. (1967). The Unbiasedness of Zellner's Seemingly Unrelated Regression Equations Estimators. Journal of the American Statistical Association, 62, 141-142.

[24] LIU, L.M. and D.M. HANSSENS (1981). A Bayesian Approach to Time Varying Cross Sectional Regression Models. Journal of Econometrics, 15, 341-356.

[25] LIU, L.M. and G.C. TIAO (1980). Random Coefficient First-Order Autoregressive Models. Journal of Econometrics, 13, 305-326.

[26] MADDALA, G.S. and T.D. MOUNT (1973). A Comparative Study of Alternate Estimators for Variance Components Models Use in Econometric Applications. Journal of the American Statistical Association, 68, 324-328.

[27] MAGNUS, J.R. (1982). Multivariate Error Components Analysis of Linear and Non-Linear Regression Models by Maximum Likelihood. Journal of Econometrics, 19, 239-285.

[28] MAGNUS, J.R. and H. NEUDECKER (1980). The Elimination Matrix : Some Theorems and Applications. SIAM Journal on Algebraic and Discrete Methods, 1, 422-449.

[29] MAZODIER, P. and A. TROGNON (1977). Données en Coupes Répétées et Modèles à Double Indice, in Modèles Régionaux et Régionaux-Nationaux, Published by Raymond Courbis, Editions Cujas, Collection GAMA, No. 1, 287-308.

[30] MUNDLAK, Y. (1978). On the Pooling of Time Series and Cross-Section Data. Econometrica, 46, 69-85.

[31] MUNDLAK, Y. (1978). Models with Variable Coefficients: Integration and Extension. Annales de l'INSEE, No. 30-31, 483-509.

[32] NAGAR, A.L. (1959). The Bias and Moment Matrix of the General k-class Estimators of the Parameters in Simultaneous Equations. Econometrica, 27, 575-595.

[33] NAGAR, A.L. (1961). A Note on the Residual Variance Estimation in Simultaneous Equations. Econometrica, 29, 238-243.

[34] NERLOVE, M. (1967). Experimental Evidence on the Estimation of Dynamic Economic Relations from a Time Series of Cross Sections. Economic Studies Quarterly, 18, 42-74.

[35] NERLOVE, M. (1971). Further Evidence on the Estimation of Dynamic Relations from a Time Series of Cross Sections. Econometrica, 39, 359-382.

[36] NERLOVE, M. (1971). A Note on Error Components Models. Econometrica, 39, 383-396.

[37] POLLOCK, D.S.G. (1979). The Algebra of Econometrics. John Wiley and Sons, Chichester.

[38] PRUCHA, I.R. (1984). On the Asymptotic Efficiency of Feasible Aitken Estimator for Seemingly Unrelated Regression Models with Error Components. Econometrica, 52, 203-207.

[39] PRUCHA, I.R. (1985). Maximum Likelihood and Instrumental Variable Estimation in Simultaneous Equation Systems with Error Components. International Economic Review, 26, 491-506.

[40] RAO, C.R. (1970). Estimation of Heteroscedastic Variances in Linear Models. Journal of the American Statistical Association, 65, 161-172.

[41] RAO, C.R. (1972). Estimation of Variance and Covariance Components in Linear Models. Journal of the American Statistical Association, 67, 112-115.

[42] RESIDENTIAL ENERGY CONSUMPTION SURVEY : Housing Characteristics 1982. Published : August 1984. U.S. Energy Information Administration Publication.

[43] RESIDENTIAL ENERGY CONSUMPTION SURVEY : Consumption and Expenditures, April 1982 through March 1983. Part 1 : National Data. Published : November 1984. U.S. Energy Information Administration Publication.

[44] RESIDENTIAL ENERGY CONSUMPTION SURVEY : Consumption and Expenditures, April 1982 through March 1983. Part 2 : Regional Data. Published : December 1984. U.S. Energy Information Administration Publication.

[45] ROSENBERG, B. (1973). The Analysis of a Cross Section of Time Series by Stochastically Convergent Parameter Regression. Annals of Economic and Social Measurement, 2, 399-428.

[46] SWAMY, P.A.V.B. (1970). Efficient Inference in a Random Coefficient Regression Model. Econometrica, 38, 311-323.

[47] SWAMY, P.A.V.B. and S.S. ARORA (1972). The Finite Sample Properties of the Estimators of Coefficients in the Error Components Regressions Models. Econometrica, 40, 253-260.

[48] SWAMY, P.A.V.B. and J. S. MEHTA (1973). Bayesian Analysis of Error Components Regression Models. Journal of the American Statistical Association, 68, 645-658.

[49] SWAMY, P.A.V.B. and J. S. MEHTA (1975). Bayesian and non-Bayesian Analysis of Switching Regression and of Random Coefficient Regression Models. Journal of the American Statistical Association, 70, 593-602.

[50] TAYLOR, W.E. (1980). Small Sample Considerations in Estimations from Panel Data. Journal of Econometrics, 13, 203-223.

[51] THEIL, H. (1971). Principles of Econometrics, North-Holland Publishing Company, Amsterdam.

[52] TROGNON, A. (1978). Miscellaneous Asymptotic Properties of Ordinary Least Squares and Maximum Likelihood Estimators in Dynamic Error Components Models. Annales de l'INSEE, No. 30-31, 631-657.

[53] VARADHARAJAN, J. (1981). Estimation of Simultaneous Linear Equation Models with Error Component Structure. Cahiers du Département d'économétrie, 81.06, Université de Genève, Switzerland.

[54] VARADHARAJAN, J. (1981). Note on the Identity of Ordinary Least Squares Estimator and Generalised Least Squares Estimator in a General Stratified Effect Component Model. Cahiers du Département d'économétrie, 81.07, Université de Genève, Switzerland.

[55] WALLACE, T.D. and A. HUSSAIN (1969). The Use of Error Components Models in Combining Cross-Section with Time-Series Data. Econometrica, 37, 55-72.

[56] WANSBEEK, T.J. (1980). Quantitative Effects in Panel Data Modelling. Doctoral Thesis, Leyden University, The Netherlands.

[57] ZELLNER, A. and H. THEIL (1962). Three-Stage Least Squares : Simultaneous Estimation of Simultaneous Equations. Econometrica, 30, 54-78.

Vol. 211: P. van den Heuvel, The Stability of a Macroeconomic System with Quantity Constraints. VII, 169 pages. 1983.

Vol. 212: R. Sato and T. Nôno, Invariance Principles and the Structure of Technology. V, 94 pages. 1983.

Vol. 213: Aspiration Levels in Bargaining and Economic Decision Making. Proceedings, 1982. Edited by R. Tietz. VIII, 406 pages. 1983.

Vol. 214: M. Faber, H. Niemes und G. Stephan, Entropie, Umweltschutz und Rohstoffverbrauch. IX, 181 Seiten. 1983.

Vol. 215: Semi-Infinite Programming and Applications. Proceedings, 1981. Edited by A.V. Fiacco and K.O. Kortanek. XI, 322 pages. 1983.

Vol. 216: H.H. Müller, Fiscal Policies in a General Equilibrium Model with Persistent Unemployment. VI, 92 pages. 1983.

Vol. 217: Ch. Grootaert, The Relation Between Final Demand and Income Distribution. XIV, 105 pages. 1983.

Vol. 218: P. van Loon, A Dynamic Theory of the Firm: Production, Finance and Investment. VII, 191 pages. 1983.

Vol. 219: E. van Damme, Refinements of the Nash Equilibrium Concept. VI, 151 pages. 1983.

Vol. 220: M. Aoki, Notes on Economic Time Series Analysis: System Theoretic Perspectives. IX, 249 pages. 1983.

Vol. 221: S. Nakamura, An Inter-Industry Translog Model of Prices and Technical Change for the West German Economy. XIV, 290 pages. 1984.

Vol. 222: P. Meier, Energy Systems Analysis for Developing Countries. VI, 344 pages. 1984.

Vol. 223: W. Trockel, Market Demand. VIII, 205 pages. 1984.

Vol. 224: M. Kiy, Ein disaggregiertes Prognosesystem für die Bundesrepublik Deutschland. XVIII, 276 Seiten. 1984.

Vol. 225: T.R. von Ungern-Sternberg, Zur Analyse von Märkten mit unvollständiger Nachfragerinformation. IX, 125 Seiten. 1984

Vol. 226: Selected Topics in Operations Research and Mathematical Economics. Proceedings, 1983. Edited by G. Hammer and D. Pallaschke. IX, 478 pages. 1984.

Vol. 227: Risk and Capital. Proceedings, 1983. Edited by G. Bamberg and K. Spremann. VII, 306 pages. 1984.

Vol. 228: Nonlinear Models of Fluctuating Growth. Proceedings, 1983. Edited by R.M. Goodwin, M. Krüger and A. Vercelli. XVII, 277 pages. 1984.

Vol. 229: Interactive Decision Analysis. Proceedings, 1983. Edited by M. Grauer and A.P. Wierzbicki. VIII, 269 pages. 1984.

Vol. 230: Macro-Economic Planning with Conflicting Goals. Proceedings, 1982. Edited by M. Despontin, P. Nijkamp and J. Spronk. VI, 297 pages. 1984.

Vol. 231: G.F. Newell, The M/M/∞ Service System with Ranked Servers in Heavy Traffic. XI, 126 pages. 1984.

Vol. 232: L. Bauwens, Bayesian Full Information Analysis of Simultaneous Equation Models Using Integration by Monte Carlo. VI, 114 pages. 1984.

Vol. 233: G. Wagenhals, The World Copper Market. XI, 190 pages. 1984.

Vol. 234: B.C. Eaves, A Course in Triangulations for Solving Equations with Deformations. III, 302 pages. 1984.

Vol. 235: Stochastic Models in Reliability Theory. Proceedings, 1984. Edited by S. Osaki and Y. Hatoyama. VII, 212 pages. 1984.

Vol. 236: G. Gandolfo, P.C. Padoan, A Disequilibrium Model of Real and Financial Accumulation in an Open Economy. VI, 172 pages. 1984.

Vol. 237: Misspecification Analysis. Proceedings, 1983. Edited by T.K. Dijkstra. V, 129 pages. 1984.

Vol. 238: W. Domschke, A. Drexl, Location and Layout Planning. IV, 134 pages. 1985.

Vol. 239: Microeconomic Models of Housing Markets. Edited by K. Stahl. VII, 197 pages. 1985.

Vol. 240: Contributions to Operations Research. Proceedings, 1984. Edited by K. Neumann and D. Pallaschke. V, 190 pages. 1985.

Vol. 241: U. Wittmann, Das Konzept rationaler Preiserwartungen. XI, 310 Seiten. 1985.

Vol. 242: Decision Making with Multiple Objectives. Proceedings, 1984. Edited by Y.Y. Haimes and V. Chankong. XI, 571 pages. 1985.

Vol. 243: Integer Programming and Related Areas. A Classified Bibliography 1981–1984. Edited by R. von Randow. XX, 386 pages. 1985.

Vol. 244: Advances in Equilibrium Theory. Proceedings, 1984. Edited by C.D. Aliprantis, O. Burkinshaw and N.J. Rothman. II, 235 pages. 1985.

Vol. 245: J.E.M. Wilhelm, Arbitrage Theory. VII, 114 pages. 1985.

Vol. 246: P.W. Otter, Dynamic Feature Space Modelling, Filtering and Self-Tuning Control of Stochastic Systems. XIV, 177 pages. 1985.

Vol. 247: Optimization and Discrete Choice in Urban Systems. Proceedings, 1983. Edited by B.G. Hutchinson, P. Nijkamp and M. Batty. VI, 371 pages. 1985.

Vol. 248: Plural Rationality and Interactive Decision Processes. Proceedings, 1984. Edited by M. Grauer, M. Thompson and A.P. Wierzbicki. VI, 354 pages. 1985.

Vol. 249: Spatial Price Equilibrium: Advances in Theory, Computation and Application. Proceedings, 1984. Edited by P. T. Harker. VII, 277 pages. 1985.

Vol. 250: M. Roubens, Ph. Vincke, Preference Modelling. VIII, 94 pages. 1985.

Vol. 251: Input-Output Modeling. Proceedings, 1984. Edited by A. Smyshlyaev. VI, 261 pages. 1985.

Vol. 252: A. Birolini, On the Use of Stochastic Processes in Modeling Reliability Problems. VI, 105 pages. 1985.

Vol. 253: C. Withagen, Economic Theory and International Trade in Natural Exhaustible Resources. VI, 172 pages. 1985.

Vol. 254: S. Müller, Arbitrage Pricing of Contingent Claims. VIII, 151 pages. 1985.

Vol. 255: Nondifferentiable Optimization: Motivations and Applications. Proceedings, 1984. Edited by V.F. Demyanov and D. Pallaschke. VI, 350 pages. 1985.

Vol. 256: Convexity and Duality in Optimization. Proceedings, 1984. Edited by J. Ponstein. V, 142 pages. 1985.

Vol. 257: Dynamics of Macrosystems. Proceedings, 1984. Edited by J.-P. Aubin, D. Saari and K. Sigmund. VI, 280 pages. 1985.

Vol. 258: H. Funke, Eine allgemeine Theorie der Polypol- und Oligopolpreisbildung. III, 237 pages. 1985.

Vol. 259: Infinite Programming. Proceedings, 1984. Edited by E.J. Anderson and A.B. Philpott. XIV, 244 pages. 1985.

Vol. 260: H.-J. Kruse, Degeneracy Graphs and the Neighbourhood Problem. VIII, 128 pages. 1986.

Vol. 261: Th.R. Gulledge, Jr., N.K. Womer, The Economics of Made-to-Order Production. VI, 134 pages. 1986.

Vol. 262: H.U. Buhl, A Neo-Classical Theory of Distribution and Wealth. V, 146 pages. 1986.

Vol. 263: M. Schäfer, Resource Extraction and Market Structure. XI, 154 pages. 1986.

Vol. 264: Models of Economic Dynamics. Proceedings, 1983. Edited by H.F. Sonnenschein. VII, 212 pages. 1986.

Vol. 265: Dynamic Games and Applications in Economics. Edited by T. Başar. IX, 288 pages. 1986.

Vol. 266: Multi-Stage Production Planning and Inventory Control. Edited by S. Axsäter, Ch. Schneeweiss and E. Silver. V, 264 pages. 1986.

Vol. 267: R. Bemelmans, The Capacity Aspect of Inventories. IX, 165 pages. 1986.

Vol. 268: V. Firchau, Information Evaluation in Capital Markets. VII, 103 pages. 1986.

Vol. 269: A. Borglin, H. Keiding, Optimality in Infinite Horizon Economies. VI, 180 pages. 1986.

Vol. 270: Technological Change, Employment and Spatial Dynamics. Proceedings 1985. Edited by P. Nijkamp. VII, 466 pages. 1986.

Vol. 271: C. Hildreth, The Cowles Commission in Chicago, 1939–1955. V, 176 pages. 1986.

Vol. 272: G. Clemenz, Credit Markets with Asymmetric Information. VIII, 212 pages. 1986.

Vol. 273: Large-Scale Modelling and Interactive Decision Analysis. Proceedings, 1985. Edited by G. Fandel, M. Grauer, A. Kurzhanski and A.P. Wierzbicki. VII, 363 pages. 1986.

Vol. 274: W.K. Klein Haneveld, Duality in Stochastic Linear and Dynamic Programming. VII, 295 pages. 1986.

Vol. 275: Competition, Instability, and Nonlinear Cycles. Proceedings, 1985. Edited by W. Semmler. XII, 340 pages. 1986.

Vol. 276: M.R. Baye, D.A. Black, Consumer Behavior, Cost of Living Measures, and the Income Tax. VII, 119 pages. 1986.

Vol. 277: Studies in Austrian Capital Theory, Investment and Time. Edited by M. Faber. VI, 317 pages. 1986.

Vol. 278: W.E. Diewert, The Measurement of the Economic Benefits of Infrastructure Services. V, 202 pages. 1986.

Vol. 279: H.-J. Büttler, G. Frei and B. Schips, Estimation of Disequilibrium Models. VI, 114 pages. 1986.

Vol. 280: H.T. Lau, Combinatorial Heuristic Algorithms with FORTRAN. VII, 126 pages. 1986.

Vol. 281: Ch.-L. Hwang, M.-J. Lin, Group Decision Making under Multiple Criteria. XI, 400 pages. 1987.

Vol. 282: K. Schittkowski, More Test Examples for Nonlinear Programming Codes. V, 261 pages. 1987.

Vol. 283: G. Gabisch, H.-W. Lorenz, Business Cycle Theory. VII, 229 pages. 1987.

Vol. 284: H. Lütkepohl, Forecasting Aggregated Vector ARMA Processes. X, 323 pages. 1987.

Vol. 285: Toward Interactive and Intelligent Decision Support Systems. Volume 1. Proceedings, 1986. Edited by Y. Sawaragi, K. Inoue and H. Nakayama. XII, 445 pages. 1987.

Vol. 286: Toward Interactive and Intelligent Decision Support Systems. Volume 2. Proceedings, 1986. Edited by Y. Sawaragi, K. Inoue and H. Nakayama. XII, 450 pages. 1987.

Vol. 287: Dynamical Systems. Proceedings, 1985. Edited by A.B. Kurzhanski and K. Sigmund. VI, 215 pages. 1987.

Vol. 288: G.D. Rudebusch, The Estimation of Macroeconomic Disequilibrium Models with Regime Classification Information. VII, 128 pages. 1987.

Vol. 289: B.R. Meijboom, Planning in Decentralized Firms. X, 168 pages. 1987.

Vol. 290: D.A. Carlson, A. Haurie, Infinite Horizon Optimal Control. XI, 254 pages. 1987.

Vol. 291: N. Takahashi, Design of Adaptive Organizations. VI, 140 pages. 1987.

Vol. 292: I. Tchijov, L. Tomaszewicz (Eds.), Input-Output Modeling. Proceedings, 1985. VI, 195 pages. 1987.

Vol. 293: D. Batten, J. Casti, B. Johansson (Eds.), Economic Evolution and Structural Adjustment. Proceedings, 1985. VI, 382 pages. 1987.

Vol. 294: J. Jahn, W. Krabs (Eds.), Recent Advances and Historical Development of Vector Optimization. VII, 405 pages. 1987.

Vol. 295: H. Meister, The Purification Problem for Constrained Games with Incomplete Information. X, 127 pages. 1987.

Vol. 296: A. Börsch-Supan, Econometric Analysis of Discrete Choice. VIII, 211 pages. 1987.

Vol. 297: V. Fedorov, H. Läuter (Eds.), Model-Oriented Data Analysis. Proceedings, 1987. VI, 239 pages. 1988.

Vol. 298: S.H. Chew, Q. Zheng, Integral Global Optimization. VII, 179 pages. 1988.

Vol. 299: K. Marti, Descent Directions and Efficient Solutions in Discretely Distributed Stochastic Programs. XIV, 178 pages. 1988.

Vol. 300: U. Derigs, Programming in Networks and Graphs. XI, 315 pages. 1988.

Vol. 301: J. Kacprzyk, M. Roubens (Eds.), Non-Conventional Preference Relations in Decision Making. VII, 155 pages. 1988.

Vol. 302: H.A. Eiselt, G. Pederzoli (Eds.), Advances in Optimization and Control. Proceedings, 1986. VIII, 372 pages. 1988.

Vol. 303: F.X. Diebold, Empirical Modeling of Exchange Rate Dynamics. VII, 143 pages. 1988.

Vol. 304: A. Kurzhanski, K. Neumann, D. Pallaschke (Eds.), Optimization, Parallel Processing and Applications. Proceedings, 1987. VI, 292 pages. 1988.

Vol. 305: G.-J.C.Th. van Schijndel, Dynamic Firm and Investor Behaviour under Progressive Personal Taxation. X, 215 pages. 1988.

Vol. 306: Ch. Klein, A Static Microeconomic Model of Pure Competition. VIII, 139 pages. 1988.

Vol. 307: T.K. Dijkstra (Ed.), On Model Uncertainty and its Statistical Implications. VII, 138 pages. 1988.

Vol. 308: J.R. Daduna, A. Wren (Eds.), Computer-Aided Transit Scheduling. VIII, 339 pages. 1988.

Vol. 309: G. Ricci, K. Velupillai (Eds.), Growth Cycles and Multisectoral Economics: the Goodwin Tradition. III, 126 pages. 1988.

Vol. 310: J. Kacprzyk, M. Fedrizzi (Eds.), Combining Fuzzy Imprecision with Probabilistic Uncertainty in Decision Making. IX, 399 pages. 1988.

Vol. 311: R. Färe, Fundamentals of Production Theory. IX, 163 pages. 1988.

Vol. 312: J. Krishnakumar, Estimation of Simultaneous Equation Models with Error Components Structure. X, 357 pages. 1988.